Optimal Adaptive Cooperative Control for Nonlinear Multiagent Systems

Xin Wang · Huaqing Li · Tingwen Huang

Optimal Adaptive Cooperative Control for Nonlinear Multiagent Systems

Springer

Xin Wang
College of Electronic and Information Engineering
Southwest University
Chongqing, China

Huaqing Li
College of Electronic and Information Engineering
Southwest University
Chongqing, China

Tingwen Huang
Faculty of Computer Science and Control Engineering
Shenzhen University of Advanced Technology
Shenzhen, Guangdong, China

ISBN 978-981-95-2963-6 ISBN 978-981-95-2964-3 (eBook)
https://doi.org/10.1007/978-981-95-2964-3

This Springer imprint is published by the registered company Springer Nature Singapore Pte Ltd.
The registered company address is: 152 Beach Road, #21-01/04 Gateway East, Singapore 189721, Singapore

Preface

The cooperative control of multi-agent systems has attracted significant attention in recent years, emerging as a pivotal research area within new generation artificial intelligence. While substantial progress has been achieved in theoretical frameworks, practical implementations face three challenges: (i) non-ideal work conditions arising from environmental disturbances and device limitations, (ii) escalating communication cost in large-scale networked systems, and (iii) the imperative for enhanced system reliability and security. To address these critical issues and bridge the gap between theories and applications, this book has an deep investigation on advanced control methodologies for the cooperation of multiagent systems under constrained environments. The book provides a systematic and in-depth introduction to the cooperative control for multiagent systems from a theoretical perspective. The proposed cooperative control methods are aiming at four aspects: communication cost reduction, system state measurability, inner unknown dynamics and uncertainties as well as external disturbances, cyber attacks and system faults. The book is organized as follows:

Chapter 1 begins with the background of multiagent systems. The primary problems challenges are discussed and the main works are summarized briefly.

Chapter 2 discusses the neural network-based adaptive tracking control issue for a class of high-order nonlinear multiagent systems both subjected to the immeasurable state variables and unknown external disturbance. Combining with the radial basis function neural networks (RBFNNs), the composite disturbance observer and state observer are established. Furthermore, to cope with the sparsity of the control resources, the proposed method is extended to the event-triggered case.

Chapter 3 proposes an event-triggered adaptive control strategy to achieve the leaderless-following consensus for a class of cyberphysical systems under the direct topology. To tackle these knotty problems, the Nussbaum gain functions are employed to replace the control gains, and the available compromised system variables are applied in the controllers. In addition, the proposed disturbance observer based on compromised state further improves the robustness of the system.

Chapter 4 investigates the leader-following consensus problem for a class of nonlinear multiagent systems subject to sensor and actuator faults in a fixed directed graph. First, a fault compensation mechanism is proposed and the command filtering method is employed to avoid the burst of complexity. Furthermore, the neural network-based state observers are designed to reconstruct the unmeasurable states of the nonlinear multiagent systems.

Chapter 5 is concerned with the event-triggered adaptive finite-time control problem for a class of cyber-physical systems (CPSs) subject to uncertain deception attacks. The finite time command filter to alleviate the explosion of complexity is deployed. To stabilize the system with a guaranteed performance index, a disturbance observer is constructed by fuzzy logic systems (FLS). Furthermore, the Nussbaum gain is utilized to resist the unknown control efficiency generated by uncertain deception attacks.

Chapter 6 is intended to solve the adaptive control issue for uncertain nonlinear systems with time-varying full-state constraints. A novel Integral Barrier Lyapunov functions (IBLFs) based neural backstepping control approach is designed, and the sliding-mode disturbance observers (SMDOs) are established to deal with the immeasurable disturbances in each order of the state-constrained uncertain nonlinear systems. Besides, the dynamic threshold-based event-sampling mechanism is constructed to deal with the sparsity of resources and system controlling burden.

Chapter 7 considers the consensus tracking problem for nonlinear semi-strict-feedback multi-agent systems with unknown states and disturbances, based on radial basis function neural networks (RBFNNs) and back-stepping techniques. The adaptive event-triggered control scheme is introduced to decrease the update times of the controller so as to save the limited communication resources. The state observer and disturbance observer as well as the first-order filter are first jointly constructed.

Chapter 8 studies the optimized adaptive finite-time consensus control issue for stochastic nonlinear multiagent systems subject to non-affine non-linear faults. Under the architecture of the adaptive optimized back-stepping method, the neural-network-based simplified reinforcement learning algorithm with an identifier-critic-actor structure is developed, where the identifier, critic and actor are put forward to estimate unknown dynamics, evaluate system performance and implement control behavior, respectively. Then, the Butterworth low-pass filter is introduced to compensate for the adverse effects brought by non-affine nonlinear faults.

Chongqing, China Xin Wang
Chongqing, China Huaqing Li
Shenzhen, China Tingwen Huang

Contents

Chapter 1
Introduction

1.1 Background

With the increasing progress of science and technology, it is an inevitable trend for artificial intelligence to penetrate into all aspects of daily life. A single artificial intelligence entity may be able to replace humans in performing certain simple tasks, but if it is required to complete larger and more complex tasks, multiple artificial intelligence entities need cooperation. This demand has promoted the development of swarm intelligence. Although it was first used to describe the incredible collective behaviors exhibited by large biological groups such as flocks of birds, schools of fish, and swarms of bees, nowadays, swarm intelligence has been extended to fields such as robotics, computers, and industrial manufacturing, and it has gradually become one of the major sustainable development directions in the field of artificial intelligence. Swarm intelligence in artificial intelligence refers to the intelligent characteristics autonomously demonstrated by multiple artificial intelligence entities that follow certain rules of interaction and collaboration. If these characteristics are utilized, it may greatly exceed the work efficiency of humans themselves in some applications. Therefore, exploring and developing the field of swarm intelligence in artificial intelligence has an urgent practical significance.

In the process of studying biology, computers, and complex social networks in recent decades, the concept of multiagent system has gradually emerged. Currently, as an effective method and tool, the theory of multiagent systems is widely applied in various fields [1–6] such as intelligent transportation, industrial manufacturing, power system management, unmanned autonomous systems, etc. An individual agent is usually simple and limited in function, but through the cooperation among multiple agents, it can break through the limitations of the individual agent's own capabilities, thus demonstrating powerful functions in terms of overall information processing and task execution capabilities. They can also complete complex objectives [7–10] such as swarming, formation, aggregation, and rendezvous. This characteristic of

X. Wang et al., *Optimal Adaptive Cooperative Control for Nonlinear Multiagent Systems*,
https://doi.org/10.1007/978-981-95-2964-3_1

multiagent systems is in line with the requirements of swarm intelligence in artificial intelligence. Therefore, studying the cooperative control problem of multiagent systems can promote the development of swarm intelligence.

The research on the cooperative control of non-ideal multiagent systems has been rapidly heating up recently, focusing on the working conditions of multiagent systems considering a large number of actual environments and various physical constraints. Since ideal multiagent systems take into account very few adverse influencing factors, relatively complex tasks can be completed by designing simple cooperative rules and control methods. However, in the actual working environment, there are a large number of unstable factors at any time and anywhere, which leads to the fact that multiagent systems always operate in a non-ideal state. These non-ideal factors arising from multiple aspects often bring adverse effects[11–16], such as model uncertainty caused by disturbances, time delays caused by communication interference, and performance losses caused by machine failures. Moreover, security control [17–21] for the multiagent systems embodied as multiple cyber-physical systems has also become an important area. Generally, it is difficult to overcome these problems with simple theories and methods, thus giving rise to many practical issues. Therefore, in order to solve these problems, it is necessary not only to develop the cooperative control theory of multiagent systems considering non-ideal factors, but also to propose specific solutions to the practical problems in the cooperation process in a targeted manner.

1.2 Research Challenges

Considering the problem of cooperative control of multiagent systems in a non-ideal environment, the content of this book mainly focuses on the following types of challenges:

Communication Cost Reduction: The key to the cooperative control of multiagent systems lies in how to achieve the self-organizing intelligent behavior of the overall system by designing proper control protocols. Since swarm intelligence benefits from a large number of individuals, in order to improve the overall performance of multiagent systems, there is often a certain requirement for large enough number of agents, which leads to high energy consumption in both control and communication processes. In order to obtain a better system energy efficiency ratio, non-real-time control methods [22–25] can be considered such as sampling control, intermittent control and impulsive control, etc. Among them, event-triggered control [26–29], due to its discontinuous sampling characteristics, is currently one of the important tools for reducing the energy consumption of control systems. As an intelligent sampling control method, the event-triggered control method can naturally achieve good results. But because it is necessary to compensate for the adverse effects brought about by non-ideal factors, a simple event-triggered control method may not be applicable. Therefore, taking the event-triggered control method as the main body, designing appropriate hybrid control method to deal with the issues yielded from applications becomes challenging.

System State Measurability: In theoretical analysis, each agent in a multiagent system has a certain dynamic model [30–33], and its input-output relationship is also observable, which makes any information in the system are easy to access. But in practical applications, due to the limitations of sensors or hardware constraints, it is difficult to accurately measure every state of the dynamic system. In fact, by considering the cost of devices, usually only the input and output can be obtained, and then the internal states of the system can be reconstructed through communication and computational algorithms. Although using state observers [34–38] is a common solution nowadays, it has limitation that the performance highly depends on the certain system model. Therefore, for large-scale multiagent systems in non-ideal environments with multiple uncertainties, designing a high-performance state observer leads to another challenge.

Inner Unknown Dynamics and Uncertainties: In an ideal scenario, the dynamics of the multiagent systems are clear and their evolving behaviors can be precisely characterized. However in practice, the dynamics of agents often exhibit uncertainties or may even remain entirely unknown. Unlike traditional robust control approaches, neural network-based adaptive control methods [39–48] developed in recent years have demonstrated significant potential in addressing such issues. By extensive real-time state data from the system to train neural networks by either online or offline learning rules, these methods can identify the dynamic uncertainties or the system dynamical models. This capability thereby facilitates the design of subsequent control protocols. However, the actual behavior of the neural networks lacks stability for online learning scheme, which depends on the sampled data, tremendous parameters and network structures, which is still a challenge.

External Disturbances, Cyber Attacks and Faults: In the actual operation of the system, external disturbances, information attacks, and failures are several common factors that frequently occur and seriously affect the system. External disturbances [37, 38, 49] generally originate from the limitations of complex environments and physical conditions. Information attacks [50–55] are an important factor to be considered in the security control of current cyber-physical systems. Failures [56–62] often include mechanical structure failures, communication module failures, and computer failures. Existing methods adopt active countermeasure strategies. For example, control parameters with strong robustness are designed so that even if the above situations occur, the system can still maintain normal performance. With the improvement of the methods, the current methods tend to identify and separate these influencing factors online, and then use the identified information to design an adaptive control strategy to compensate for or reduce their negative impacts to a certain extent, thus maintaining the original performance of the system.

1.3 Content Categorization

This book will carry out a series of corresponding work around the above-mentioned types of challenges in the cooperative control of multiagent systems. Each piece of work may not only involve a single problem, but there may also be situations where several types of problems are coupled.

Event Triggered Control Method: The event-triggered control scheme is applied to multiagent systems widely. When the system encounters unknown state time delays, external disturbances, deception attacks, and actuator failures, how to design a reasonable framework for event-triggered control that can not only maintain low communication costs in multiagent systems but also ensure the convergence of the system? These questions are addressed in Chaps. 3–5 and 7.

Dynamic Observer Techniques: Multiple observer techniques are used to address issues such as partial state unmeasurability, external disturbances, and various unknown factors that multiagent systems encounter in complex environments. When the system faces specific environmental adverse factors, how to online approximate system states or unknown nonlinear functions based on observer techniques and design corresponding efficient cooperative control algorithms? The details can be seen in Chaps. 2–4, 6 and 7.

Advanced Adaptive Control Scheme: To address the impacts of unknown dynamics and external disturbances encountered by multiagent systems, advanced intelligent control techniques have been introduced, including online/offline learning of neural networks and fuzzy logic systems. Not limited to single nonlinear function approximation and identification, when the system is affected by multiple adverse factors, how to design online learning rules for neural networks and fuzzy logic systems while integrating multiple techniques to achieve compensation for factors such as unknown dynamics, uncertainties, and external disturbances in multiagent systems? Chapters 2–8 will all give answers to these problems accordingly.

Optimized Control Framework: While standard optimal control design methodologies rely on system models and ideal environments, reinforcement learning approaches can be employed to achieve approximate optimal control in practical non-ideal scenarios. In the presence of multiple uncertainties, how to design rapid and efficient reinforcement learning rules and neural network adaptive control frameworks that not only counteract the effects of non-ideal environments but also enable online optimization of both controller performance and system behavior? Chapter 8 presents an alternative solution.

1.4 Cooperative Control Problems in Multiagent Systems

Multiagent systems (MAS) cooperative control aims to design distributed control protocols that enable multiple agents to achieve a desired collective behavior through local information exchange. These collective behaviors can typically be categorized into several core control problems, which form the basis for the research presented in the subsequent chapters of this book.

- **Consensus Control**: This is the most fundamental problem in MAS cooperative control. The objective is to design a control protocol such that all agents (or followers) in the system reach an agreement on certain quantities of interest, meaning their states converge to a common value. This common value can be a predetermined constant or a function dependent on the initial states of all agents.
- **Tracking Control (Leader-Follower Consensus)**: Tracking control can be viewed as a special form of consensus. In this paradigm, one or more special agents, referred to as leaders, exist within the system, and their state or trajectory serves as the reference for the entire group. The goal for the other agents, known as followers, is to track the leader's state.
  We denote the leader node as v_0. The connections between followers and this leader, termed leader adjacency, can be described by a diagonal matrix $B = \text{diag}(b_1, b_2, \ldots, b_N) \in \mathbb{R}^{N\times N}$. In this matrix, $b_i > 0$ indicates that follower agent v_i can directly receive information from the leader v_0 (b_i can be a binary value, such as 0 or 1, or can represent the weight of the connection). If $b_i = 0$, follower v_i does not directly receive information from the leader. This representation is particularly suitable when the leader's signal is considered an external input or a virtual signal.
  Alternatively, in some analyses, it is more convenient to consider the leader as an integral node in the network. This leads to the concept of an augmented graph, $\bar{\mathcal{G}} = (\bar{\mathcal{V}}, \bar{\mathcal{E}})$, which includes both the leader and all followers. The vertex set for this graph is $\bar{\mathcal{V}} = \mathcal{V} \cup \{v_0\} = \{v_0, v_1, \ldots, v_N\}$, and its edge set $\bar{\mathcal{E}}$ encompasses edges among followers as well as edges from the leader to the followers. The adjacency and Laplacian matrices of this augmented graph can be defined accordingly.
- **Formation Control**: Formation control requires a group of agents to maintain a predefined geometric pattern while moving. Unlike consensus control, which focuses on state agreement, formation control emphasizes maintaining specific relative states among agents. This is essential for applications like drone swarms and satellite constellations.
- **Containment Control**: The containment control problem typically involves multiple leaders and multiple followers. The control objective is to drive all followers to enter and remain within the convex hull formed by the leaders. This problem is significant in applications such as cooperative search, area surveillance, and resource protection.

1.5 Graph Theory Fundamentals

This section introduces the basic graph theory concepts required to describe the interaction topology in multi-agent systems. Consider a system composed of N agents (also called followers), whose information interaction topology can be represented by a directed graph $\mathcal{G} = (\mathcal{V}, \mathcal{E}, A)$.

The fundamental components of this graph include the vertex set, denoted as $\mathcal{V} = \{v_1, v_2, \ldots, v_N\}$, which is the collection of vertices representing the N follower agents in the system. Interactions are defined by the edge set $\mathcal{E} \subseteq \mathcal{V} \times \mathcal{V}$, consisting of directed edges. An edge $(v_j, v_i) \in \mathcal{E}$ indicates that information can be directly transmitted from agent v_j (information sender) to agent v_i (information receiver).

To mathematically capture these connections, we use the adjacency matrix $A = [a_{ij}] \in \mathbb{R}^{N \times N}$. Its elements a_{ij} are defined as follows: if there is an information flow from agent v_j to agent v_i (i.e., $(v_j, v_i) \in \mathcal{E}$), then $a_{ij} > 0$; typically, in unweighted graphs $a_{ij} = 1$, while in weighted graphs a_{ij} represents the connection weight. Conversely, if there is no information flow from v_j to v_i, then $a_{ij} = 0$. In this monograph, we assume that agents do not have self-loops, i.e., $a_{ii} = 0$ for all $i \in \{1, \ldots, N\}$. If $a_{ij} > 0$, agent v_j is termed a neighbor of agent v_i, and the set of such neighbors pointing to v_i is denoted as $\mathcal{N}_i = \{v_j \in \mathcal{V} | (v_j, v_i) \in \mathcal{E}\}$.

For directed graphs, the degree matrix typically concerns the in-degree of nodes. The in-degree of an agent v_i (denoted d_i) is defined as the sum of weights of edges pointing to it: $d_i = \sum_{j=1}^{N} a_{ij}$. Then, the in-degree matrix $D = \mathrm{diag}(d_1, d_2, \ldots, d_N) \in \mathbb{R}^{N \times N}$ is a diagonal matrix whose diagonal elements are these in-degrees. Based on these, the Laplacian matrix of the graph $\mathcal{G}$ is defined as $L = D - A$. Its elements l_{ij} are specifically given by:

$$l_{ij} = \begin{cases} \sum_{k=1, k \neq i}^{N} a_{ik} & \text{if } i = j \\ -a_{ij} & \text{if } i \neq j \text{ and } (v_j, v_i) \in \mathcal{E} \\ 0 & \text{otherwise} \end{cases}$$

Note that $\sum_{k=1, k \neq i}^{N} a_{ik}$ in this expression is d_i. In some literature, if a_{ij} represents a connection from i to j, then L might be defined based on out-degrees, or $L_{ij} = -a_{ij}$ $(i \neq j)$ and $L_{ii} = \sum_{k \neq i} a_{ik}$ (ensuring row sums are zero). Here, we adopt the standard definition based on in-degrees, which is more common in multi-agent cooperative control.

1.6 Important Graph Properties

Several graph properties are crucial for understanding information flow and system behavior. A **directed path** from a node v_j to a node v_i is a finite sequence of distinct vertices $(v_j, v_{p1}, v_{p2}, \ldots, v_{pk}, v_i)$ such that any two consecutive vertices (v_a, v_b) in the sequence form an edge $(v_a, v_b) \in \mathcal{E}$. Furthermore, a graph is said to contain a

directed spanning tree if there exists a node (called a **root**) from which there is a directed path to all other nodes in the graph. In leader-follower systems, if at least one follower can receive information from the leader (i.e., at least one $b_i > 0$), and if directed paths exist from these leader-connected followers to all other followers, then information can propagate throughout the entire follower network. A directed graph $\mathcal{G}$ exhibits **strong connectivity** if for every ordered pair of distinct nodes (v_i, v_j), there is a directed path from v_i to v_j.

The Laplacian matrix L possesses several fundamental properties that are central to the analysis of consensus and stability in multi-agent systems. These properties are frequently used in the stability analysis throughout this book.

- **Zero Eigenvalue and Connectivity**: The Laplacian matrix L is always positive semi-definite. It has at least one eigenvalue equal to zero, with the corresponding eigenvector being the vector of all ones, $\mathbf{1}_N = [1, 1, \ldots, 1]^T$. For a directed graph $\mathcal{G}$, the matrix L has a simple zero eigenvalue if and only if the graph has a directed spanning tree. This property is fundamental for reaching consensus, as the zero eigenvalue corresponds to the steady-state consensus value.
- **Rank of Laplacian**: For an undirected graph, the algebraic multiplicity of the zero eigenvalue equals the number of connected components in the graph. For a graph with a single connected component (i.e., a connected graph), the rank of L is $N - 1$.

1.7 Useful Lemmas

This section provides several fundamental lemmas and inequalities that will be frequently used in the stability analysis and controller design throughout the book.

Lemma 1.1 *(Young's Inequality)*

1. *For any scalars $X, Y \in \mathbb{R}$ and any positive constants $a > 0$, $b > 0$ and $\varepsilon > 0$, the following inequality holds:*

$$|X|^a |Y|^b \leq \frac{a}{a+b} \varepsilon |X|^{a+b} + \frac{b}{a+b} \varepsilon^{-a/b} |Y|^{a+b}. \tag{1.1}$$

2. *For any vectors $x, y \in \mathbb{R}^n$, the inequality has the following useful forms. For any positive constant $\varepsilon > 0$:*

$$x^T y \leq \frac{\varepsilon}{2} \|x\|^2 + \frac{1}{2\varepsilon} \|y\|^2. \tag{1.2}$$

For any positive definite matrix $P \in \mathbb{R}^{n \times n}$:

$$x^T y + y^T x \leq x^T P x + y^T P^{-1} y. \tag{1.3}$$

This inequality, in both its scalar and vector forms, is fundamental for decoupling and bounding terms in Lyapunov function analysis.

Lemma 1.2 *(Property of Hyperbolic Tangent Function) For any given constant $x \in \mathbb{R}$ and any $\epsilon > 0$, the following inequality always holds:*

$$0 \leq |x| - x \tanh\left(\frac{x}{\epsilon}\right) \leq 0.2785\epsilon. \tag{1.4}$$

This specific inequality is useful for handling saturation nonlinearities and event-triggered control design, as seen in later chapters.

Lemma 1.3 *(Uniform Ultimate Boundedness) Consider a continuous positive definite function $V(t)$. If $V(t)$ satisfies the following differential inequality:*

$$\dot{V}(t) \leq -c_1 V(t) + c_2, \tag{1.5}$$

where c_1 and c_2 are positive constants, then for any given $V(t_0)$, $V(t)$ is uniformly ultimately bounded (UUB) and satisfies:

$$V(t) \leq V(t_0)e^{-c_1(t-t_0)} + \frac{c_2}{c_1}. \tag{1.6}$$

This implies that $V(t)$ will eventually converge to the compact set $\Omega_V = \{V \in \mathbb{R} \mid V \leq c_2/c_1\}$. This lemma is a cornerstone for proving the boundedness of all signals in the closed-loop adaptive control systems presented in this book.

Lemma 1.4 *(Properties of Matrix Trace and Kronecker Product) The following properties hold for any matrices A, B, C of appropriate dimensions:*

1. *Trace Property: $Tr(ABC) = Tr(BCA) = Tr(CAB)$.*
2. *Kronecker Product Property: $(A \otimes B)(C \otimes D) = (AC \otimes BD)$.*

These properties are essential for the analysis of stochastic systems (Chap. 8) and systems with observer-based designs (Chap. 7) , respectively.

1.8 Notaitions

Sets and Spaces

$\mathbb{R}$ Set of real numbers
$\mathbb{R}^n$ n-dimensional real Euclidean space
$\mathbb{R}^{n \times m}$ Set of $n \times m$ real matrices
$\mathbb{N}$ Set of non-negative integers (assuming $0, 1, 2, \ldots$)
$\mathbb{N}^+$ Set of positive integers (assuming $1, 2, 3, \ldots$)

Matrix and Vector Operations

A^T, x^T Transpose of matrix A or vector x
A^{-1} Inverse of matrix A (if A is invertible)
I, I_n Identity matrix (of dimension n if specified)
0 Zero vector or matrix (dimensions clear from context)
$\text{diag}(d_1, \ldots, d_n)$ Diagonal matrix with diagonal entries $d_1, \ldots, d_n$
$|| \cdot ||$ Norm of a vector (typically Euclidean norm) or a matrix (e.g., Frobenius or induced norm)
$\lambda_{min}(A)$ Minimum eigenvalue of a symmetric matrix A
$\lambda_{max}(A)$ Maximum eigenvalue of a symmetric matrix A
$A > (\geq) 0$ Matrix A is positive (semi-)definite
$\text{col}\{v_1, \ldots, v_n\}$ Column vector formed by stacking elements $v_1, \ldots, v_n$

Graph Theory

$\mathcal{G} = (\mathcal{V}, \mathcal{E}, A)$ Directed graph with vertex set $\mathcal{V}$, edge set $\mathcal{E}$, and adjacency matrix A
$\mathcal{V} = \{v_1, \ldots, v_N\}$ Set of vertices (representing agents)
$\mathcal{E} \subseteq \mathcal{V} \times \mathcal{V}$ Set of directed edges
$A = [a_{ij}]$ Adjacency matrix of $\mathcal{G}$
$D = \text{diag}(d_1, \ldots, d_N)$ In-degree matrix of $\mathcal{G}$ ($d_i = \sum_{j=1}^{N} a_{ij}$)
$L = D - A$ Laplacian matrix of $\mathcal{G}$
$\mathcal{N}_i = \{v_j \in \mathcal{V} | (v_j, v_i) \in \mathcal{E}\}$ Set of neighbors of agent v_i (information senders to v_i)
v_0 Leader node (in leader-follower systems)
$B = \text{diag}(b_1, \ldots, b_N)$ Leader adjacency matrix; $b_i > 0$ if follower i receives from leader

System Variables and Dynamics

N Number of follower agents
$x_i \in \mathbb{R}^{n_i}$ State vector of the i-th agent
$\bar{x}_{i,k} = [x_{i,1}^T, \ldots, x_{i,k}^T]^T$ State vector of agent i up to its k-th component/subsystem
u_i Control input to the i-th agent
y_i Output of the i-th agent
y_r Desired reference trajectory or leader's output
$f_{i,k}(\cdot), h_{i,j}(\cdot), q_r(\cdot)$ Unknown or known smooth nonlinear functions in system dynamics
$d_{i,k}(t), \Upsilon_{i,k}(t), r_i(t)$ External disturbances, compounded disturbances, or fault signals
$\tau_{i,j}(t)$ Time-varying state time-delay
$\varrho_{i,j}, \varrho_i^u$ Deception attack signals on sensors or actuators

Estimation, Errors, NNs and FLSs

$\hat{z}$ Estimate of a variable z (e.g., $\hat{x}_i, \hat{W}, \hat{\vartheta}_i, \hat{\Upsilon}_i$)
$\tilde{z} = z - \hat{z}$ or $\hat{z} - z$ Estimation error of z (e.g., $\tilde{x}_i, \tilde{W}, \tilde{\vartheta}_i$)

$e_{i,k}, z_{i,k}, \nu_{i,k}, \Xi_k, \varsigma_i, \zeta_{i,j}$ Error variables or transformed error variables in control design

W^*, Θ^* Ideal (optimal) weight vector/matrix for NNs or FLSs

$S(Z), \Lambda(Q), \psi(Z), \phi(X)$ Basis function vector for NNs or FLSs

$\epsilon(Z), \delta(Z), \omega_{i,k}$ Approximation error of NNs or FLSs

$\vartheta_i, \phi_i, \Theta_i$ Parameters related to NN weight bounds or adaptive laws in general

Control Design and Lyapunov Theory

$V(\cdot), V_{i,k}(\cdot)$ Lyapunov function or Lyapunov function candidate

$\alpha_{i,k}$ Virtual control law

$N(\Phi)$ Nussbaum-type function

P_i, Q_i Positive definite matrices in Lyapunov equations (e.g., $A_i^T P_i + P_i A_i \leq -2Q_i$)

Event-Triggered Control (ETC)

t_k^i k-th triggering instant for agent i

$\epsilon_i(t)$ or $e_i(t)$ or $D(t)$ Measurement error for event-triggering condition

$m_{1,i}, m_{2,i}, b_i, \delta_i, \mu_i, \bar{m}_i, \aleph_i$ Positive design parameters for event-triggered mechanisms

Other Common Mathematical Symbols

$|\cdot|$ Absolute value of a scalar

$\mathrm{sign}(x)$ Sign function of x: 1 if $x > 0$, -1 if $x < 0$, 0 if $x = 0$

$\tanh(x)$ Hyperbolic tangent function of x

$x^{(r)}$ r-th order time derivative of x

$\mathcal{L}V$ Infinitesimal generator of $V(x)$ (for stochastic systems)

$\triangleq$ or $:=$ Is defined as

$\forall$ For all

$\exists$ There exists

$\in$ Belongs to

∞ Infinity

$\sum$ Summation operator

$\int$ Integral operator

lim Limit operator

$\sup(\cdot)$ Supremum (least upper bound)

$\inf(\cdot)$ Infimum (greatest lower bound)

$E[\cdot]$ Expectation of a random variable.

References

1. Chen B, Cheng H (2010) A review of the applications of agent technology in traffic and transportation systems. IEEE Trans Intell Transp Syst 22(2):485–497
2. Singh V, Sharma SK (2024) Applying hybrid approach of AHP, DEMATEL, and TOPSIS to study the impact of multiagent technology on the performance of manufacturing organizations. IEEE Trans Eng Manage 71(5):9299–9313
3. Zhou H, Aral A, Brandić I, Erol-Kantarci M (2022) Multiagent Bayesian deep reinforcement learning for microgrid energy management under communication failures. IEEE Internet Things J 9(14):11685–11698
4. Rao S, Ghose D (2014) Sliding mode control-based autopilots for leaderless consensus of unmanned aerial vehicles. IEEE Trans Control Syst Technol 22(5):1964–1972
5. Jin X, Wang S, Qin J, Zheng W, Kang Y (2018) Adaptive fault-tolerant consensus for a class of uncertain nonlinear second-order multi-agent systems with circuit implementation. IEEE Trans Circuits Syst I Regul Pap 65(7):2243–2255
6. Liang H, Chang Z, Pan Y (2024) Dual-event-triggered intelligence security control for multiagent systems against DoS attacks with applications in mobile robot systems. IEEE Trans Artif Intell 5(2):916–924
7. Tanner HG, Jadbabaie A, Pappas GJ (2007) Flocking in fixed and switching networks. IEEE Trans Autom Control 52(5):863–868
8. Meng Z, Dimarogonas DV, Johansson KH (2014) Leader-follower coordinated tracking of multiple heterogeneous Lagrange systems using continuous control. IEEE Trans Rob 30(3):739–745
9. Xiao Q, Lewis FL, Zeng Z (2019) Containment control for multiagent systems under two intermittent control schemes. IEEE Trans Autom Control 64(3):1236–1243
10. Wang Y, Song Y, Ren W (2018) Distributed adaptive finite-time approach for formation-containment control of networked nonlinear systems under directed topology. IEEE Trans Neural Netw Learn Syst 29(7):3164–3175
11. Wang X, Li S, Yu X, Yang J (2017) Distributed active anti-disturbance consensus for leader-follower higher-order multi-agent systems with mismatched disturbances. IEEE Trans Autom Control 62(11):5795–5801
12. Deng C, Wen C (2020) Distributed resilient observer-based fault-tolerant control for heterogeneous multiagent systems under actuator faults and DoS attacks. IEEE Trans Control Netw Syst 7(3):1308–1318
13. Li W, Sader M, Zhu Z, Liu Z, Chen Z (2023) Event triggered fault-tolerant secure containment control of multi-agent systems through impulsive scheme. Inf Sci 622:1128–1140
14. You L, Jiang X, Zhang X, Yan H, Huang T (2022) Distributed edge event-triggered control of nonlinear fuzzy multiagent systems with saturation constraint hybrid impulsive protocols. IEEE Trans Fuzzy Syst 30(10):4142–4151
15. Zhu F, Du W (2024) Observer-based consensus of multi-agent systems under odd distributed impulsive control protocol. Appl Math Comput 466:128474
16. Yang Z, Ling G, Ge M (2024) Secure impulsive tracking of multi-agent systems with directed hypergraph topologies against hybrid deception attacks. Neural Netw 180:106691
17. Li Y, Shi L, Chen T (2017) Detection against linear deception attacks on multi-sensor remote state estimation. IEEE Trans Control Netw Syst 5:846–856
18. Jin X, Haddad WM, Yucelen T (2017) An adaptive control architecture for mitigating sensor and actuator attacks in cyber-physical systems. IEEE Trans Autom Control 62:6058–6064
19. Yoo SJ (2019) Neural-network-based adaptive resilient dynamic surface control against unknown deception attacks of uncertain nonlinear time-delay cyberphysical systems. IEEE Trans Neural Netw Learn Syst 31:4341–4353
20. Gao R, Huang J, Su X, Zhao L (2022) K-filter-based adaptive output feedback control for high-order nonlinear systems subject to actuator and sensor attacks. Int J Robust Nonlinear Control 32:3469–3484

21. Meng M, Xiao G, Li B (2020) Adaptive consensus for heterogeneous multi-agent systems under sensor and actuator attacks. Automatica 122:109242
22. Wang F, Liu Z, Chen Z (2022) Sampled-hold-based consensus control for second-order multiagent systems under aperiodically intermittent communication. IEEE Trans Circuits Syst I Regul Pap 69(9):3794–3803
23. Wu W, Tong S (2023) Fuzzy adaptive consensus control for nonlinear multiagent systems with intermittent actuator faults. IEEE Trans Cybern 53(5):2969–2979
24. Han Y, Xiao Q, Zeng Z (2023) Neuroadaptive impulsive control on consensus of uncertain multiagent systems using continuous and sampled information. IEEE Trans Neural Netw Learn Syst 34(8):5086–5098
25. Guan Z, Han G, Li J, He D, Feng G (2015) Impulsive multiconsensus of second-order multiagent networks using sampled position data. IEEE Trans Neural Netw Learn Syst 26(11):2678–2688
26. Wang A, Liu L, Qiu J, Feng G (2019) Event-triggered robust adaptive fuzzy control for a class of nonlinear systems. IEEE Trans Fuzzy Syst 27(8):1648–1658
27. Li Y, Yang G (2018) Adaptive neural control of pure-feedback nonlinear systems with event-triggered communications. IEEE Trans Neural Netw Learn Syst 29(12):6242–6251
28. Zhou Q, Wang W, Ma H, Li H (2021) Event-triggered fuzzy adaptive containment control for nonlinear multiagent systems with unknown Bouc-Wen hysteresis input. IEEE Trans Fuzzy Syst 29(3):731–741
29. Sun K, Qiu J, Karimi HR, Fu Y (2021) Event-triggered robust fuzzy adaptive finite-time control of nonlinear systems with prescribed performance. IEEE Trans Fuzzy Syst 29(5):1460–1471
30. Lu KF, Xia YQ (2013) Adaptive attitude tracking control for rigid spacecraft with finite-time convergence. Automatica 49(12):3591–3599
31. Wang CL, Guo L, Wen CY, Hu QL, Qiao JZ (2020) Event-triggered adaptive attitude tracking control for spacecraft with unknown actuator faults. IEEE Trans Ind Electron 67(3):2241–2250
32. Luo B, Wu HN, Li HX (2015) Adaptive optimal control of highly dissipative nonlinear spatially distributed processes with neuro-dynamic programming. IEEE Trans Neural Netw Learn Syst 26(4):684–696
33. Xu Y, Lu RQ, Shi P, Li HY, Xie SL (2018) Finite-time distributed state estimation over sensor networks with round-robin protocol and fading channels. IEEE Trans Cybern 48(1):336–345
34. Chen CLP, Wen GX, Liu YJ, Liu Z (2016) Observer-based adaptive backstepping consensus tracking control for high-order nonlinear semi-strict-feedback multiagent systems. IEEE Trans Cybern 46(7):1591–1601
35. Yang Y, Xu HW, Yue D (2019) Observer-based distributed secure consensus control of a class of linear multi-agent systems subject to random attacks. IEEE Trans Circuits Syst I Regul Pap 66(8):3089–3099
36. Chen B, Lin C, Liu XP, Liu KF (2016) Observer-based adaptive fuzzy control for a class of nonlinear delayed systems. IEEE Trans Syst Man Cybern Syst 46(1):27–36
37. Chen M, Ge SS (2015) Adaptive neural output feedback control of uncertain nonlinear systems with unknown hysteresis using disturbance observer. IEEE Trans Ind Electron 62(12):7706–7716
38. Min HF, Xu SY, Ma Q, Zhang BY, Zhang ZQ (2017) Composite-observer-based output-feedback control for nonlinear time-delay systems with input saturation and its application. IEEE Trans Ind Electron 65(7):5856–5863
39. Chen CLP, Wen G, Liu Y, Wang F (2014) Adaptive consensus control for a class of nonlinear multiagent time-delay systems using neural networks. IEEE Trans Neural Netw Learn Syst 25(6):1217–1226
40. Wen G, Chen CLP, Liu Y, Liu Z (2015) Neural-network-based adaptive leader-following consensus control for second-order nonlinear multi-agent systems. IET Control Theory Appl 9(13):1927–1934
41. Chen B, Zhang H, Liu X, Lin C (2018) Neural observer and adaptive neural control design for a class of nonlinear systems. IEEE Trans Neural Netw Learn Syst 29(9):4261–4271
42. Liu Y, Ma L, Liu L, Tong S, Chen CLP (2020) Adaptive neural network learning controller design for a class of nonlinear systems with time-varying state constraints. IEEE Trans Neural Netw Learn Syst 31(1):66–75

43. Hou Z, Cheng L, Tan M (2009) Decentralized robust adaptive control for the multiagent system consensus problem using neural networks. IEEE Trans Syst Man Cybern B Cybern 39(3):636–647
44. Chen G, Song Y (2014) Cooperative tracking control of nonlinear multiagent systems using self-structuring neural networks. IEEE Trans Neural Netw Learn Syst 25(8):1496–1507
45. Wen G, Yu W, Li Z, Yu X, Cao J (2017) Neuro-adaptive consensus tracking of multiagent systems with a high-dimensional leader. IEEE Trans Cybern 47(7):1730–1742
46. Qin J, Zhang G, Zheng W, Kang Y (2019) Neural network-based adaptive consensus control for a class of nonaffine nonlinear multiagent systems with actuator faults. IEEE Trans Neural Netw Learn Syst 30(12):3633–3644
47. Zhao K, Song Y, Shen Z (2018) Neuroadaptive fault-tolerant control of nonlinear systems under output constraints and actuation faults. IEEE Trans Neural Netw Learn Syst 29(2):286–298
48. Ding D, Wang Z, Han Q (2020) Neural-network-based consensus control for multiagent systems with input constraints: the event-triggered case. IEEE Trans Cybern 50(8):3719–3730
49. Ma H, Liang HJ, Zhou Q, Ahn CK (2019) Adaptive dynamic surface control design for uncertain nonlinear strict-feedback systems with unknown control direction and disturbance. IEEE Trans Syst Man Cybern Syst 49(3):506–515
50. An L, Yang G (2018) Improved adaptive resilient control against sensor and actuator attacks. Inf Sci 423:145–156
51. Wang W, Han Z, Liu K, Li J (2021) Distributed adaptive resilient formation control of uncertain nonholonomic mobile robots under deception attacks. IEEE Trans Circuits Syst I Regul Pap 68(9):3822–3835
52. Yoo SJ (2020) Neural-network-based adaptive resilient dynamic surface control against unknown deception attacks of uncertain nonlinear time-delay cyberphysical systems. IEEE Trans Neural Netw Learn Syst 31(11):4341–4353
53. Zhu Y, Zheng WX (2020) Observer-based control for cyber-physical systems with periodic DoS attacks via a cyclic switching strategy. IEEE Trans Autom Control 65(9):3714–3721
54. Zhu M, Martínez S (2014) On the performance analysis of resilient networked control systems under replay attacks. IEEE Trans Autom Control 59(3):804–808
55. Jin X, Haddad WM, Yucelen T (2017) An adaptive control architecture for mitigating sensor and actuator attacks in cyber-physical systems. IEEE Trans Autom Control 62(11):6058–6064
56. Jin X, Wang S, Qin J, Zheng WX, Kang Y (2018) Adaptive fault-tolerant consensus for a class of uncertain nonlinear second-order multi-agent systems with circuit implementation. IEEE Trans Circuits Syst I Regul Pap 65(7):2243–2255
57. Xu Y, Sun J, Wang G, Wu Z (2021) Dynamic triggering mechanisms for distributed adaptive synchronization control and its application to circuit systems. IEEE Trans Circuits Syst I Regul Pap 68(5):2246–2256
58. Liu J, Yang M, Tian E, Cao J, Fei S (2019) Event-based security control for state-dependent uncertain systems under hybrid-attacks and its application to electronic circuits. IEEE Trans Circuits Syst I Regul Pap 66(12):4817–4828
59. Chen S, Ho DWC, Li L, Liu M (2015) Fault-tolerant consensus of multi-agent system with distributed adaptive protocol. IEEE Trans Cybern 45(10):2142–2155
60. Shen Q, Yue C, Goh CH, Wang D (2019) Active fault-tolerant control system design for spacecraft attitude maneuvers with actuator saturation and faults. IEEE Trans Ind Electron 66(5):3763–3772
61. Wang X, Yang GH (2020) Fault-tolerant consensus tracking control for linear multiagent systems under switching directed network. IEEE Trans Cybern 50(5):1921–1930
62. Ma HJ, Yang GH (2016) Adaptive fault-tolerant control of cooperative heterogeneous systems with actuator faults and unreliable interconnections. IEEE Trans Autom Control 61(11):3240–3255

Chapter 2
Neural Network-Based Adaptive Tracking Control for Nonlinear Multiagent Systems: The Observer Case

Traditional periodic control updates may result in unnecessary resource consumption, while event-triggered schemes improve efficiency by activating controllers only upon specific conditions, thereby minimizing communication load. Nevertheless, trajectory tracking control where follower agents must match the state of leader remains a key challenge for nonlinear systems with unknown dynamics, even with the integration of neural networks and backstepping methods [1, 2].

Current approaches typically require complete state information [3, 4], ignoring key practical issues such as partial observability and unknown disturbances [5, 6]. Although event-triggered control reduces resource usage [7, 8], its effectiveness in nonlinear multiagent systems under such constraints remains unverified, thus motivating our work.

Extensive research has been conducted on state observer design for nonlinear systems with immeasurable states, particularly through neural network-based observers and adaptive tracking strategies for multiagent and output-only systems. The backstepping technique, widely adopted for controller synthesis [9, 10], has been extended to accommodate practical challenges such as actuator failures, input quantization, state-dependent constraints, discrete-time dynamics, and disturbance attenuation. Recent developments include low-order system tracking control [11, 12], reinforcement learning-enhanced backstepping, and adaptive disturbance compensation methods [13]. Additionally, observer-based frameworks have advanced by unifying state and disturbance estimation for broader nonlinear system applications.

To alleviate communication burdens in disturbed nonlinear multiagent systems, event-triggered mechanisms have been implemented, paralleling approaches in general nonlinear and networked systems. Although fuzzy neural networks have been utilized in constrained robotic systems, the problem of tracking control for nonlinear multi-agent systems subject to both immeasurable states and unknown disturbances has received limited attention.

X. Wang et al., *Optimal Adaptive Cooperative Control for Nonlinear Multiagent Systems*,
https://doi.org/10.1007/978-981-95-2964-3_2

2.1 Preliminaries

2.1.1 Preliminaries and Problem Formulation

This section introduces key graph-theoretic terminologies, presents the system model, and describes the RBF neural network framework. Furthermore, we develop both state observers and composite disturbance observers for each follower agent.

2.1.2 System Formulation

Consider a nonlinear multiagent system comprising N follower agents (denoted as agents 1 through N). The dynamics of each follower agent follow an n_i-order nonstrict-feedback structure described by:

$$\begin{aligned}\dot{x}_{i,k} &= f_{i,k}(x_i) + x_{i,k+1} + d_{i,k}(t),\\ \dot{x}_{i,n_i} &= u_i + f_{i,n_i}(x_i) + d_{i,n_i}(t),\\ y_i &= x_{i,1},\end{aligned} \tag{2.1}$$

where $i = 1, 2, \ldots, N$, $k = 1, 2, \ldots, n_i - 1$, the state vector of the i-th agent is denoted by $x_i = \mathrm{col}\{x_{i,1}, s_{i,2}, \ldots, s_{i,n_i}\} \in \mathbb{R}^{n_i}$, t stands for time, $y_i \in \mathbb{R}$ and $u_i \in \mathbb{R}$ refer to as the output and control input of the i-th follower, respectively. The intrinsic nonlinear feature is represented by $f_{i,k}(x_i)$, which is an unknown smooth function. The external disturbance is expressed as the unknown bounded function $d_{i,k}(t)$.

2.1.3 Radial Basis Function Neural Networks

It is clear that given smooth function $h(\mathcal{Z}) : \mathbb{R}^q \to \mathbb{R}$ defined on a compact set $\Omega_{\mathcal{Z}}$ and a positive constant $\delta > 0$, there exist the RBF NNs $W^{\mathrm{T}} S(\mathcal{Z})$ satisfying

$$h(\mathcal{Z}) = W^{\mathrm{T}} S(\mathcal{Z}) + \delta(\mathcal{Z}) \ \forall \mathcal{Z} \in \Omega_{\mathcal{Z}} \subset \mathbb{R}^q, \tag{2.2}$$

where $S(\mathcal{Z}) = [S_1(\mathcal{Z}), S_2(\mathcal{Z}), \ldots, S_l(\mathcal{Z})]^{\mathrm{T}}$ denotes the Gaussian basis function vector, $l > 1$ represents the number of neuron nodes, $\delta(\mathcal{Z})$ is the estimate error satisfying $|\delta(\mathcal{Z})| < \delta$. Particularly, the gaussian basis function is selected as

$$S(\mathcal{Z}) = \exp\left[\frac{(\mathcal{Z} - v_i)^{\mathrm{T}}(\mathcal{Z} - v_i)}{\omega_i^2}\right], \tag{2.3}$$

where $\nu_i = [\nu_{i1}, \nu_{i2}, \ldots, c_{iq}]^{\mathrm{T}} \in \mathbb{R}^q$, $i = 1, 2, \ldots, l$ is the center of the receptive field, ω_i denotes the width of the Gaussian function. $W^* \in \mathbb{R}^l$ is referred to as an ideal weight vector defined as follows

$$W^* = \arg\min_{W \in \mathbb{R}^l} \left\{ \sup_{\mathcal{Z} \in \Omega} |h(\mathcal{Z}) - W^{\mathrm{T}} S(\mathcal{Z})| \right\}. \tag{2.4}$$

2.1.4 State Observer and Disturbance Observer

In this sequel, to address the impacts of both immeasurable state variables and unknown external disturbances, we design a state observer and a composite disturbance observer. Prior to their design, the following notation is established.

Originated from the propositions in [14, 15], the following adaptive control parameters are proposed as

$$\vartheta_i = \max\left\{ \|W_{i,\xi}^*\|^2, \xi = 1, \ldots, n_i \right\}, i = 1, \ldots, N,$$

in which $W_{i,\xi}^*$ stands for the ideal constant weight vector and the estimate of ϑ_i is denoted by $\hat{\vartheta}_i$.

Define $\Delta f_{i,k}(\hat{x}_i) = f_{i,k}(x_i) - f_{i,k}(\hat{x}_i)$, where $\hat{x}_i$ is the estimate of x_i. In this context, the RBF NNs $W_{i,k,1}^{*\mathrm{T}} S_{i,k}(\hat{x}_i)$ will be deployed to approximate the unknown nonlinear function $\lambda_{i,k} f_{i,k}(\hat{x}_i)$ with $\lambda_{i,k} > 0$, one has

$$\begin{aligned}
\dot{x}_{i,k} &= \frac{1}{\lambda_{i,k}} W_{i,k,1}^{*\mathrm{T}} S_{i,k}(\hat{x}_i) + x_{i,k+1} + \Upsilon_{i,k}(t), \\
\dot{x}_{i,n_i} &= u_i + \frac{1}{\lambda_{i,n_i}} W_{i,n_i,1}^{*\mathrm{T}} S_{i,n_i}(\hat{x}_i) + \Upsilon_{i,n_i}(t), \\
y_i &= x_{i,1},
\end{aligned} \tag{2.5}$$

where $\Upsilon_{i,k}(t) = \Delta f_{i,k}(\hat{x}_i) + \frac{1}{\lambda_{i,k}} \delta_{i,k}(\hat{x}_i) + d_{i,k}$ with $\delta_{i,k}(\hat{s}_i)$ satisfying $|\delta_{i,k}(\hat{x}_i)| < \bar{\delta}_{i,k}$, $k = 1, 2, \ldots, n_i$. Actually, the term $\Upsilon_{i,k}$ can be regarded as a composite disturbance.

Inspired by the propositions in [5, 6], the state observer of i-th agent is

$$\begin{aligned}
\dot{\hat{x}}_{i,k} &= \frac{1}{\lambda_{i,k}} \hat{W}_{i,k,1}^{*\mathrm{T}} S_{i,k}(\hat{x}_i) + g_{i,k}(y_i - \hat{x}_{i,1}) + \hat{x}_{i,k+1} + \hat{\Upsilon}_{i,k}, \\
\dot{\hat{x}}_{i,n_i} &= u_i + \frac{1}{\lambda_{i,n_i}} \hat{W}_{i,n_i,1}^{*\mathrm{T}} S_{i,n_i}(\hat{x}_i) + g_{i,n_i}(y_i - \hat{x}_{i,1}) + \hat{\Upsilon}_{i,n_i}, \\
\hat{y}_i &= \hat{x}_{i,1},
\end{aligned} \tag{2.6}$$

where $g_{i,k}$ refers to as the state observer parameter, $\hat{W}_{i,k,1}$ and $\hat{\Upsilon}_{i,k}$ are interpreted as the estimate of the $W_{i,k,1}$ and $\Upsilon_{i,k}$, respectively.

For ease of description, let

$$\tilde{x}_i = \left[\tilde{x}_{i,1}, \ldots, \tilde{x}_{i,n_i}\right]^{\mathrm{T}} = \left[\left(x_{i,1} - \hat{x}_{i,1}\right), \ldots, \left(x_{i,n_i} - \hat{x}_{i,n_i}\right)\right]^{\mathrm{T}},$$
$$\tilde{\eta}_{i,j} = \frac{W_{i,j,1}^{*\mathrm{T}} S_{i,j}(\hat{x}_i) - \hat{W}_{i,j,1}^{*\mathrm{T}} S_{i,j}(\hat{x}_i)}{\lambda_{i,j}}, \tilde{\eta}_i = \left[\tilde{\eta}_{i,1}, \ldots, \tilde{\eta}_{i,n_i}\right]^{\mathrm{T}},$$
$$\tilde{\Upsilon}_i = \left[\left(\Upsilon_{i,1} - \hat{\Upsilon}_{i,1}\right), \ldots, \left(\Upsilon_{i,n_i} - \hat{\Upsilon}_{i,n_i}\right)\right]^{\mathrm{T}},$$

From (2.5) and (2.6), one obtains

$$\dot{\tilde{x}}_i(t) = A_i \tilde{x}_i(t) + \tilde{\eta}_i + \tilde{\Upsilon}_i, \tag{2.7}$$

where $A_i = \begin{bmatrix} -g_{i,1} & & \\ -g_{i,2} & I_{(n_i-1)\times(n_i-1)} & \\ \vdots & & \\ -g_{i,n_i} & 0 & \cdots 0 \end{bmatrix}$, and $\Lambda_i = \left[g_{i,1}, \ldots, g_{i,n_i}\right]^{\mathrm{T}}$ should be properly chosen such that A_i is a Hurwitz matrix. In other words, considering a symmetric positive matrix $P_i = P_i^{\mathrm{T}} > 0$, there exists matrix $Q_i = Q_i^{\mathrm{T}} > 0$ satisfying

$$A_i^{\mathrm{T}} P_i + P_i A_i \leq -2Q_i. \tag{2.8}$$

Meanwhile, this section proposes a composite disturbance observer to estimate the unknown compounded disturbance $\Upsilon_{i,k}$, compensating for the mismatched disturbance.

$$\begin{aligned} \hat{\Upsilon}_i &= \hat{s}_i + \lambda_i \hat{x}_i, \\ \dot{\hat{s}}_i &= \hat{W}_i^{*\mathrm{T}} S_i(\hat{x}_i) - \lambda_i \left(\hat{x}_{i+1} + \hat{\Upsilon}_i\right), \end{aligned} \tag{2.9}$$

where $i = 1, 2, \ldots, n_i$, $\hat{x}_{i,n_i+1} = u_i$.

Define the disturbance estimate error $\tilde{s}_{i,k} = s_{i,k} - \hat{s}_{i,k}$ with $s_{i,k} = \Upsilon_{i,k} + \lambda_{i,k} x_{i,k}$, one obtains

$$\begin{aligned} \dot{\tilde{s}}_{i,k} = \dot{\Upsilon}_{i,k} - \lambda_{i,k} &\left(\frac{\tilde{W}_{i,k,1}^{*\mathrm{T}} S_{i,k}(\hat{x}_i)}{\lambda_{i,k}} + \tilde{x}_{i,k+1}\right) \\ &- \lambda_{i,k} \left(\tilde{s}_{i,k} + \lambda_{i,k} \tilde{x}_{i,k}\right), \end{aligned} \tag{2.10}$$

where $\tilde{W}_{i,k,1}^{*} = W_{i,k,1}^{*} - \hat{W}_{i,k,1}^{*}$, $\tilde{x}_{i,n_i+1} = 0$.

Remark 2.1 Numerous adaptive tracking protocols have been proposed for nonlinear multi-agent systems in recent literature. Nevertheless, the majority of these results implicitly assume prior knowledge of either the system states or external disturbances, significantly constraining their real-world applicability. Departing from these existing approaches, this chapter considers the more realistic scenario where both state variables and external disturbances are immeasurable and unknown. To address this, we develop a state observer alongside a composite disturbance observer. A key advantage of the proposed composite observer is its simultaneous consideration of estimation errors $\frac{1}{\lambda_{i,k}}\delta_{i,k}(x_i)$ and $\Delta f_{i,k}(\hat{x}_i)$, which enhances tracking performance compared to conventional methods.

Remark 2.2 The construction of a disturbance observer represents the primary approach for handling unknown disturbances, as demonstrated by extensive existing research. Compared with the methods proposed in [14, 16], this work advances the disturbance observer design by considering scenarios where state variables are also unmeasurable. Furthermore, the proposed scheme is extended to an event-triggered framework, significantly enhancing the novelty and applicability of our results.

A neural network-based distributed consensus protocol u_i is developed in this chapter for multi-agent systems with immeasurable state variables and unknown external disturbances, utilizing the backstepping method. Based on [3], the graph-based error surface $e_{i,1}$ is defined as

$$\begin{aligned} e_{i,1} &= \sum_{j\in\mathcal{N}_i} a_{i,j}\left(\hat{y}_i - \hat{y}_j\right) + b_i(\hat{y}_i - y_r), \\ e_{i,\xi} &= \hat{x}_{i,\xi} - \alpha_{i,\xi-1}, \end{aligned} \tag{2.11}$$

where $i = 1, 2, \ldots, N$, $\xi = 2, \ldots, n_i$, and $\alpha_{i,\xi-1}$ denotes the virtual control law to be specified subsequently and y_r stands for the desired tracking target. To end this section, some necessary assumptions about the graph topology and the disturbance as well as the lemma shall be presented.

Assumption 2.1 Consider the nonlinear multiagent systems (2.1), there exists a positive constant $\bar{d}_{i,k}$ such that $|\dot{\Upsilon}_{i,k}| < \bar{d}_{i,k}$.

Assumption 2.2 The augmented graph $\bar{\mathcal{G}}$ contains a spanning tree with the root node being the leader.

Remark 2.3 Recently, there has been a trend in regarding the nonlinear multiagent systems with external disturbance as a control target owing to their vast applications, where the natural idea is to suppose that external disturbance satisfies some assumptions. For example, pioneering yet excellent results have been pursued in [16, 17]. The control protocols were conducted based on the assumption that the disturbance was bounded in [18]. The disturbance and its derivative are both required bounded in [16, 17]. Assumption 2.1 in this chapter, however, only requires that the derivative of the disturbance is bounded, which is much milder and easy to be satisfied.

Remark 2.4 It should be pointed out that if the assumption 2.2 is not satisfied, there exist a set $\mathcal{V}_k$ containing k $(1 \leq k \leq N)$ followers which do not have directed paths from the leader. That is to say, the set of k followers can not receive information from the followers in $\bar{\mathcal{V}}_k = \mathcal{V} \setminus \mathcal{V}_k$. Obviously, there must be some isolated followers or isolated group in the set $\mathcal{V}_k$. The tracking performance can not achieved no matter what case.

Lemma 2.1 *Sun et al. [19] Consider* $S(\chi_h) = \left[S_1(\chi_h), \ldots, S_p(\chi_h)\right]^{\mathrm{T}}$, *where* $\chi_h = [\chi_1, \ldots, \chi_h]^{\mathrm{T}}$ *denotes the basis function vector of RBF NNs. The following inequality is well established for any positive scalar* $q \leq h$.

$$\|S(\chi_h)\|^2 \leq \|S(\chi_q)\|^2.$$

2.2 Distributed Consensus Protocols Designing

In this section, a distributed consensus protocol u_i is proposed to synchronize the follower outputs y_i with the leader's output y_r via the back-stepping method.

Step 1: Taking the derivative of $e_{i,1}$ along (2.6), one has

$$\begin{aligned}
\dot{e}_{i,1} &= (d_i + b_i)\,\dot{\hat{y}}_i - \sum_{j\in\mathcal{N}_i} a_{ij}\dot{\hat{y}}_j - b_i\dot{y}_r \\
&= (d_i + b_i)\left(\frac{\hat{W}_{i,1,1}^{*\mathrm{T}} S_{i,1}(\hat{x}_i)}{\lambda_{i,1}} + g_{i,1}\tilde{x}_{i,1} + e_{i,2} + \alpha_{i,1} + \hat{\Upsilon}_{i,1}\right) \\
&- \sum_{j\in\mathcal{N}_i} a_{ij}\left(\frac{\hat{W}_{j,1,1}^{*\mathrm{T}} S_{j,1}(\hat{x}_j)}{\lambda_{j,1}} + g_{j,1}\tilde{x}_{j,1} + \hat{x}_{j,2} + \hat{\Upsilon}_{j,1}\right) - b_i\dot{y}_r.
\end{aligned} \tag{2.12}$$

Define the Lyapunov function candidate $V_{i,1}$:

$$V_{i,1} = \frac{1}{2}\left[V_{i,0} + e_{i,1}^2 + \tilde{s}_{i,1}^2 + \frac{\tilde{W}_{i,1,1}^{*\mathrm{T}}\tilde{W}_{i,1,1}^{*}}{\gamma_{i,1}}\right], \tag{2.13}$$

where $V_{i,0} = \tilde{x}_i^{\mathrm{T}} P_i \tilde{x}_i + \frac{\tilde{\vartheta}_i^2}{\varrho_i}$ with $\tilde{\vartheta}_i = \vartheta_i - \hat{\vartheta}_i$, ϱ_i and $\gamma_{i,1}$ are the corresponding design constants.

The derivative of $V_{i,1}$ along (2.7), (2.10) and (2.12) is expressed as

$$\begin{aligned}
\dot{V}_{i,1} &= \tilde{x}_i^{\mathrm{T}} P_i \dot{\tilde{x}}_i + e_{i1}\dot{e}_{i1} + \tilde{s}_{i,1}\dot{\tilde{s}}_{i,1} - \frac{\tilde{\vartheta}_i}{\varrho_i}\dot{\hat{\vartheta}}_i - \frac{\tilde{W}_{i,1,1}^{*}}{\gamma_{i,1}}\dot{\hat{W}}_{i,1,1}^{*} \\
&= \frac{1}{2}\tilde{x}_i^{\mathrm{T}}\left[A_i^T P_i + P_i A_i\right]\tilde{x}_i + \tilde{x}_i^{\mathrm{T}} P_i \tilde{\eta}_i + \tilde{x}_i^{\mathrm{T}} P_i \tilde{\Upsilon}_i + \\
&\quad e_{i1}\left[g_{i,1}(\chi_{i,1}) + (d_i + b_i)\left(g_{i,1}\tilde{x}_{i,1} + e_{i,2} + \alpha_{i,1}\right)\right] \\
&\quad + \tilde{s}_{i,1}\left[\dot{\Upsilon}_{i,1} - \lambda_{i,1}\left(\frac{\tilde{W}_{i,1,1}^{*\mathrm{T}}}{\lambda_{i,1}} S_{i,1}(\hat{x}_i) + \tilde{x}_{i,2}\right) - \lambda_{i,1}\tilde{\Upsilon}_{i,1}\right] \\
&\quad - \frac{\tilde{\vartheta}_i}{\varrho_i}\dot{\hat{\vartheta}}_i - \frac{\tilde{W}_{i,1,1}^{*}}{\gamma_{i,1}}\dot{\hat{W}}_{i,1,1}^{*} - e_{i,1}\sum_{j\in\mathcal{N}_i} a_{ij} g_{j,1}\tilde{x}_{j,1} - \frac{d_i}{2}e_{i,1}^2,
\end{aligned} \tag{2.14}$$

where

$$g_{i,1}(\chi_{i,1}) = (d_i + b_i)\left(\frac{\hat{W}^{*\mathrm{T}}_{i,1,1}S_{i,1}(\hat{x}_i)}{\lambda_{i,1}} + \hat{\Upsilon}_{i,1}\right) - b_i\dot{y}_r$$
$$- \sum_{j\in\mathcal{N}_i} a_{ij}\left(\frac{W^{*\mathrm{T}}_{j,1,1}S_{j,1}(\hat{x}_j)}{\lambda_{j,1}} + \hat{x}_{j,2} + \hat{\Upsilon}_{j,1}\right) + \frac{d_i}{2}e_{i,1}$$

The vector a approximates the unknown function b on a compact set, meaning âc > 0 such that: The RBF NNs $W^{*\mathrm{T}}_{i,1}S_{i,1}(\chi_{i,1})$ are utilized to approximate the unknown function $g_{i,1}(\chi_{i,1})$ on a compact set, meaning $\exists\varepsilon_{i,1}$ such that

$$g_{i,1}(\chi_{i,1}) = W^{*\mathrm{T}}_{i,1}S_{i,1}(\chi_{i,1}) + \omega_{i,1}(\chi_{i,1}), \tag{2.15}$$

where $\chi_{i,1} = \left[\hat{x}_i^{\mathrm{T}}, \hat{x}_j^{\mathrm{T}}, \hat{\Upsilon}^{\mathrm{T}}_{i,1}, \hat{\Upsilon}^{\mathrm{T}}_{j,1}, y_r, \dot{y}_r, a_{ij}\right]^{\mathrm{T}}$ $(j \in \mathcal{N}_i)$, $\omega_{i,1}$ denotes the approximate error satisfying $|\omega_{i,1}(\chi_{i,1})| \leq \varepsilon_{i,1}$.

In view of the Young's inequality and Lemma 2.1, for a positive constant $q_{i,1}$, one obtains

$$\begin{aligned} e_{i1}g_{i,1}(\chi_{i,1}) &\leq \frac{\|W^*_{i,1}\|^2}{2q^2_{i,1}}e^2_{i,1}S^{\mathrm{T}}_{i,1}(\chi_{i,1})S_{i,1}(\chi_{i,1}) \\ &\quad + \frac{q^2_{i,1}}{2} + \frac{e^2_{i,1}}{2} + \frac{\varepsilon^2_{i,1}}{2} \\ &\leq \frac{\vartheta_i}{2q^2_{i,1}}e^2_{i,1}S^{\mathrm{T}}_{i,1}(Z_{i,1})S_{i,1}(Z_{i,1}) \\ &\quad + \frac{q^2_{i,1}}{2} + \frac{e^2_{i,1}}{2} + \frac{\varepsilon^2_{i,1}}{2}, \end{aligned} \tag{2.16}$$

where $Z_{i,1} = \left[\hat{x}^{\mathrm{T}}_{i,1}, \hat{x}^{\mathrm{T}}_{j,1}, \hat{\Upsilon}^{\mathrm{T}}_{i,1}, \hat{\Upsilon}^{\mathrm{T}}_{j,1}, y_r, b_i\dot{y}_r, a_{ij}\right]^{\mathrm{T}}$ $(j \in \mathcal{N}_i)$.

By applying the Young's inequality, the following inequalities can be obtained:

$$\tilde{x}_i^{\mathrm{T}}P_i\tilde{\eta}_i \leq \frac{\beta_{i0}\rho_i^2}{\underline{\lambda}_i}\|P_i\|^2\|\tilde{x}_i\|^2 + \frac{\|\tilde{W}_i^*\|^2}{\beta_{i0}}, \tag{2.17}$$

$$\begin{aligned} \tilde{x}_i^{\mathrm{T}}P_i\tilde{\Upsilon}_i &\leq \frac{1}{2}\|P_i\|^2\|\tilde{x}_i\|^2 + \frac{1}{2}\left\|\tilde{\Upsilon}_i\right\|^2 \\ &\leq \frac{1}{2}\|P_i\|^2\|\tilde{x}_i\|^2 + \|\tilde{s}_i\|^2 + \bar{\lambda}_i^2\|\tilde{x}_i\|^2, \end{aligned} \tag{2.18}$$

$$-\lambda^2_{i,1}s_{i,1}\tilde{x}_{i,1} \leq \frac{1}{2}\lambda^4_{i,1}\tilde{x}^2_{i,1} + \frac{1}{2}\tilde{s}^2_{i,1}, \tag{2.19}$$

$$\tilde{s}_{i,1}\dot{\Upsilon}_{i,1} \leq \frac{1}{2}\tilde{s}^2_{i,1} + \frac{1}{2}\bar{d}^2_{i1}, \tag{2.20}$$

$$- e_{i,1}\sum_{j\in\mathcal{N}_i} a_{ij} g_{j,1}\tilde{x}_{j,1} \le \frac{d_i}{2}e_{i,1}^2 + \sum_{j\in\mathcal{N}_i} a_{ij}\frac{g_{j,1}^2}{2}\tilde{x}_{j,1}^2. \tag{2.21}$$

where $\underline{\lambda}_i = \min\left\{\|\lambda_{i,j}\|, j = 1, 2, \ldots, n_i\right\}$, $\bar{\lambda}_i = \max\left\{\|\lambda_{i,j}\|, j = 1, 2, \ldots, n_i\right\}$, $\rho_i = \|S_i(\tilde{x}_i)\|$, $\beta_{i0} > 0$ is a adaptive control design parameter.

The distributed virtual control law $\alpha_{i,1}$ and the adaptation law $\hat{M}_{i,1}^*$ are designed as

$$\alpha_{i,1} = -g_{i,1}(y_i - \hat{x}_{i,1}) + \frac{1}{b_i + d_i}\Bigg(- (\kappa_{i,1} + \frac{1}{2})e_{i,1}$$
$$- \frac{\hat{\vartheta}_i}{2q_{i,1}^2} e_{i,1} S_{i,1}^{\mathrm{T}}(Z_{i,1}) S_{i,1}(Z_{i,1})\Bigg), \tag{2.22}$$
$$\dot{\hat{W}}_{i,1,1}^* = -\gamma_{i,1}\tilde{s}_{i,1} S_{i,1}(\hat{x}_i) - \sigma_{i,1}\hat{W}_{i,1,1}^*. \tag{2.23}$$

Utilizing the adaptation law (2.23) and substituting (2.16)–(2.22) into (2.14) yields

$$\dot{V}_{i,1} \le -\tilde{x}_i^{\mathrm{T}}\left[Q_i - \Pi_i \otimes I_{n_i\times n_i}\right]\tilde{x}_i + (d_i + b_i)\, e_{i,1}e_{i,2}$$
$$- \left(\lambda_{i,1} - \frac{3}{2}\right)\tilde{s}_{i,1}^2 + \frac{1}{2}\lambda_{i,1}^4\tilde{x}_{i,1}^2 + \frac{1}{2}\lambda_{i,1}^2 x_{i,2}^2 + \frac{\|\tilde{W}_i^*\|^2}{\beta_{i0}}$$
$$+ \frac{\sigma_{i,1}}{\gamma_{i,1}}\tilde{W}_{i,1,1}^*\hat{W}_{i,1,1}^* - \kappa_{i,1}e_{i,1}^2 + \sum_{j\in\mathcal{N}_i} a_{ij}\frac{l_{j,1}^2}{2}\tilde{x}_{j,1}^2 + \upsilon_{i,1}$$
$$+ \tilde{\vartheta}_i\left(\frac{1}{2q_{i,1}^2}e_{i,1}^2 S_{i,1}^{\mathrm{T}}(Z_{i,1}) S_{i,1}(Z_{i,1}) - \frac{1}{\varrho_i}\dot{\hat{\vartheta}}_i\right), \tag{2.24}$$

where $\Pi_i = \frac{\beta_{i0}\rho_i^2}{\underline{\lambda}_i}\|P_i\|^2 + \frac{1}{2}\|P_i\|^2 + \bar{\lambda}_i^2$, $\upsilon_{i,1} = \frac{\bar{d}_{i,1}^2}{2} + \frac{q_{i,1}^2}{2} + \frac{\varepsilon_{i,1}^2}{2}$.

Step $\xi\,(\xi = 2, \ldots, n_i - 1)$: After differentiating $e_{i,\xi}$ along (2.6) and (2.11), one has

$$\dot{e}_{i,\xi} = \frac{\hat{W}_{i,\xi,1}^{*\mathrm{T}} S_{i,\xi}(\hat{x}_i)}{\lambda_{i,\xi}} + g_{i,\xi}\tilde{x}_{i,1} + e_{i,\xi+1} + \alpha_{i,\xi} + \hat{\Upsilon}_{i,\xi} - \dot{\alpha}_{i,\xi-1}, \tag{2.25}$$

where

$$\dot{\alpha}_{i,\xi-1} = \sum_{\tau=0}^{\xi-1}\frac{\partial\alpha_{i,\xi-1}}{\partial y_r^{\tau}} y_r^{\tau+1} + \frac{\partial\alpha_{i,\xi-1}}{\partial\hat{\vartheta}_i}\dot{\hat{\vartheta}}_i + \frac{\partial\alpha_{i,\xi-1}}{\partial x_{i,1}} x_{i,2}$$
$$+ \sum_{\tau=1}^{\xi-1}\frac{\partial\alpha_{i,\tau-1}}{\partial\hat{s}_{i,\tau}}\left[-\lambda_{i,\tau}\left(\frac{\hat{W}_{i,\tau,1}^{*\mathrm{T}} S_{i,\tau}(\hat{x}_i)}{\lambda_{i,\tau}} + \hat{x}_{i,\tau+1} + \hat{s}_{i,\tau}\right)\right]$$
$$+ \sum_{\tau=1}^{\xi-1}\sum_{j\in\mathcal{N}_i}\frac{\partial\alpha_{i,\tau-1}}{\partial\hat{s}_{j,\tau}}\left[-\lambda_{j,\tau}\left(\frac{\hat{W}_{j,\tau,1}^{*\mathrm{T}} S_{j,\tau}(\hat{x}_j)}{\lambda_{j,\tau}} + \hat{x}_{j,\tau+1}\right)\right]$$

$$+\frac{\partial \alpha_{i,\xi-1}}{\partial x_{i,1}}\left[f_{i,1}(x_i)+d_{i,1}(t)\right]-\sum_{\tau=1}^{\xi-1}\sum_{j\in\mathcal{N}_i}\frac{\partial \alpha_{i,\tau-1}}{\partial \hat{s}_{j,\tau}}\lambda_{j,\tau}\hat{s}_{j,\tau}$$
$$-\sum_{\tau=1}^{\xi-1}\frac{\partial \alpha_{i,\tau-1}}{\partial \hat{s}_{i,\tau}}\lambda_{i,\tau}^2\hat{x}_{i,\tau}-\sum_{\tau=1}^{\xi-1}\sum_{j\in\mathcal{N}_i}\frac{\partial \alpha_{i,\tau-1}}{\partial \hat{s}_{j,\tau}}\lambda_{j,\tau}^2\hat{x}_{j,\tau}.$$

Select the following Lyapunov candidate function as

$$V_{i,\xi}=V_{i,\xi-1}+\frac{1}{2}\left[e_{i,\xi}^2+\tilde{s}_{i,\xi}^2+\frac{\tilde{W}_{i,\xi,1}^{*\mathrm{T}}\tilde{W}_{i,\xi,1}^{*}}{\gamma_{i,\xi}}\right]. \tag{2.26}$$

Proceeding similarly, calculate the derivative of (2.26). Then, we get

$$\begin{aligned}\dot{V}_{i,\xi}&=\dot{V}_{i,\xi-1}+e_{i,\xi}\dot{e}_{i,\xi}+\tilde{s}_{i,\xi}\dot{\tilde{s}}_{i,\xi}-\frac{\tilde{W}_{i,\xi,1}}{\gamma_{i,\xi}}\dot{\hat{W}}_{i,\xi,1}^{*}\\
&=\dot{V}_{i,\xi-1}+e_{i,\xi}\left(\frac{\hat{W}_{i,\xi,1}^{*\mathrm{T}}S_{i,\xi}(\hat{x}_i)}{\lambda_{i,\xi}}+g_{i,\xi}(y_i-\hat{x}_{i,1})+e_{i,\xi+1}\right)\\
&-\tilde{s}_{i,\xi}\left[\lambda_{i,\xi}\left(\frac{\tilde{W}_{i,\xi,1}^{*\mathrm{T}}S_{i,\xi}(\hat{x}_i)}{\lambda_{i,\xi}}+\tilde{x}_{i,\xi+1}+\tilde{s}_{i,\xi}+\lambda_{i,\xi}^2\tilde{x}_{i,\xi}\right)\right]\\
&+e_{i,\xi}\left(\alpha_{i,\xi}+\hat{\Upsilon}_{i,\xi}-\dot{\alpha}_{i,\xi-1}\right)+\tilde{s}_{i,\xi}\dot{\Upsilon}_{i,\xi}-\frac{\tilde{W}_{i,\xi,1}}{\gamma_{i,\xi}}\dot{\hat{W}}_{i,\xi,1}^{*}.\end{aligned} \tag{2.27}$$

Define

$$\begin{aligned}g_{i,\xi}(\chi_{i,\xi})&=\frac{\hat{W}_{i,\xi,1}^{*\mathrm{T}}S_{i,\xi}(\hat{x}_i)}{\lambda_{i,\xi}}-\sum_{\tau=0}^{\xi-1}\frac{\partial \alpha_{i,\xi-1}}{\partial y_r^{\tau}}y_r^{\tau+1}-\psi_{i,\xi}(\chi_{i,\xi})\\
&-\frac{\partial \alpha_{i,\xi-1}}{\partial x_{i,1}}\left[\hat{s}_{i,1}+\lambda_{i,1}\hat{x}_{i,1}+\frac{W_{i,1,1}^{*\mathrm{T}}S_{i,1}(x_i)}{\lambda_{i,1}}+x_{i,2}\right]\\
&+\sum_{\tau=1}^{\xi-1}\frac{\partial \alpha_{i,\tau-1}}{\partial \hat{s}_{i,\tau}}\left[\lambda_{i,\tau}\left(\frac{\hat{W}_{i,\tau,1}^{*\mathrm{T}}S_{i,\tau}(\hat{x}_i)}{\lambda_{i,\tau}}+\hat{x}_{i,\tau+1}+s_{i,\tau}\right)\right]\\
&+\sum_{\tau=1}^{\xi-1}\sum_{j\in\mathcal{N}_i}\frac{\partial \alpha_{i,\tau-1}}{\partial \hat{s}_{j,\tau}}\left[\lambda_{j,\tau}\left(\frac{\hat{W}_{j,\tau,1}^{*\mathrm{T}}S_{j,\tau}(\hat{x}_j)}{\lambda_{j,\tau}}+\hat{x}_{j,\tau+1}+s_{j,\tau}\right)\right]\\
&+(r_{i,\xi}+\bar{r}_{i,\xi}+\hat{r}_{i,\xi}+\frac{1}{2})e_{i,\xi}+(\tilde{d}_i+\tilde{b}_i)e_{i,\xi-1}+\hat{\Upsilon}_{i,\xi}\\
&+\sum_{\tau=1}^{\xi-1}\frac{\partial \alpha_{i,\tau-1}}{\partial \hat{s}_{i,\tau}}\lambda_{i,\tau}^2\hat{x}_{i,\tau}+\sum_{\tau=1}^{\xi-1}\sum_{j\in\mathcal{N}_i}\frac{\partial \alpha_{i,\tau-1}}{\partial \hat{s}_{j,\tau}}\lambda_{j,\tau}^2\hat{x}_{j,\tau}.\end{aligned}$$

Without loss of generality, let $\tilde{d}_i+\tilde{b}_i=d_i+b_i$ if $\xi=2$, and $\tilde{d}_i+\tilde{b}_i=1$ if $\xi>2$.

Following the same methodology employed for the unknown nonlinear function $g_{i,r}(\chi_{i,r})$,$r = 1, \ldots, \xi - 1$, the function $g_{i,\xi}(\chi_{i,\xi})$ can similarly be approximated as

$$g_{i,\xi}(\chi_{i,\xi}) = W_{i,\xi}^{*\mathrm{T}} S_{i,\xi}(\chi_{i,\xi}) + \omega_{i,\xi}, \tag{2.28}$$

where $\chi_{i,\xi} = \left[\hat{x}_i^{\mathrm{T}}, \hat{x}_j^{\mathrm{T}}, \hat{\Upsilon}_{i,\xi}^{\mathrm{T}}, \hat{\Upsilon}_{j,\xi}^{\mathrm{T}}, y_r, \hat{\vartheta}_i, a_{ij}\right]^{\mathrm{T}}$ $(j \in \mathcal{N}_i)$, and $\omega_{i,\xi}$ satisfies $|\omega_{i,\xi}| \leq \varepsilon_{i,\xi}$.

Building upon the compensation technique introduced in [20], we employ function $\psi_{i,\xi}(\chi_{i,\xi})$ to compensate for term $\frac{\partial \alpha_{i,\xi-1}}{\partial \hat{\vartheta}_i}\dot{\hat{\vartheta}}_i$. Following the approaches in [14, 15], $\psi_{i,\xi}(\chi_{i,\xi})$ is characterized as

$$\begin{aligned} \psi_{i,\xi}(\chi_{i,\xi}) = & -\epsilon_i \hat{\vartheta}_i \frac{\partial \alpha_{i,\xi-1}}{\partial \hat{\vartheta}_i} - \sum_{\tau=2}^{\xi} \frac{e_{i,\xi} \varrho_i}{2q_{i,1}^2} \left| e_{i,\tau} \frac{\partial \alpha_{i,\tau-1}}{\partial \hat{\vartheta}_i} \right| \\ & + \sum_{\tau=2}^{\xi-1} \frac{\partial \alpha_{i,\xi-1}}{\partial \hat{\vartheta}_i} \frac{\varrho_i}{2q_{i,\tau}^2} e_{i,\tau}^2 S_{i,\tau}^{\mathrm{T}}(Z_{i,\tau}) S_{i,\tau}(Z_{i,\tau}). \end{aligned} \tag{2.29}$$

In view of the Young's inequality and Lemma 2.1, we also have

$$\begin{aligned} e_{i,\xi} g_{i,\xi}(\chi_{i,\xi}) \leq & \frac{\vartheta_i}{2q_{i,\xi}^2} e_{i,\xi}^2 S_{i,\xi}^{\mathrm{T}}(Z_{i,\xi}) S_{i,\xi}(Z_{i,\xi}) \\ & + \frac{q_{i,\xi}^2}{2} + \frac{e_{i,\xi}^2}{2} + \frac{\varepsilon_{i,\xi}^2}{2}, \\ - e_{i,\xi} \sum_{\tau=1}^{\xi-1} \frac{\partial \alpha_{i,\tau-1}}{\partial \hat{s}_{i,\tau}} \lambda_{i,\tau} \tilde{s}_{i,\tau} \leq & \, r_{i,\xi} e_{i,\xi}^2 + \frac{1}{2} \sum_{\tau=1}^{\xi-1} \tilde{s}_{i,\tau}^2, \\ - e_{i,\xi} \sum_{\tau=1}^{\xi-1} \sum_{j \in \mathcal{N}_i} \frac{\partial \alpha_{i,\tau-1}}{\partial \hat{s}_{j,\tau}} \lambda_{j,\tau} \tilde{s}_{j,\tau} \leq & \, \bar{r}_{i,\xi} e_{i,\xi}^2 + \frac{1}{2} \sum_{\tau=1}^{\xi-1} \sum_{j \in \mathcal{N}_i} h_i \tilde{s}_{j,\tau}^2, \\ - e_{i,\xi} \frac{\partial \alpha_{i,\xi-1}}{\partial x_{i,1}} (\tilde{s}_{i,1} + \lambda_{i,1} \tilde{x}_{i,1}) \leq & \, \hat{r}_{i,\xi} e_{i,\xi}^2 + \frac{1}{2} \tilde{s}_{i,1}^2 + \frac{1}{2} \tilde{x}_{i,1}^2. \end{aligned} \tag{2.30}$$

where $Z_{i,\xi} = \left[\hat{x}_{i,\xi}^{\mathrm{T}}, \hat{x}_{j,\xi}^{\mathrm{T}}, \hat{\Upsilon}_{i,\xi}^{\mathrm{T}}, \hat{\Upsilon}_{j,\xi}^{\mathrm{T}}, y_r, \hat{\vartheta}_i, a_{ij}\right]^{\mathrm{T}}$, $r_{i,\xi} = \frac{1}{2} \sum_{\tau=1}^{\xi-1} \left(\frac{\partial \alpha_{i,\tau-1}}{\partial \hat{s}_{i,\tau}} \lambda_{i,\tau}\right)^2$, $\bar{r}_{i,\xi} = \frac{1}{2h_i} \sum_{\tau=1}^{\xi-1} \left(\frac{\partial \alpha_{i,\tau-1}}{\partial \hat{s}_{j,\tau}} \lambda_{j,\tau}\right)^2$, $\hat{r}_{i,\xi} = \frac{1}{2} \left(\left(\frac{\partial \alpha_{i,\xi-1}}{\partial x_{i,1}}\right)^2 + \left(\lambda_{i,1} \frac{\partial \alpha_{i,\xi-1}}{\partial x_{i,1}}\right)^2\right)$.

Choose the distributed virtual control law $\alpha_{i,\xi}$ and adaptation law $\hat{W}_{i,\xi,1}^*$ as follows:

$$\begin{aligned} \alpha_{i,\xi} = & -g_{i,\xi}(y_i - \hat{x}_{i,1}) - \frac{\hat{\vartheta}_i}{2q_{i,\xi}^2} e_{i,\xi} S_{i,\xi}^{\mathrm{T}}(Z_{i,\xi}) S_{i,\xi}(Z_{i,\xi}) \\ & - (\kappa_{i,\xi} + \frac{1}{2}) e_{i,\xi}, \end{aligned} \tag{2.31}$$

$$\dot{\hat{W}}_{i,\xi,1}^* = -\gamma_{i,\xi} \tilde{s}_{i,\xi} S_{i,\xi}(\hat{x}_i) - \sigma_{i,\xi} \hat{W}_{i,\xi,1}^*. \tag{2.32}$$

According to (2.27) and (2.29)-(2.32), it yields

$$\begin{aligned}
\dot{V}_{i,\xi} \leq & -\tilde{x}_i^{\mathrm{T}}\left[Q_i - \Pi_i \otimes I_{n_i\times n_i}\right]\tilde{x}_i + e_{i,\xi}e_{i,\xi+1} \\
& + \tilde{\vartheta}_i\left(\sum_{\tau=1}^{\xi}\frac{1}{2q_{i,\xi}^2}e_{i,\xi}^2 S_{i,\xi}^{\mathrm{T}}(Z_{i,\xi})S_{i,\xi}(Z_{i,\xi}) - \frac{1}{\varrho_i}\dot{\hat{\vartheta}}_i\right) \\
& - \sum_{\tau=1}^{\xi}\left[\left(\lambda_{i,\xi} - \frac{3}{2}\right)\tilde{s}_{i,\xi}^2 - \frac{1}{2}\lambda_{i,\tau}^4\tilde{x}_{i,\tau}^2 - \frac{1}{2}\lambda_{i,\tau}^2 x_{i,\tau+1}^2\right] \\
& + \frac{\|\tilde{W}_i^*\|^2}{\beta_{i0}} + \sum_{\tau=1}^{\xi}\left(\frac{\sigma_{i,\tau}}{\gamma_{i,\tau}}\tilde{W}_{i,\tau,1}^*\hat{W}_{i,\tau,1}^* - \kappa_{i,\tau}e_{i,\tau}^2\right) + \frac{1}{2}\sum_{\tau=1}^{\xi-1}\tilde{s}_{i,\tau}^2 \\
& + \sum_{\tau=2}^{\xi}e_{i,\tau}\left(\psi_{i,\tau}(\chi_{i,\tau}) - \frac{\partial\alpha_{i,\tau-1}}{\partial\hat{\vartheta}_i}\dot{\hat{\vartheta}}_i\right) + \frac{1}{2}\sum_{\tau=1}^{\xi-1}\sum_{j\in\mathcal{N}_i}h_i\tilde{s}_{j,\tau}^2 \\
& + \sum_{j\in\mathcal{N}_i}a_{ij}\frac{l_{j,1}^2}{2}\tilde{x}_{j,1}^2 + \frac{1}{2}\sum_{\tau=1}^{\xi-1}(\tilde{s}_{i,1}^2 + \tilde{x}_{i,1}^2) + \upsilon_{i,\xi},
\end{aligned} \tag{2.33}$$

where $\upsilon_{i,\xi} = \upsilon_{i,\xi-1} + \frac{\bar{d}_{i,\xi}^2}{2} + \frac{q_{i,\xi}^2}{2} + \frac{\varepsilon_{i,\xi}^2}{2}$.

Step n_i: The derivative of e_{i,n_i} along (2.11) is described by

$$\dot{e}_{i,n_i} = \frac{\hat{W}_{i,n_i,1}^{*\mathrm{T}}S_{i,n_i}(\hat{x}_i)}{\lambda_{i,n_i}} + g_{i,n_i}\tilde{x}_{i,1} + u_i + \hat{\Upsilon}_{i,n_i} - \dot{\alpha}_{i,n_i-1}, \tag{2.34}$$

where

$$\begin{aligned}
\dot{\alpha}_{i,n_i-1} = & \sum_{\tau=0}^{n_i-1}\frac{\partial\alpha_{i,n_i-1}}{\partial y_r^{\tau}}y_r^{\tau+1} + \frac{\partial\alpha_{i,n_i-1}}{\partial\hat{\vartheta}_i}\dot{\hat{\vartheta}}_i + \frac{\partial\alpha_{i,n_i-1}}{\partial x_{i,1}}x_{i,2} \\
& + \sum_{\tau=1}^{n_i-1}\frac{\partial\alpha_{i,n_i-1}}{\partial\hat{s}_{i,\tau}}\left[-\lambda_{i,\tau}\left(\frac{\hat{W}_{i,\tau,1}^{*\mathrm{T}}S_{i,\tau}(\hat{x}_i)}{\lambda_{i,\tau}} + \hat{x}_{i,\tau+1} + \hat{s}_{i,\tau}\right)\right] \\
& + \sum_{\tau=1}^{n_i-1}\sum_{j\in\mathcal{N}_i}\frac{\partial\alpha_{i,\tau-1}}{\partial\hat{s}_{j,\tau}}\left[-\lambda_{j,\tau}\left(\frac{\hat{W}_{j,\tau,1}^{*\mathrm{T}}S_{j,\tau}(\hat{x}_j)}{\lambda_{j,\tau}} + \hat{x}_{j,\tau+1}\right)\right] \\
& + \frac{\partial\alpha_{i,n_i-1}}{\partial x_{i,1}}\left[f_{i,1}(x_i) + d_{i,1}(t)\right] - \sum_{\tau=1}^{n_i-1}\sum_{j\in\mathcal{N}_i}\frac{\partial\alpha_{i,\tau-1}}{\partial\hat{s}_{j,\tau}}\lambda_{j,\tau}\hat{s}_{j,\tau} \\
& - \sum_{\tau=1}^{n_i-1}\frac{\partial\alpha_{i,\tau-1}}{\partial\hat{s}_{i,\tau}}\lambda_{i,\tau}^2\hat{x}_{i,\tau} - \sum_{\tau=1}^{n_i-1}\sum_{j\in\mathcal{N}_i}\frac{\partial\alpha_{i,\tau-1}}{\partial\hat{s}_{j,\tau}}\lambda_{j,\tau}^2\hat{x}_{j,\tau}.
\end{aligned}$$

Construct the following Lyapunov candidate function

$$V_{i,n_i} = V_{i,n_i-1} + \frac{1}{2}\left[e_{i,n_i}^2 + \tilde{s}_{i,n_i}^2 + \frac{\tilde{W}_{i,n_i,1}^{*\mathrm{T}}\tilde{W}_{i,n_i,1}^*}{\gamma_{i,n_i}}\right]. \tag{2.35}$$

Define

$$g_{i,n_i}(\chi_{i,n_i}) = (r_{i,n_i} + \bar{r}_{i,n_i} + \hat{r}_{i,n_i} + \frac{1}{2})e_{i,n_i} - \psi_{i,n_i}(\chi_{i,n_i})$$
$$+ \sum_{\tau=1}^{n_i-1} \sum_{j\in\mathcal{N}_i} \frac{\partial\alpha_{i,n_i-1}}{\partial\hat{s}_{j,\tau}} \left[\lambda_{j,\tau}\left(\frac{\hat{W}_{j,\tau,1}^{*\mathrm{T}} S_{j,\tau}(\hat{x}_j)}{\lambda_{j,\tau}} + \hat{x}_{j,\tau+1} + s_{j,\tau}\right)\right]$$
$$+ \sum_{\tau=1}^{n_i-1} \frac{\partial\alpha_{i,n_i-1}}{\partial\hat{s}_{i,\tau}} \left[\lambda_{i,\tau}\left(\frac{\hat{W}_{i,\tau,1}^{*\mathrm{T}} S_{i,\tau}(\hat{x}_i)}{\lambda_{i,\tau}} + \hat{x}_{i,\tau+1} + s_{i,\tau}\right)\right]$$
$$- \frac{\partial\alpha_{i,n_i-1}}{\partial x_{i,1}} \left[\hat{s}_{i,1} + \lambda_{i,1}\hat{x}_{i,1} + \frac{W_{i,1,1}^{*\mathrm{T}} S_{i,1}(x_i)}{\lambda_{i,1}} + x_{i,2}\right] + \hat{\Upsilon}_{i,n_i}$$
$$+ \frac{\hat{W}_{i,n_i,1}^{*\mathrm{T}} S_{i,n_i}(\hat{x}_i)}{\lambda_{i,n_i}} - \sum_{\tau=0}^{n_i-1} \frac{\partial\alpha_{i,n_i-1}}{\partial y_r^{\tau}} y_r^{\tau+1} + (\tilde{d}_i + \tilde{b}_i)e_{i,n_i-1}$$
$$+ \sum_{\tau=1}^{n_i-1} \frac{\partial\alpha_{i,n_i-1}}{\partial\hat{s}_{i,\tau}} \lambda_{i,\tau}^2 \hat{x}_{i,\tau} + \sum_{\tau=1}^{n_i-1} \sum_{j\in\mathcal{N}_i} \frac{\partial\alpha_{i,n_i-1}}{\partial\hat{s}_{j,\tau}} \lambda_{j,\tau}^2 \hat{x}_{j,\tau}.$$

Hence, one also obtains

$$g_{i,n_i}(\chi_{i,n_i}) = W_{i,n_i}^{*\mathrm{T}} S_{i,n_i}(\chi_{i,n_i}) + \omega_{i,n_i}, \tag{2.36}$$

where $\chi_{i,n_i} = \left[\hat{x}_i^{\mathrm{T}}, \hat{x}_j^{\mathrm{T}}, \hat{\Upsilon}_{i,n_i}^{\mathrm{T}}, \hat{\Upsilon}_{j,n_i}^{\mathrm{T}}, y_r, \hat{\vartheta}_i, a_{ij}\right]^{\mathrm{T}}$ $(j \in \mathcal{N}_i)$, and ω_{i,n_i} satisfies $|\omega_{i,n_i}| \leq \varepsilon_{i,n_i}$.

The uncertain nonlinear function $\psi_{i,n_i}(\chi_{i,n_i})$ is also introduced to compensate the term $\frac{\partial\alpha_{i,n_i-1}}{\partial\hat{\vartheta}_i}\dot{\hat{\vartheta}}_i$, which is defined as

$$\psi_{i,n_i}(\chi_{i,n_i}) = -\epsilon_i\hat{\vartheta}_i \frac{\partial\alpha_{i,n_i-1}}{\partial\hat{\vartheta}_i} - \sum_{\tau=2}^{n_i} \frac{e_{i,n_i}\varrho_i}{2q_{i,\tau}^2} \left| e_{i,\tau}\frac{\partial\alpha_{i,\tau-1}}{\partial\hat{\vartheta}_i} \right|$$
$$+ \sum_{\tau=2}^{n_i-1} \frac{\partial\alpha_{i,n_i-1}}{\partial\hat{\vartheta}_i} \frac{\varrho_i}{2q_{i,n_i}^2} e_{i,\tau}^2 S_{i,\tau}^{\mathrm{T}}(Z_{i,\tau}) S_{i,\tau}(Z_{i,\tau}). \tag{2.37}$$

where $Z_{i,n_i} = \left[\hat{x}_{i,n_i}^{\mathrm{T}}, \hat{x}_{j,n_i}^{\mathrm{T}}, \hat{\Upsilon}_{i,n_i}^{\mathrm{T}}, \hat{\Upsilon}_{j,n_i}^{\mathrm{T}}, y_r, \hat{\vartheta}_i, a_{ij}\right]^{\mathrm{T}}$.

In what follows, the distributed practical control input of ith follower and the adaption laws $\hat{\vartheta}_i$ and $\hat{M}_{i,n_i}^*$ are designed as

$$u_i = -g_{i,n_i}(y_i - \hat{x}_{i,1}) - \frac{\hat{\vartheta}_i}{2q_{i,n_i}^2} e_{i,n_i} S_{i,n_i}^{\mathrm{T}}(Z_{i,n_i}) S_{i,n_i}(Z_{i,n_i})$$
$$- (k_{i,n_i} + \frac{1}{2})e_{i,n_i}, \tag{2.38}$$

$$\dot{\hat{\vartheta}}_i = \sum_{\tau=1}^{n_i} \frac{\varrho_i}{2q_{i,\tau}^2} e_{i,\tau}^2 S_{i,n_i}^{\mathrm{T}}(Z_{i,\tau}) S_{i,\tau}(Z_{i,\tau}) - \mu_i \hat{\vartheta}_i, \tag{2.39}$$

$$\dot{\hat{W}}_{i,n_i,1}^* = -\gamma_{i,n_i} \tilde{s}_{i,n_i} S_{i,n_i}(\hat{x}_i) - \sigma_{i,n_i} \hat{W}_{i,n_i,1}^*. \tag{2.40}$$

Proceeding similarly, take the derivative of (2.35) and substitute (2.37)-(2.40) into it. Then, one can obtain

$$\begin{aligned}
\dot{V}_{i,n_i} \leq & -\tilde{x}_i^{\mathrm{T}} \left[Q_i - \Pi_i \otimes I_{n_i \times n_i} \right] \tilde{x}_i + \frac{1}{2} \sum_{\tau=1}^{n_i-1} \sum_{j \in \mathcal{N}_i} \tilde{s}_{j,\tau}^2 \\
& + \frac{\mu_i}{\varrho_i} \tilde{\vartheta}_i \hat{\vartheta}_i + \frac{(n_i - 1)}{2} \left(\sum_{j \in \mathcal{N}_i} a_{ij} l_{j,1}^2 \tilde{x}_{j,1}^2 + \tilde{s}_{i,1}^2 + \tilde{x}_{i,1}^2 \right) \\
& - \sum_{\tau=1}^{n_i} \left[\left(\lambda_{i,n_i} - \frac{3}{2} \right) \tilde{s}_{i,n_i}^2 - \frac{1}{2} \lambda_{i,\tau}^4 \tilde{x}_{i,\tau}^2 - \frac{1}{2} \lambda_{i,\tau}^2 x_{i,\tau+1}^2 \right] \\
& + \sum_{\tau=1}^{n_i} \left(\frac{\sigma_{i,\tau}}{\gamma_{i,\tau}} \tilde{W}_{i,\tau,1}^* \hat{W}_{i,\tau,1}^* - \kappa_{i,\tau} e_{i,\tau}^2 \right) + \frac{1}{2} \sum_{\tau=1}^{n_i-1} \tilde{s}_{i,\tau}^2 + \upsilon_{i,n_i} \\
& + \frac{\| \tilde{W}_i^* \|^2}{\beta_{i0}},
\end{aligned} \tag{2.41}$$

where $\upsilon_{i,n_i} = \sum_{l=1}^{n_i} (\frac{\bar{d}_{i,l}^2}{2} + \frac{q_{i,l}^2}{2} + \frac{\varepsilon_{i,l}^2}{2})$.

Building upon the preceding analysis, we now present our main contribution: a neural network-based adaptive tracking control protocol for nonlinear multiagent systems subject to simultaneous unknown disturbances and immeasurable state variables. This principal result is formally stated in Theorem 2.1.

Theorem 2.1 *Under assumption 2.1, consider the nonlinear multiagent systems (2.1), whose interaction topology is described by a digraph or undirected graph. If the actual control input (2.38) and the distributed virtual controls (2.22) and (2.31), as well as the adaption laws (2.23), (2.32), (2.39) and (2.40), are both taken into account, it can be guaranteed that all the signals of the system (2.1) are semiglobally bound and the global graph-based synchronization error located in a bounded compact set.*

Proof. For the stability analysis, the Lyapunov function candidate is chosen as

$$V = \sum_{i=1}^{N} V_{i,n_i}. \tag{2.42}$$

By utilizing the Young's inequality, one obtains

$$\frac{\mu_i}{\varrho_i} \tilde{\vartheta}_i \hat{\vartheta}_i \leq -\frac{\mu_i}{2\varrho_i} \tilde{\vartheta}_i^2 + \frac{\mu_i}{2\varrho_i} \vartheta_i^2, \tag{2.43}$$

$$\frac{\sigma_{i,\tau}}{\gamma_{i,\tau}}\tilde{W}_{i,\tau,1}^{*\mathrm{T}}\hat{W}_{i,\tau,1}^{*} \leq -\frac{\sigma_{i,\tau}}{2\gamma_{i,\tau}}\tilde{W}_{i,\tau,1}^{*\mathrm{T}}\tilde{W}_{i,\tau,1}^{*} + \frac{\sigma_{i,\tau}}{2\gamma_{i,\tau}}W_{i,\tau,1}^{*\mathrm{T}}W_{i,\tau,1}^{*}. \tag{2.44}$$

Let

$$\begin{aligned} K_i^1 &:= \max\left\{\frac{(n_i-1)}{2}\left(\sum_{j\in\mathcal{N}_i} a_{ji}l_{i,1}^2+1\right), \frac{1}{2}\lambda_{i,1}^2, \ldots,\right. \\ &\left.\frac{1}{2}\lambda_{i,n_i-1}^2\right\}, \\ K_i^2 &:= \max\left\{\frac{1}{2}\lambda_{i,1}^4, \ldots, \frac{1}{2}\lambda_{i,n_i}^4\right\}. \end{aligned}$$

Differentiating (2.42) and combining with (2.43)–(2.44), it implies,

$$\begin{aligned} \dot{V} \leq &-\sum_{i=1}^{N}\left[\tilde{x}_i^{\mathrm{T}}\left[Q_i-(\Pi_i-K_i^1+K_i^2)\otimes I_{n_i\times n_i}\right]\tilde{x}_i\right] \\ &-\sum_{i=1}^{N}\sum_{\tau=1}^{n_i}\kappa_{i,\tau}e_{i,\tau}^2-\sum_{i=1}^{N}\sum_{\tau=1}^{n_i}\left(\frac{\sigma_{i,\tau}}{2}-\frac{\gamma_{i,\tau}}{\beta_{i0,\tau}}\right)\frac{\tilde{W}_{i,\tau,1}^{*\mathrm{T}}\tilde{W}_{i,\tau,1}^{*}}{\gamma_{i,\tau}} \\ &-\sum_{i=1}^{N}\left\{\left[\left(\lambda_{i,1}-\frac{3}{2}\right)-\frac{(n_i-1)}{2}-\frac{\sum_{j\in\mathcal{N}_i}}{2}\right]\tilde{s}_{i,1}^2\right. \\ &+\sum_{\tau=1}^{n_i-1}\left(\lambda_{i,\tau}-\frac{3}{2}-\frac{n_i-\tau}{2}\left(1+\sum_{j\in\mathcal{N}_i}h_j\right)\right)\tilde{s}_{i,\tau}^2 \\ &\left.+\left(\lambda_{i,n_i}-\frac{3}{2}\right)\tilde{s}_{i,n_i}^2\right\}-\sum_{i=1}^{N}\frac{\mu_i}{2\varrho_i}\tilde{\vartheta}_i^2 \\ &+\upsilon_{i,n_i}+\sum_{i=1}^{N}\left[\frac{\mu_i}{2\varrho_i}\vartheta_i^2+\sum_{\tau=1}^{n_i}\left(\frac{\sigma_{i,\tau}}{2\gamma_{i,\tau}}W_{i,\tau,1}^{*\mathrm{T}}W_{i,\tau,1}^{*}\right)\right]. \end{aligned} \tag{2.45}$$

Define

$$\begin{aligned} K_i = \min&\left\{\left(\lambda_{i,1}-\frac{3}{2}\right)-\frac{(n_i-1)}{2}-\frac{\sum_{j\in\mathcal{N}_i}}{2},\right. \\ &\left.\left(\lambda_{i,\tau}-\frac{3}{2}-\frac{n_i-\tau}{2}\left(1+\sum_{j\in\mathcal{N}_i}h_j\right)\right), \lambda_{i,n_i}-\frac{3}{2}\right\}, \end{aligned}$$

where $\tau = 2, 3, \ldots, n_i - 1$, and

$$K := \min_{1\le i\le N}\left\{\frac{\lambda_{\min}\left\{Q_i - (\Pi_i - K_i^1 + K_i^2)\otimes I_{n_i\times n_i}\right\}}{\lambda_{\max}\{P_i\}}, 2K_i,\right.$$
$$\left.\mu_i, \sigma_{i,\tau} - \frac{2\gamma_{i,\tau}}{\beta_{i0,\tau}}\right\},$$
$$C := \upsilon_{i,n_i} + \sum_{i=1}^{N}\left[\frac{\mu_i}{2\varrho_i}\vartheta_i^2 + \sum_{\tau=1}^{n_i}\left(\frac{\sigma_{i,\tau}}{2\gamma_{i,\tau}}W_{i,\tau,1}^{*\mathrm{T}}W_{i,\tau,1}^{*}\right)\right].$$

To ensure closed-loop stability, the controller parameters must be selected according to the following constraints

$$Q_i - (\Pi_i - K_i^1 + K_i^2)\otimes I_{n_i\times n_i} > 0, \tag{2.46}$$
$$\frac{\sigma_{i,\tau}}{2} - \frac{\gamma_{i,\tau}}{\beta_{i0,\tau}} > 0, \tag{2.47}$$
$$\left(\lambda_{i,1} - \frac{3}{2}\right) - \frac{(n_i - 1)}{2} - \frac{\sum_{j\in\mathcal{N}_i}}{2} > 0, \tag{2.48}$$
$$\lambda_{i,\tau} - \frac{3}{2} - \frac{n_i - \tau}{2}\left(1 + \sum_{j\in\mathcal{N}_i} h_j\right) > 0, \tau = 2, \ldots, n_i. \tag{2.49}$$

Hence, the equation (2.45) can be reorganized as

$$\dot{V}(t) \le -KV(t) + C, \tag{2.50}$$

which implies

$$V(t) \le \left[V(t_0) - \frac{C}{K}\right]\exp(-K(t - t_0)) + \frac{C}{K}. \tag{2.51}$$

Thus, one can conclude that all the signals of the system are bounded. Further, one has $\frac{1}{2}e_{i,1}^2 \le V(t) \le \left[V(t_0) - \frac{C}{K}\right]\exp(-K(t - t_0)) + \frac{C}{K}$. On the other hand, the equation (2.11) implies

$$\|y - \bar{y}_r\| \le \frac{\|e_1\|}{\sigma(L + B)} \tag{2.52}$$

where $y = [y_1, y_2, \ldots, y_N]^{\mathrm{T}}$, $e_1 = \left[e_{1,1}, e_{2,1}, \ldots, e_{N,1}\right]^{\mathrm{T}}$, $\bar{y}_r = [y_r, y_r, \ldots, y_r]^{\mathrm{T}} \in \mathbb{R}^N$. Hence, one has

$$\lim_{t\to\infty}\|y - y_r\| \le \frac{1}{\sigma(L + B)}\sqrt{\frac{2C}{K}}. \tag{2.53}$$

If the assumption 2.2 holds, then the matrix $L + B$ is nonsingular. Thus, one has $\frac{1}{\sigma(L+B)}\sqrt{\frac{2C}{K}} \leq \bar{\pi}$ holds for any positive $\bar{\pi}$ by specifying the designed parameters based on the definitions of K and C. Hence, one can conclude that the outputs of all followers track the desired reference is well proved, which completes the proof.

The above description is sufficient for implementation, but a key issue remains: the adaptive control law relies on real-time signal transmission to actuators, increasing resource consumption and network congestion. Fortunately, a general solution exists. As noted in the introduction, event-triggered mechanisms can significantly reduce controller transmissions by executing only when specific conditions are met. Thus, we incorporate this approach into multiagent tracking control protocols. Here, we adopt a sample-and-hold strategy: the control input u_i remains constant until the triggering condition is violated, defined via a sequence $(t_k^i, r_i(t_k))_{k\in N}$ as follows:

$$u_i(t) = r_i(t_k), t \in [t_k, t_{k+1}), \tag{2.54}$$

$$t_{k+1}^i = \inf\left\{t > t_k^i \mid |e_i(t)| \geq m_{1,i}|r_i(t_k^i)| + m_{2,i}\right\}, \tag{2.55}$$

where $i = 1, 2, \ldots, N$, $k = 0, 1, 2, \ldots,$, $e_i(t) = r_i(t) - u_i(t)$, $0 < m_{1,i} < 1$, $m_{2,i} > 0$ refer to as the design control parameters, $r(t)$ denotes the intermediate control vector and will be specified subsequently, t_k^i is the controller update instant of ith follower. Clearly, the controller $u_i(t)$ holds as $r_i(t_k^i)$ for $t \in [t_k^i, t_{k+1}^i)$. Hence, one notes that there exist functions $\rho_{i,1}(t)$ and $\rho_{i,2}(t)$ satisfying $|\rho_{i,1}(t)| < 1$ and $|\rho_{i,1}(t) < 1|$ such that $r_i(t) = (1 + \rho_{i,1}(t)m_{1,i})u_i(t) + \rho_{i,2}(t)m_{i,2}$, which yields

$$u_i(t) = \frac{r_i(t)}{1 + \rho_{i,1}(t)m_{1,i}} - \frac{\rho_{i,2}(t)m_{i,2}}{1 + \rho_{i,1}(t)m_{1,i}} \tag{2.56}$$

The adaptive controller r_i is presented as

$$\begin{aligned} r_i(t) = -(1 + m_{i,1})\bigg[&\alpha_{i,n_i}\tanh\left(\frac{\alpha_{i,n_i}e_{i,n_i}}{\tau_i}\right) \\ &+ \bar{m}_i\tanh\left(\frac{\bar{m}_i e_{i,n_i}}{\tau_i}\right)\bigg] \end{aligned} \tag{2.57}$$

where the distributed virtual control law α_{i,n_i} is designed as

$$\begin{aligned} \alpha_{i,n_i} = &-l_{i,n_i}(y_i - \hat{x}_{i,1}) - \frac{\hat{\vartheta}_i}{2q_{i,n_i}^2}e_{i,n_i}S_{i,n_i}^{\mathrm{T}}(Z_{i,n_i})S_{i,n_i}(Z_{i,n_i}) \\ &- (\kappa_{i,n_i} + \frac{1}{2})e_{i,n_i}. \end{aligned} \tag{2.58}$$

and the adaption laws $\hat{\vartheta}_i$, $\hat{W}_{i,n_i,1}^*$ are given as (2.39) and (2.40).

Based on the proposed design approach, the following theorem establishes the main result for the event-triggered control scheme, ensuring both tracking performance and the exclusion of Zeno behavior.

Theorem 2.2 *If the assumption 2.1 is satisfied, under the state observer (2.6), composite disturbance observer (2.9), the event-triggered control law (2.53), and the adaption laws (2.39), (2.40), (2.57), the following claims hold.*

- *All the signals of the closed-loop system are uniformly bounded.*
- *The tracking error exponentially converges to compact set.*
- *The Zeno-behavior can be precluded. That is, there exists a positive constant $\bar{t}$ such that the inter-sampling intervals of ith follower satisfies $t_{k+1}^i - t_k^i \geq \bar{t}$.*

Proof. Step n_i: Proceeding similarly, differentiating (2.35) yields

$$\begin{aligned}
\dot{V}_{i,n_i} \leq & -\tilde{x}_i^{\mathrm{T}}\left[Q_i - \Pi_i \otimes I_{n_i \times n_i}\right]\tilde{x}_i + \frac{1}{2}\sum_{\tau=1}^{n_i-1}\sum_{j\in\mathcal{N}_i} \tilde{s}_{j,\tau}^2 \\
& + \frac{\mu_i}{\varrho_i}\tilde{\vartheta}_i\hat{\vartheta}_i + \frac{(n_i-1)}{2}\left(\sum_{j\in\mathcal{N}_i} a_{ij} l_{j,1}^2 \tilde{x}_{j,1}^2 + \tilde{s}_{i,1}^2 + \tilde{x}_{i,1}^2\right) \\
& - \sum_{\tau=1}^{n_i}\left[\left(\lambda_{i,n_i} - \frac{3}{2}\right)\tilde{s}_{i,n_i}^2 - \frac{1}{2}\lambda_{i,\tau}^4\tilde{x}_{i,\tau}^2 - \frac{1}{2}\lambda_{i,\tau}^2 x_{i,\tau+1}^2\right] \\
& + \sum_{\tau=1}^{n_i}\left(\frac{\sigma_{i,\tau}}{\gamma_{i,\tau}}\tilde{W}_{i,\tau,1}^*\hat{W}_{i,\tau,1}^* - \kappa_{i,\tau}e_{i,\tau}^2\right) + \frac{1}{2}\sum_{\tau=1}^{n_i-1}\tilde{s}_{i,\tau}^2 + \upsilon_{i,n_i} \\
& + \frac{\|\tilde{W}_i^*\|^2}{\beta_{i0}} + 0.557\tau_i - |e_{i,n_i}\bar{m}_i| - |\frac{e_{i,n_i}m_{i,2}}{1-m_{i,1}}|,
\end{aligned} \tag{2.59}$$

The overall Lyapunov candidate function is chosen as

$$V = \sum_{i=1}^{N} V_{i,n_i}. \tag{2.60}$$

Proceeding similarly, calculating the derivative of (2.59) and letting $C_1 := \upsilon_{i,n_i} + \sum_{i=1}^{N}\left[\frac{\mu_i}{2\varrho_i}\vartheta_i^2 + \sum_{\tau=1}^{n_i}\left(\frac{\sigma_{i,\tau}}{2\gamma_{i,\tau}}W_{i,\tau,1}^{*\mathrm{T}}W_{i,\tau,1}^*\right)\right] + 0.557\tau_i$, one obtains the following.

$$V(t) \leq -KV(t) + C_1 \tag{2.61}$$

The rest proof is similar as proof in theorem 2.1, which are omitted here.

In what follows, we shall show that there exists $\bar{t}$ such that the inter-sampling intervals $t_{k+1}^i - t_k^i \geq \bar{t}$.

One notes that

$$\begin{aligned}
\frac{d}{dt}|e_i(t)| &= \frac{d}{dt}(e_i(t) \times e_i(t))^{\frac{1}{2}} = \operatorname{sign}(e_i(t))\dot{e}_i(t) \\
&\leq |\dot{r}_i(t)|
\end{aligned} \tag{2.62}$$

It is clear that $r_i(t)$ is differentiable and $\dot{r}_i(t)$ is a function of the bounded signals. Hence, one has $|\dot{r}_i(t)| \leq \bar{r}$. On the other hand, $e_i(t_k^i) = 0$ and $\lim_{t\to t_{k+1}^i} e_i(t) =$

$m_{1,i}|r_i(t_k^i)| + m_{2,i}$. Thus, the lower bound of inter-sampling intervals $\bar{t}$ satisfies $\bar{t} \geq \frac{m_{1,i}|r_i(t_k^i)|+m_{2,i}}{\bar{r}}$ for $t \in [t_k^i, t_{k+1}^i)$, $i = 1, 2, \ldots, N$. That is, the zeno-phenomenon is excluded. The proof is completed.

2.3 Simulation Example

This section presents a numerical simulation to validate the efficacy of the proposed tracking control strategy. Through comparative analysis with the approach in [21], we demonstrate the superior performance of our method. Furthermore, we conduct simulations under event-triggered conditions to evaluate its practical implementation.

Consider the nonlinear multiagent systems with three followers and one leader, each agent is governed by

$$\begin{aligned} \dot{x}_{i1} &= x_{i2} + \sin(x_{i2})x_{i1}^2 + d_{i1}, \\ \dot{x}_{i2} &= u_i + x_{i2}\sin(x_{i1}) + d_{i2}, \\ y_i &= x_{i1}, i = 1, 2, 3, \end{aligned} \tag{2.63}$$

with $d_{i1} = \sin(x_{i2})\cos(t)x_{i1}$, $d_{i2} = \sin^2(t)x_{i1}x_{i2} + 1$. The dynamics of the leader is modeled

$$\begin{aligned} \dot{x}_r &= 3t^2\exp(-t) + 0.5\left(\cos\left(\frac{t}{2} + 3\right) - 1\right) - 1.2x_r, \\ y_r &= x_r. \end{aligned} \tag{2.64}$$

The communication interaction topology is given by a directed graph, whose adjacency matrix is chosen as follows

$$A = \begin{bmatrix} 0 & 0.1 & 0.1 \\ 0.1 & 0 & 0.1 \\ 0.1 & 0.1 & 0 \end{bmatrix}, \tag{2.65}$$

and the weight matrix between the leader and the followers is $B = \text{diag}\{0.8, 0.8, 0.8\}$. Here, the RBF NNs are applied to design the state observer and the disturbance observer, which contain 36 nodes and centers are evenly located in $[-6, 6]$, the width is $\omega_i = 2$. The initial value of this simulation is chosen as $x_r(0) = [-2]$, $x_1(0) = [1, 3]^{\mathrm{T}}$, $x_2(0) = [2, 3]^{\mathrm{T}}$, $x_3(0) = [3, 1]^{\mathrm{T}}$. Besides, the other initial conditions of some variables, including the RBF NNs weights $\hat{M}_{ij}^*$, the estimation of the ϑ_i, and the disturbance observer $\Upsilon_{i,j}$, $i = 1, 2, 3$, $j = 1, 2$, all are chosen as $[0]$, where $[0]$ stands for the zero matrix with suitable dimension. The state observer's initial condition is given by $\hat{x}_r(0) = [-2]$, $\hat{x}_1(0) = [1, 3]^{\mathrm{T}}$, $\hat{x}_2(0) = [2, 3]^{\mathrm{T}}$, $\hat{x}_3(0) = [3, 1]^{\mathrm{T}}$. In what follows, the effectiveness of the tracking protocol shall be verified. Before proceeding, some control design parameters are chosen as: $l_{i1} = 18$, $\kappa_{i,1} = 22$, $q_{i1} = 10$,

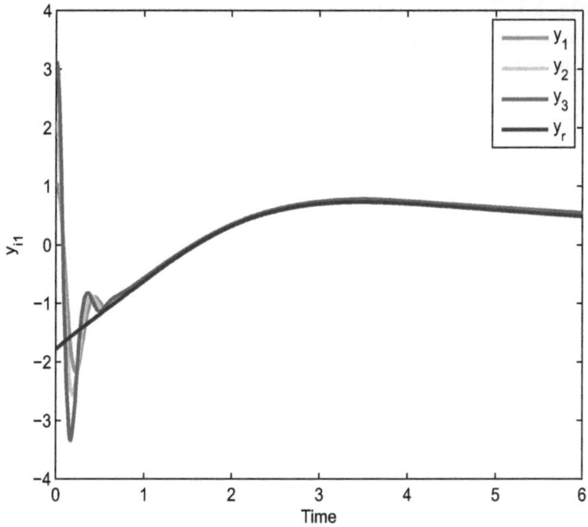

Fig. 2.1 The outputs of the multiagent systems (1)

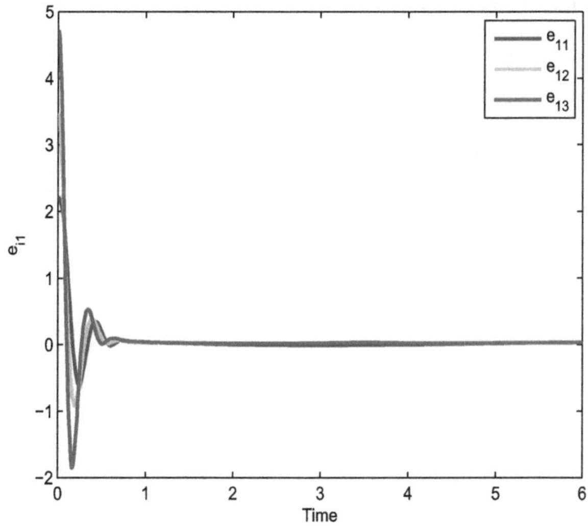

Fig. 2.2 The tracking performance of the multiagent systems (1)

$\gamma_{i,1} = 3$, $\sigma_{i1} = 14$, $l_{i2} = 22.5$, $l_{i2} = 22.5$, $\kappa_{i,2} = 14$, $q_{i1} = 5$, $\gamma_{i,2} = 18$, $\sigma_{i2} = 4$, $\varrho_i = 1$, $\mu_i = 1$, $i = 1, 2, 3$. The simulation results are depicted in Figs. 2.1, 2.2, 2.3 2.4 and 2.5. From Fig. 2.1, it is well verified that the adaptive control protocol can guarantee that the output of the followers track that of the leader. The response curve of the graph-based synchronization error e_{i1} is depicted in Fig. 2.2. The state curves x_{i2} of the followers are shown in Fig. 2.3, it also demonstrates the tracking control scheme is with better performance. The estimation of the adaptive control law ϑ_i and the control input signal are drawn in Figs. 2.4 and 2.5, respectively.

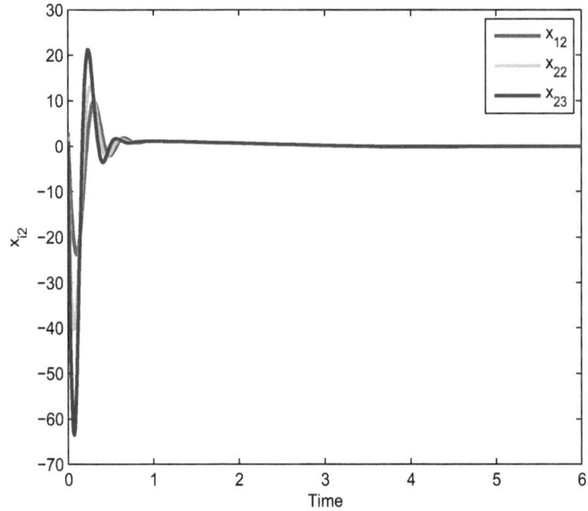

Fig. 2.3 The state response curves x_{i2} of the three followers

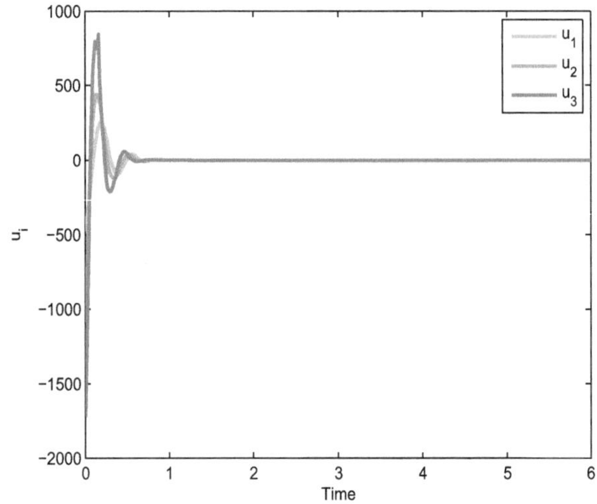

Fig. 2.4 The control input signal u_i of the three followers

So as to further show the merit of the proposed approach, we compare the tracking performance of the proposed method with the one of the methods proposed in [21], and the comparison result is depicted in Fig. 2.6. From which, one notes that our method can achieve a faster tracking performance when compared with the result obtained from [21].

In what follows, we will further verify the effectiveness of the theorem 2.2, choosing the event-triggered control parameters as $m_{1,i} = 0.26, m_{2,i} = 0.18, \tau_i = 2.6$. The corresponding simulation results are depicted in Figs. 2.7 and 2.8, where Fig. 2.7 shows the output response curves of the agents and Fig. 2.8 depicts invent triggering intervals of three agents. One notes that a slower tracking performance is obtained

Fig. 2.5 The estimation of the adaptive variables ϑ_i of the three followers

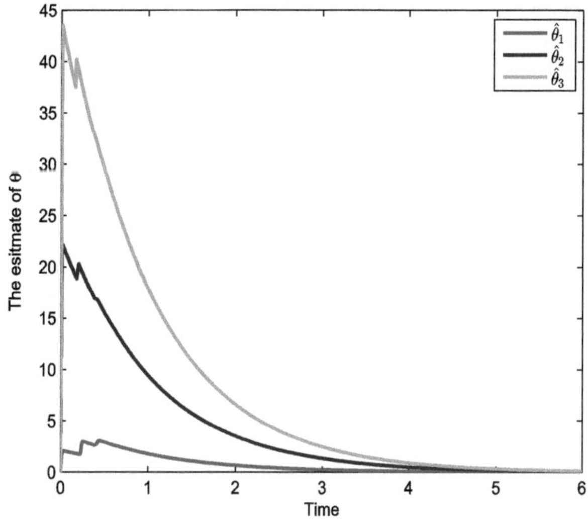

Fig. 2.6 The outputs of the multiagent systems from the method developed in the event-triggered manner

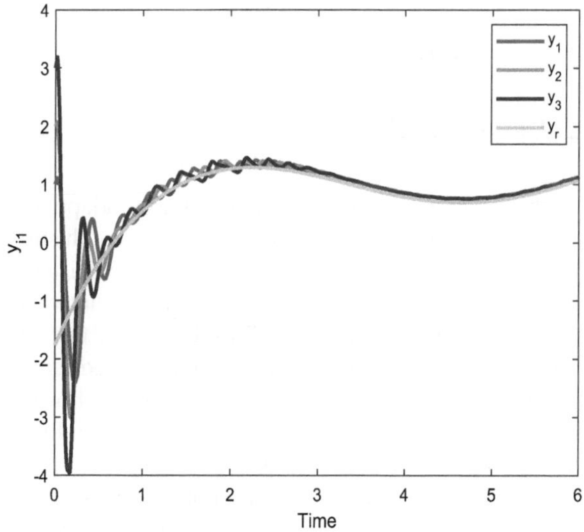

while the control signals are sent to the actuator aperiodically in a piecewise manner, which reduces the number of the communication from the controller to the actuator. Thus, the validity of the theorem 2.2 is verified.

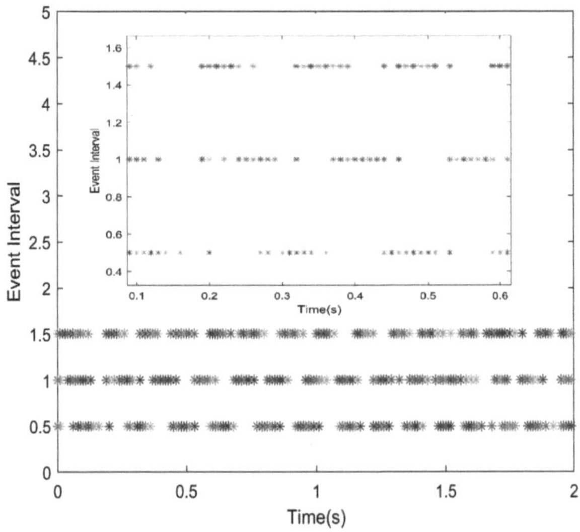

Fig. 2.7 Event triggering intervals of the three followers

2.4 Conclusion

2.4.1 Contribution

In this chapter, the neural network-based adaptive tracking problem was explored for a class of nonlinear multiagent systems both subjected to the immeasurable state variables and unknown external disturbance. Distinguished from existing literatures, we mainly focused on designing the neural network-based state observer and composite disturbance observer. Based on this, the adaptive tracking control scheme which drive the followers to track the desired target has been developed by applying the back-stepping technique for each follower, whose communication topology can be modeled as a directed graph or undirected graph. The scheme ensures semiglobal boundedness of all system signals and achieves synchronization within a bounded compact set. Furthermore, so as to reduce the communication burden, the proposed method is then extended into the event-triggered case, and the adaptive event-triggered tracking control scheme is also established for the nonlinear multiagent systems. Finally, a simulation example was carried out to verify the validity and effectiveness of our theoretical results.

2.4.2 *Limitations of the Current Work*

Despite the promising results, several limitations remain:

Computational Complexity: The reliance on radial basis function (RBF) neural networks and multiple observers increases computational demands, which may hinder real-time implementation for large-scale systems.

Assumption Dependency: The stability analysis hinges on restrictive assumptions, such as bounded disturbances (Assumption 2.1) and the existence of a spanning tree in the communication graph (Assumption 2.2). These may not hold in real-world scenarios with unbounded disturbances or dynamic topologies.

Parameter Sensitivity: The control performance is highly sensitive to the choice of design parameters(e.g., $\kappa_{i,\tau}$, $\lambda_{i,\tau}$). Suboptimal tuning could degrade tracking accuracy or even destabilize the system.

Event-Triggered Trade-offs: While event-triggered mechanisms reduce communication, they introduce delays and slower convergence (as seen in Fig. 2.7), which may compromise performance in time-critical applications.

2.4.3 *Future Directions*

To address these limitations, future research could focus on:

Reduced-Order Observers: Developing lightweight observer designs to lower computational overhead while maintaining estimation accuracy.

Robustness Enhancements: Incorporating adaptive mechanisms to handle unbounded disturbances or relax topology assumptions, such as using switching graph theories or hybrid control strategies.

Machine Learning Integration: Leveraging reinforcement learning or metaheuristic algorithms to automate parameter tuning and improve adaptability.

Advanced Event-Triggering Schemes: Exploring dynamic triggering thresholds or hybrid time-event triggers to balance communication efficiency and control performance.

Experimental Validation: Testing the proposed methods on physical platforms (e.g., robotic swarms or smart grids) to evaluate scalability and robustness under real-world constraints.

By addressing these challenges, the framework could be extended to broader applications, such as autonomous vehicles, distributed energy systems, or bio-inspired collective behaviors, where nonlinear dynamics and resource constraints are critical.

This work lays a foundation for adaptive control in complex multiagent systems, but bridging the gap between theoretical guarantees and practical deployment remains an open frontier. Future advancements should prioritize scalability, robustness, and real-world feasibility to unlock the full potential of neural network-based control in distributed systems.

References

1. Luo B, Wu HN, Li HX (2015) Adaptive optimal control of highly dissipative nonlinear spatially distributed processes with neuro-dynamic programming. IEEE Trans Neural Netw Learn Syst 26(4):684–696
2. Xu Y, Lu RQ, Shi P, Li HY, Xie SL (2018) Finite-time distributed state estimation over sensor networks with round-robin protocol and fading channels. IEEE Trans Cybern 48(1):336–345
3. Yang Y, Xu HW, Yue D (2019) Observer-based distributed secure consensus control of a class of linear multi-agent systems subject to random attacks. IEEE Trans Circuits Syst I Regular Pap 66(8):3089–3099
4. Chen B, Lin C, Liu XP, Liu KF (2016) Observer-based adaptive fuzzy control for a class of nonlinear delayed systems. IEEE Trans Syst Man Cybern Syst 46(1):27–36
5. Chen M, Ge SS (2015) Adaptive neural output feedback control of uncertain nonlinear systems with unknown hysteresis using disturbance observer. IEEE Trans Ind Electron 62(12):7706–7716
6. Min HF, Xu SY, Ma Q, Zhang BY, Zhang ZQ (2017) Composite-observer-based output-feedback control for nonlinear time-delay systems with input saturation and its application. IEEE Trans Ind Electron 65(7):5856–5863
7. Wang X, Lemmon MD (2011) Event-triggering in distributed networked control systems. IEEE Trans Autom Control 56(3):586–601
8. Xing LT, Wen CY, Liu ZT, Su HY, Cai JP (2017) Event-triggered adaptive control for a class of uncertain nonlinear systems. IEEE Trans Autom Control 62(4):2071–2076
9. Zhang WB, Tang Y, Miao QY, Du W (2013) Exponential synchronization of coupled switched neural networks with mode-dependent impulsive effects. IEEE Trans Neural Netw Learn Syst 24(8):1316–1326
10. Li HY, Wang LJ, Du HP, Boulkroune A (2017) Adaptive fuzzy backstepping tracking control for strict-feedback systems with input delay. IEEE Trans Fuzzy Syst 25(3):642–652
11. Mei J, Ren W, Li B, Ma GF (2015) Distributed containment control for multiple unknown second-order nonlinear systems with application to networked Lagrangian systems. IEEE Trans Neural Netw Learn Syst 26(9):1885–1899
12. Wang F, Chen B, Lin C, Zhang J, Meng XZ (2018) Adaptive neural network finite-time output feedback control of quantized nonlinear systems. IEEE Trans Cybern 48(6):1839–1848
13. Ma H, Liang HJ, Zhou Q, Ahn CK (2019) Adaptive dynamic surface control design for uncertain nonlinear strict-feedback systems with unknown control direction and disturbances. IEEE Trans Syst Man Cybern Syst 49(3):506–515
14. Zhang YH, Sun J, Liang HJ, Li HY (2020) Event-triggered adaptive tracking control for multiagent systems with unknown disturbances. IEEE Trans Cybern 50(3):890–901
15. Shang Y, Chen B, Lin C (2020) Consensus tracking control for distributed nonlinear multiagent systems via adaptive neural backstepping approach. IEEE Trans Syst Man Cybern Syst 50(7):2436–2444
16. Liu WH, Li P (2019) Disturbance observer-based fault-tolerant adaptive control for nonlinearly parameterized systems. IEEE Trans Ind Electron 66(11):8681–8691
17. Huang Y, Wang JZ, Shi DW, Wang JF, Shi L (2019) Event-triggered sampled-data control: an active disturbance rejection approach. IEEE/ASME Trans Mechatron 24(5):2052–2063
18. Wang AQ, Liu L, Qiu JB, Feng G (2019) Event-triggered robust adaptive fuzzy control for a class of nonlinear systems. IEEE Trans Fuzzy Syst 27(8):1648–1658

19. Sun YM, Chen B, Lin C, Wang HH, Zhou SW (2016) Adaptive neural control for a class of stochastic nonlinear systems by backstepping approach. Inf Sci 369:748–764
20. Chen B, Liu XP, Liu KF, Lin C (2009) Direct adaptive fuzzy control of nonlinear strict-feedback systems. Automatica 45(6):1530–1535
21. Chen CLP, Wen GX, Liu YJ, Liu Z (2016) Observer-based adaptive backstepping consensus tracking control for high-order nonlinear semi-strict-feedback multiagent systems. IEEE Trans Cybern 46(7):1591–1601

Chapter 3
Neural Network-Based Event-Triggered Cooperative Adaptive Neural Control for Cyber-Physical Systems with Unknown State Time-Delays and Deception Attacks

In recent years, research on cooperative control of cyber-physical systems has become a focus in both academic and industrial circles, demonstrating tremendous application potential in numerous practical systems such as robotics, unmanned aerial vehicles, and spacecrafts. As a core element of cooperative control, the consensus problem has been extensively and in-depth studied, with leaderless consensus and leader-following consensus [1] forming two key research directions. It is widely recognized that the evolution of state parameters in practical systems, whether in nature or industry, is typically an interactive process, often requiring modeling as a high-order system in analysis [2–4]. As a result, significant attention has been given to addressing the consensus control problem in cyber-physical systems.

The event-triggered controllers have gained increasing research focus in the past two decades, resulting in notable advancements in consensus protocol design for cyber-physical systems [5–7]. In [8], the event-triggered robust adaptive control problem was studied for a class of uncertain nonlinear systems, using fuzzy logic systems to approximate the nonlinear uncertainties.

Notably, the network communication environment of cyber-physical systems is highly complex, and transmissions between neighboring nodes may be subject to delays due to bandwidth limitations. This can degrade system performance to a certain extent and even undermine system stability. Therefore, with the increasing popularity of cyber-physical systems, the consensus problem concerning state delays has become an active research topic [9, 10]. Although some scholars have proposed relevant control methods to compensate for time delays, most of these methods are only applicable to time-triggered control cases, and there are relatively few studies under event-triggered control strategies.

Meanwhile, the cyber-physical system is a multidimensional complex system integrating computing, network, and physical environment, requiring more communication resources to maintain its controllability. Meanwhile, the security risks caused by cyberattacks are increasing, which is why researchers are paying more and more attention to the security of the cyber-physical system [11, 12]. Deception attacks are relatively common in practical applications. Although some research achievements

X. Wang et al., *Optimal Adaptive Cooperative Control for Nonlinear Multiagent Systems*,
https://doi.org/10.1007/978-981-95-2964-3_3

have been made to address this thorny issue, cooperative control under deception attacks remains a challenging topic when communication resources are limited and system security must be ensured.

3.1 Preliminaries

Consider a nonlinear cyber-physical system described by the following model:

$$\begin{aligned}\dot{x}_{i,j} &= x_{i,j+1} + g_{i,j}(\bar{x}_{i,j}) + \delta_{i,j}(\bar{x}_{i,j}(\tau_{i,j}))\\ \dot{x}_{i,n} &= u_i^{\varrho} + g_{i,n}(\bar{x}_{i,n}) + \delta_{i,j}(\bar{x}_{i,n}(\tau_{i,n})) + l_i(t)\\ y_i &= \breve{x}_{i,1}\\ \breve{x}_{i,j} &= x_{i,j} + \varsigma_{i,j}(t, x_{i,j})\\ \bar{x}_i(t) &= \eta_i(t), t \in [-\tau_{\max}, 0],\end{aligned} \tag{3.1}$$

where $1 \le i \le N$, $1 \le j \le n-1$. $\bar{x}_i = [x_{i,1}, x_{i,2}, \ldots, x_{i,n}]^T \in \mathbb{R}^n$, $u_i \in \mathbb{R}$, $y_i \in \mathbb{R}$ are the state variable vector, control input, and output of the i-th agent, respectively. $\bar{x}_{i,j} = [x_{i,1}, x_{i,2}, \ldots, x_{i,j}]^T \in \mathbb{R}^j, \bar{x}_{i,j}(\tau_{i,j}) = [x_{i,1}(t - \tau_{i,1}(t)), x_{i,2}(t - \tau_{i,2}(t)), \ldots,$ $x_{i,j}(t - \tau_{i,j}(t))]^T \in \mathbb{R}^j$. $g_{i,j}$ and $\delta_{i,j}$ are unknown smooth functions, and the external disturbance function l_i is unknown and bounded. $\tau_{i,j}$, $\tau_{\max}$ are the unknown time-varying state time-delay and the upper bound of the time-delay $\tau_{i,j}$. $\eta_i(t) \in \mathbb{R}^n$ is the initial state vector, $u_i^{\varsigma} = u_i + \varsigma_i^u(t, \bar{x}_{i,n})$ is the control input under actuator deception attack, and $\varsigma_i^u(t, \bar{x}_{i,n})$ is the unknown attack function. Define the compounded disturbance as $\Phi_i = l_i + \varsigma_i^u(t, \bar{x}_{i,n})$, the observer $\hat{\Phi}_i$ is employed to estimate Φ_i. $\varsigma_{i,j}(t, x_{i,j})$ is the sensor deception attack signal, and $\breve{x}_{i,j}$ refers to as the state variable compromised by the sensor attacks, $\bar{\breve{x}}_{i,j} = [\breve{x}_{i,1}, \ldots, \breve{x}_{i,j}]^T$. $\tau_{i,j}$ denotes the unknown time-delay satisfying $0 < \tau_{i,j} < c_{i,j} < \infty$ and $\dot{\tau}_{i,j} < \bar{c}_{i,j} < 1$, where $c_{i,j}$ and $\bar{c}_{i,j}$ are unknown positive constants.

Control objective The primary objective of this chapter is to design an event-triggered distributed delay-independent adaptive controller that ensures asymptotic consensus of the output signals of all agents and boundedness of all signals in the closed-loop system.

Remark 3.1 The model (3.1) can represent many practical applications, such as robotic systems, flight systems [13], chemical reactors [14], and so on. Moreover, the system (3.1) takes into account the widely existing issue of state-delay, making it more representative of real-world engineering systems. When devices in the system exchange data through communication networks, security control issues cannot be overlooked. Among these safety hazards, deception attack that acts on the sensor and actuator signals by malicious attackers is one of the most common phenomena. When communication security cannot be ensured, minimizing information transmission becomes one of the most effective methods to preserve the controllability of cyber-physical systems. However, the existing literature does not address these issues. Given the above considerations, proposing an event-triggered adaptive control strategy is highly practical for a class of cyber-physical systems that face unknown deception attacks and state time delays.

In this chapter, radial basis function neural networks will be used to estimate the unknown continuous nonlinear function $S(z)$: $\mathbb{R}^q \mapsto \mathbb{R}$, which is defined on a compact set Ω_z. That is to say,

$$H(z) = W^{*T} S(z) + \epsilon(z),$$

where $S(z) = [S_1(z), S_2(z), \cdots, S_r(z)]^T$ is the basis function vector, $r \geq 1$ denotes the number of neural networks node, and $\epsilon(z)$ represents the estimation error. The basis function is chosen as $S_i(z) = \exp\left[-\frac{(z-\gamma_i)^T(z-\gamma_i)}{\xi_i^2}\right], i = 1, 2, \cdots, r$, where $\gamma_i = [\gamma_{i1}, \gamma_{i2}, \cdots, \gamma_{iq}]^T \in \mathbb{R}^q$ is the center of receptive filed and $\xi_i \in \mathbb{R}$ represents the width of this function. $W^* = [W_1^*, W_2^*, \ldots, W_r^*]^T \in \mathbb{R}^r$ denotes the ideal weight vector defined as $W^* = \arg\min_{\hat{W}}[\sup_{z\in\Omega_z} |H(z) - \hat{W}^T S(z)|]$, and $\hat{W}$ is the estimate of W^*. Without loss of generality, $W_{i,j}^*$ and $\epsilon_{i,j}$ satisfy $||W_{i,j}^*|| \leq \bar{W}_{i,j}$ and $|\epsilon_{i,j}| \leq \iota_{i,j}$, where $i = 1, 2, \ldots, N$, $j = 1, 2, \ldots, n$, and $\bar{W}_{i,j}$, $\iota_{i,j}$ are unknown positive constants, respectively.

This section concludes by presenting the necessary assumptions, definitions, and lemmas required to achieve the control objective.

Assumption 3.1 Jin et al. [15] Suppose that $\varsigma_{i,j}$ and ς_i^u satisfy $\varsigma_{i,j}(t, x_{i,j}(t)) = b_{ij}(t)x_{i,j}(t)$, and $\varsigma_i^u(t, \bar{x}_{i,n}(t)) = \varrho_i(t)\Xi_i(\bar{x}_{i,n}(t))$ with $i = 1, 2, \ldots, N, j = 1, 2, \ldots, n$, respectively, where $b_{ij}(t)$, $\varrho_i(t)$ are the unknown time-varying signals, and $\Xi_i(\bar{x}_{i,n})$ is the unknown continuous nonlinear function. There are unknown positive constants $\bar{b}_{ij}$, $\underline{b}_{ij}$ and $\bar{\varrho}_i$ satisfying $-1 < \underline{b}_{ij} \leq b_{ij}(t) \leq \bar{b}_{ij}$ and $\varrho_i(t)\Xi_i(\bar{x}_{i,n}(t)) \leq \bar{\varrho}_i$.

Remark 3.2 In this chapter, we introduce the coordinate transformation $\theta_{i,j} = 1/(1 + b_{ij}(t))$. It is straightforward to verify that $\theta_{i,j}$ remains strictly positive. Moreover, there exist some unknown positive constants $\underline{\theta}_{i,j}$, $\bar{\theta}_{i,j}$ and $\bar{\Theta}_{i,j}$ such that $\underline{\theta}_{i,j} \leq \theta_{i,j} \leq \bar{\theta}_{i,j}$, and the time derivative satisfies $|\dot{\theta}_{i,j}| \leq \bar{\Theta}_{i,j}$.

Assumption 3.2 [16] The direct graph $\mathcal{G}$ is strongly connected.

Assumption 3.3 The time-delay function $\delta_{i,j}(\bar{x}_{i,j})(i = 1, \ldots, N;\ j = 1, \ldots, n)$ in (3.1) satisfies the inequality $|\delta_{i,j}(\bar{x}_{i,j})| \leq \sum_{h=1}^{j} \bar{a}_{i,j,h}(x_{i,h})$, where $\bar{a}_{i,j,h}(x_{i,h})$ is an unknown positive function.

Lemma 3.1 *[17] It holds that* $\Upsilon_{i,j}^T(z)\Upsilon_{i,j}(z) \leq h_{i,j}$, *where* $h_{i,j}$ *is the number of neurons for* $i = 1, \cdots, N;\ j = 1, \cdots, n$.

Lemma 3.2 $\lim_{\upsilon\to 0} tanh^2(\varepsilon/l)/\varepsilon = 0$, *where* $l > 0$ *is a constant and* $\varepsilon \in \mathbb{R}$ *is a variable.*

Lemma 3.3 *Consider a compact set* $\Gamma_{z_{i,j}}$ *defined as* $\Gamma_{z_{i,j}} := \{z_{i,j} | |z_{i,j}| < 0.8814\lambda_{i,j}\}$ *for* $i = 1, \cdots, N$; $j = 1, \cdots, n$. *Then, for* $z_{i,j} \notin \Gamma_{z_{i,j}}$, *the inequality* $[1 - 2tanh^2(z_{i,j}/\lambda_{i,j})] \leq 0$ *holds, where* $\lambda_{i,j} > 0$ *is a constant.*

Lemma 3.4 *Ge et al. [18] Let* $\mathcal{V}(\cdot)$, Ψ_i *be the smooth functions defined on* $[0, t_f)$ *with* $\mathcal{V}(t) \geq 0$, *and* $\mathcal{N}(\cdot)$ *be an even smooth Nussbaum-type function. If the following inequality hold:*

$$\mathcal{V}(t) \leq d_0 + e^{-d_1 t} \int_0^t [f_{i,n_i}(\cdot)\mathcal{N}(\Psi_i) + 1]\dot{\Psi}_i e^{d_1 \tau} d\tau, \tag{3.2}$$

where d_0 *is a suitable constant,* d_1 *represents a positive constant, and* $f_{i,n_i}(\cdot)$ *is a time-varying bounded parameter which takes values in the unknown closed intervals* $I = [l-, l+]$, *with* $0 \notin I$, *then* $\mathcal{V}(t)$, $\int_0^t f_{i,n_i}(\cdot)\mathcal{N}(\Psi_i)\dot{\Psi}_i e^{d_1 \tau} d\tau$ *must be bounded on* $[0, t_f)$.

Definition 3.1 Ye et al. [19] A continuous function $\mathcal{N}(\Psi) : \mathbb{R} \rightarrow \mathbb{R}$ can be declared to be a function of Nussbaum type, if it satisfies the following conditions:

$$\limsup_{s \rightarrow +\infty} \frac{1}{s} \int_0^s \mathcal{N}(\Psi) d\Psi = +\infty \tag{3.3}$$

$$\liminf_{s \rightarrow +\infty} \frac{1}{s} \int_0^s \mathcal{N}(\Psi) d\Psi = -\infty. \tag{3.4}$$

Such as, the smooth function $\mathcal{N}(\Psi) = \Psi^2 \cos(\Psi)$ and $e^{\Psi^2} \cos(\pi/2\Psi)$ are applied in [18, 20].

Remark 3.3 Assumption 3.1 is derived from the adaptive control results for linear CPSs without state time delays [15] and is crucial for ensuring the plant's controllability under sensor deception attacks. Specifically, Assumption 3.1 prevents the situation where $1 + b_{ij}(t) = 0$, which would render the compromised state $\breve{x}_{i,j}$ unusable in the controller design process. Assumption 3.2 ensures that every agent can directly or indirectly send or receive information from other agents. This is a common assumption in MASs, as seen in [16] *et.al.* Assumption 3.3 indicates that $\delta_{i,j}(\bar{x}_{i,j}(\tau_{i,j}))$ can be decomposed into nonlinear functions based on each delayed state. Notably, the function $(\bar{a}_{i,j,h}(x_{i,h})$ remains unknown, which provides a relaxation compared to the restriction in [21]. In practical engineering, external disturbances are unavoidable and can significantly degrade the control performance of systems [22, 23]. To tackle this issue, this chapter employs an adaptive NN-based compound disturbance observer that integrates external disturbances with actuator deception attacks. We assume that the disturbance observer satisfie $\Phi_i \leq \bar{\Phi}_i$, and $\dot{\Phi}_i \leq \bar{\bar{\Phi}}_i$, where $\bar{\Phi}_i$, $\bar{\bar{\Phi}}_i$ are unknown positive constants.

Remark 3.4 This chapter addresses the unknown control coefficients resulting from deception attacks, which pose theoretical challenges in controller design. Fortunately, the Nussbaum gain technique has been introduced as a solution to address this issue [24–27]. The core idea of this technique is to analyze the system's stability by substituting the unknown control coefficient with a Nussbaum-type function. For generality, the more versatile Nussbaum-type function $\mathcal{N}(\Psi) = \Psi^2\cos(\Psi)$ is adopted in this chapter.

3.2 Event-Triggered Adaptive Neural Consensus Control Against Unknown Time-Delay and Deception Attacks

In this section, the consensus problem of cyber physical multi-agent systems with time delays and deception attacks is examined, using adaptive dynamic surface backstepping control in conjunction with the Nussbaum gain technique. Due to the deception attacks, the actual state variable $x_{i,j}$ cannot be directly accessed, and the compromised state $\breve{x}_{i,j}$ is provided instead. First, the error variables are defined as follows:

$$\begin{cases} \breve{z}_{i,1} = \breve{x}_{i,1} - \beta_{i,1} \\ \breve{z}_{i,j} = \breve{x}_{i,j} - \beta_{i,j} \\ \dot{\beta}_{i,1} = -\sum_{j=1}^{N} b_{ij}(\breve{x}_{i,1} - \breve{x}_{j,1}) \\ \mu_{i,j} = \beta_{i,j} - \alpha_{i,j-1}, \end{cases} \tag{3.5}$$

where $i = 1, \cdots, N$, $j = 2, \cdots, n$. $\omega_{i,1}$ denotes the measurement error between the outputs of the i-th agent and its neighbors, $\beta_{i,1}(0) = \breve{x}_{i,1}(0)$, and $\alpha_{i,j-1}$ represents the virtual control law to be defined later. Define $\mu_{i,j}$ as the first-order filtering error, $\beta_{i,j}$ as the output and $\alpha_{i,j-1}$ as input of the first-order low-pass filters, which are defined as follows:

$$\omega_{i,j}\dot{\beta}_{i,j} + \beta_{i,j} = \alpha_{i,j-1}, \beta_{i,j}(0) = \alpha_{i,j-1}(0), \tag{3.6}$$

where $\omega_{i,j} > 0$ is a small time constant. From the equation (3.6), we have $\dot{\beta}_{i,j} = \frac{\alpha_{i,j-1} - \beta_{i,j}}{\omega_{i,j}} = \frac{-\mu_{i,j}}{\omega_{i,j}}$. Furthermore, we can obtain:

$$\begin{aligned} \dot{\mu}_{i,j} &= \dot{\beta}_{i,j} - \dot{\alpha}_{i,j-1} \\ &= -\frac{\mu_{i,j}}{\omega_{i,j}} + \mathcal{A}_{i,j}(\dot{\breve{z}}_{i,j-1}, \dot{\breve{x}}_{i,j-1}, \hat{\ni}_{i-1}, \dot{\Psi}_{i,j-1}), \end{aligned} \tag{3.7}$$

where $\mathcal{A}_{i,j}(\cdot)$ is a continuous function, in a compact set its upper bound can be obtained [28]. In this chapter, the upper bound of $\mathcal{A}_{i,j}(\cdot)$ is expressed as $\bar{\mathcal{A}}_{i,j}$. The adaptive law $\hat{\ni}_i$ will be defined later.

To construct a more suitable Lyapunov function, let $z_{i,j} = \theta_{i,j}\check{z}_{i,j}$, $x_{i,j} = \theta_{i,j}\check{x}_{i,j}$, we have

$$\begin{cases} z_{i,1} = x_{i,1} - \theta_{i,1}\beta_{i,1} \\ z_{i,j} = x_{i,j} - \theta_{i,j}\beta_{i,j}, \end{cases} \tag{3.8}$$

where $z_{i,1}$, $z_{i,j}$ are not available, they cannot be used directly in the virtual and actual controllers.

Next, the proposed event-triggered adaptive neural control scheme will be introduced step by step based on the DSC technique.

Step $(i, 1)$: Choose the $1st$ Lyapunov candidate function:

$$\mathcal{V}_{i,1} = \frac{1}{2}z_{i,1}^2 + \mathcal{V}_{K_{i,1}}, \tag{3.9}$$

where $\mathcal{V}_{K_{i,1}}$ is defined as

$$\mathcal{V}_{K_{i,1}} = \frac{1}{2}\frac{e^{\ell_{i,1}c_{i,1}}}{1-\bar{c}_{i,1}}\int_{t-\tau_{i,1}}^{t} e^{-\ell_{i,1}(t-\xi)}\bar{a}_{i,1,1}^2(x_{i,1}(\xi))d\xi,$$

with $\ell_{i,1}$ being a positive constant.

Calculating the derivative of $V_{i,1}$, one can easily obtain:

$$\begin{aligned}\dot{\mathcal{V}}_{i,1} =& z_{i,1}(z_{i,2} + \theta_{i,2}(\alpha_{i,1} + \mu_{i,2}) + g_{i,1} + \delta_{i,1}(x_{i,1}(\tau_{i,1})) \\ &- \dot{\theta}_{i,1}\beta_{i,1} - \theta_{i,1}\dot{\beta}_{i,1}) - \ell_{i,1}\mathcal{V}_{K_{i,1}} \\ &- \frac{1}{2}\frac{e^{\ell_{i,1}c_{i,1}}}{1-\bar{c}_{i,1}}(1-\dot{\tau}_{i,1})e^{-\ell_{i,1}\tau_{i,1}}\bar{a}_{i,1,1}^2(x_{i,1}(\tau_{i,1})) \\ &+ \frac{1}{2}\frac{e^{\ell_{i,1}c_{i,1}}}{1-\bar{c}_{i,1}}\bar{a}_{i,1,1}^2(x_{i,1}(t)) \\ \le& z_{i,1}(z_{i,2} + \theta_{i,2}(\alpha_{i,1} + \mu_{i,2}) + g_{i,1} + \delta_{i,1}(x_{i,1}(\tau_{i,1})) \\ &- \dot{\theta}_{i,1}\beta_{i,1} - \theta_{i,1}\dot{\beta}_{i,1}) - \ell_{i,1}\mathcal{V}_{K_{i,1}} - \frac{1}{2}\bar{a}_{i,1,1}^2(x_{i,1}(\tau_{i,1})) \\ &+ \frac{1}{2}\frac{e^{\ell_{i,1}c_{i,1}}}{1-\bar{c}_{i,1}}\bar{a}_{i,1,1}^2(x_{i,1}(t)).\end{aligned} \tag{3.10}$$

Based on Lemma 1.1 and Assumption 3.3, we can obtain:

$$z_{i,1}\delta_{i,1}(x_{i,1}(\tau_{i,1})) \le \frac{1}{2}z_{i,1}^2 + \frac{1}{2}\bar{a}_{i,1,1}^2(x_{i,1}(\tau_{i,1})) \tag{3.11}$$

$$z_{i,1}z_{i,2} \le \frac{1}{2}z_{i,1}^2 + \frac{1}{2}z_{i,2}^2 \tag{3.12}$$

$$\theta_{i,2}z_{i,1}\mu_{i,2} \le \frac{l_{i,1}^2}{2}\mu_{i,2}^2 + \frac{1}{2l_{i,1}^2}z_{i,1}^2\overline{\theta}_{i,2}^2, \tag{3.13}$$

where $l_{i,1}$ is a positive control parameter.

Substituting the inequalities derived above into equation (3.10), we obtain:

$$\begin{aligned}\dot{\mathcal{V}}_{i,1} \le & z_{i,1}(\theta_{i,2}\alpha_{i,1} + \mathcal{J}_{i,1}(\nu_{i,1})) - d_{i,1}z_{i,1}^2 - \ell_{i,1}\mathcal{V}_{K_{i,1}} \\ & - z_{i,1}^2 + d_{i,1}\breve{z}_{i,1}^2 + \frac{1}{2}z_{i,2}^2 + \frac{r_{i,1}^2}{2}\mu_{i,2}^2 \\ & + (1 - 2\tanh^2(\frac{z_{i,1}}{\lambda_{i,1}}))\mathcal{F}_{i,1}, \end{aligned} \tag{3.14}$$

where $d_{i,1}$, $\lambda_{i,1}$ refer to as the control parameters, $\nu_{i,1} = [z_{i,1}, x_{i,1}, \theta_{i,1}]^T$, $\mathcal{J}_{i,1}(\nu_{i,1}) = g_{i,1} + (z_{i,1}\overline{\theta}_{i,2}^2)/(2r_{i,1}^2) + 2/(z_{i,1})\tanh^2(z_{i,1}/\lambda_{i,1})\mathcal{F}_{i,1} + 2z_{i,1} + d_{i,1}z_{i,1} - c_{i,1}z_{i,1}/\theta_{i,1}^2 - \dot{\theta}_{i,1}\beta_{i,1} - \theta_{i,1}\dot{\beta}_{i,1}$, and $\mathcal{F}_{i,1}(x_{i,1}) = e^{\ell_{i,1}c_{i,1}}/(2 - 2\overline{c}_{i,1})\bar{a}_{i,1,1}^2(x_{i,1}(t))$.

To proceed with the analysis, we first approximate the unknown nonlinear term $\mathcal{J}_{i,1}(\nu_{i,1})$ using the universal approximation capability of RBFNNs, which leads to the following expression for $\mathcal{J}_{i,1}(\nu_{i,1})$:

$$\mathcal{J}_{i,1}(\nu_{i,1}) = W_{i,1}^{*T}S_{i,1}(\breve{\nu}_{i,1}) + \epsilon_{i,1} + \chi_{i,1}, \tag{3.15}$$

where $\nu_{i,1}$ is not available but $\breve{\nu}_{i,1} = [\breve{z}_{i,1}, \breve{x}_{i,1}]^T$ is known, $\chi_{i,1} = W_{i,1}^{*T}(S_{i,1}(\nu_{i,1}) - S_{i,1}(\breve{\nu}_{i,1}))$.

Based on Lemma 1.1, we can obtain:

$$\begin{aligned} z_{i,1}(W_{i,1}^{*T}S_{i,1}(\breve{\nu}_{i,1}) + \epsilon_{i,1}) \le & \frac{z_{i,1}^2}{2\bar{h}_{i,1}^2}||W_{i,1}^{*T}||^2 S_{i,1}^T(\breve{\nu}_{i,1})S_{i,1}(\breve{\nu}_{i,1}) \\ & + \frac{\bar{h}_{i,1}^2}{2} + \frac{1}{2}z_{i,1}^2 + \frac{1}{2}\iota_{i,1}^2. \end{aligned} \tag{3.16}$$

Substituting (3.15) and (3.16) into (3.14), it is not difficult to obtain:

$$\begin{aligned}\dot{\mathcal{V}}_{i,1} \le & - d_{i,1}z_{i,1}^2 + z_{i,1}\chi_{i,1} + z_{i,1}\theta_{i,2}\alpha_{i,1} + \frac{l_{i,1}^2}{2}\mu_{i,2}^2 - \frac{1}{2}z_{i,1}^2 \\ & + \frac{z_{i,1}^2}{2\bar{h}_{i,1}^2}||W_{i,1}^{*T}||^2 S_{i,1}^T(\breve{\nu}_{i,1})S_{i,1}(\breve{\nu}_{i,1}) + d_{i,1}\breve{z}_{i,1}^2 + \frac{1}{2}z_{i,2}^2 \\ & + (1 - 2\tanh^2(\frac{z_{i,1}}{\lambda_{i,1}}))\mathcal{F}_{i,1} - \ell_{i,1}\mathcal{V}_{K_{i,1}} + o_{i,1}, \end{aligned} \tag{3.17}$$

where $o_{i,1} = \bar{h}_{i,1}^2/2 + \iota_{i,1}^2/2$.

The virtual control law $\alpha_{i,1}$ and the update law for the auxiliary variable $\Psi_{i,1}$ can be constructed as follows:

$$\alpha_{i,1} = \mathcal{N}(\Psi_{i,1})(d_{i,1}\breve{z}_{i,1} + \frac{\breve{z}_{i,1}}{2\bar{h}_{i,1}^2}\hat{\ni}_i S_{i,1}^T(\breve{\nu}_{i,1})S_{i,1}(\breve{\nu}_{i,1})), \tag{3.18}$$

$$\dot{\Psi}_{i,1} = d_{i,1}\breve{z}_{i,1}^2 + \frac{\breve{z}_{i,1}^2}{2\bar{h}_{i,1}^2}\hat{\ni}_i S_{i,1}^T(\breve{\nu}_{i,1})S_{i,1}(\breve{\nu}_{i,1}), \tag{3.19}$$

where $\hat{\daleth}_i$ will be defined in the last step.

Substituting (3.18) and (3.19) into (3.17), one has:

$$\begin{aligned}\dot{\mathcal{V}}_{i,1} \le & -d_{i,1}z_{i,1}^2 + z_{i,1}\chi_{i,1} + \frac{l_{i,1}^2}{2}\mu_{i,2}^2 + \frac{1}{2}z_{i,2}^2 - \frac{1}{2}z_{i,1}^2 \\ & + \frac{\breve{z}_{i,1}^2}{2\bar{h}_{i,1}^2}(\theta_{i,1}^2\|W_{i,1}^{*T}\|^2 - \hat{\daleth}_i)S_{i,1}^T(\breve{\nu}_{i,1})S_{i,1}(\breve{\nu}_{i,1}) \\ & + (1 - 2\tanh^2(\frac{z_{i,1}}{\lambda_{i,1}}))\mathcal{F}_{i,1}(x_{i,1}) - \ell_{i,1}\mathcal{V}_{K_{i,1}} + o_{i,1} \\ & + (\theta_{i,1}\theta_{i,2}\mathcal{N}(\Psi_{i,1}) + 1)\dot{\Psi}_{i,1}.\end{aligned} \tag{3.20}$$

where $\mathcal{F}_{i,1}(x_{i,1}) = (e^{\ell_{i,1}c_{i,1}}/(2(1-\bar{c}_{i,1})))\bar{a}_{i,1,1}^2(x_{i,1}(t))$.

Step $(i,2)$: The derivative of the dynamics $z_{i,2}$ can be expressed as:

$$\begin{aligned}\dot{z}_{i,2} = & x_{i,3} + g_{i,2} + \delta_{i,2}(\bar{x}_{i,2}(\tau_{i,2})) \\ & - \dot{\theta}_{i,2}\beta_{i,2} - \theta_{i,2}\dot{\beta}_{i,2}.\end{aligned} \tag{3.21}$$

Choose the $2nd$ Lyapunov candidate function as:

$$\mathcal{V}_{i,2} = \mathcal{V}_{i,1} + \frac{1}{2}z_{i,2}^2 + \frac{1}{2}\mu_{i,2}^2 + \mathcal{V}_{K_{i,2}}, \tag{3.22}$$

where $\mathcal{V}_{K_{i,2}}$ is defined as:

$$\mathcal{V}_{K_{i,2}} = \frac{1}{2}\sum_{l=1}^{2}\frac{e^{\ell_{i,2}c_{i,l}}}{1-\bar{c}_{i,l}}\int_{t-\tau_{i,l}}^{t} e^{-\ell_{i,2}(t-\xi)}\bar{a}_{i,2,l}^2(x_{i,l}(\xi))d\xi,$$

with $\ell_{i,2}$ being a positive constant.

Taking the time derivative of $\mathcal{V}_{i,2}$, and applying (3.21), we obtain:

$$\begin{aligned}\dot{\mathcal{V}}_{i,2} \le & \dot{\mathcal{V}}_{i,1} + z_{i,2}(x_{i,3} + g_{i,2} + \delta_{i,2}(\bar{x}_{i,2}(\tau_{i,2})) \\ & - \dot{\theta}_{i,2}\beta_{i,2} - \theta_{i,2}\dot{\beta}_{i,2}) + \mu_{i,2}\dot{\mu}_{i,2} - \ell_{i,2}\mathcal{V}_{K_{i,2}} \\ & - \sum_{l=1}^{2}\frac{1}{2}\bar{a}_{i,2,l}^2(x_{i,l}(\tau_{i,l})) + \frac{1}{2}\sum_{l=1}^{2}\frac{e^{\ell_{i,2}c_{i,2}}}{1-\bar{c}_{i,2}}\bar{a}_{i,l,1}^2(x_{i,l}(t)).\end{aligned} \tag{3.23}$$

Similar to (3.16), this results in:

$$z_{i,2}\theta_{i,3}\mu_{i,3} \leq \frac{\overline{\theta}_{i,3}^2 z_{i,2}^2}{2l_{i,2}^2} + \frac{l_{i,2}^2\mu_{i,3}^2}{2}, \tag{3.24}$$

$$z_{i,2}\delta_{i,2}(x_{i,2}(\tau_{i,2})) \leq \frac{1}{2}z_{i,2}^2 + \sum_{l=1}^{2}\frac{1}{2}\bar{a}_{i,2,l}^2(x_{i,l}(\tau_{i,l})), \tag{3.25}$$

$$z_{i,2}z_{i,3} \leq \frac{1}{2}z_{i,2}^2 + \frac{1}{2}z_{i,3}^2, \tag{3.26}$$

$$\mu_{i,2}\dot{\mu}_{i,2} \leq -\frac{\mu_{i,2}^2}{\omega_{i,2}} + \frac{\bar{A}_{i,2}^2\mu_{i,2}^2}{2\kappa_{i,2}^2} + \frac{\kappa_{i,2}^2}{2}, \tag{3.27}$$

where $l_{i,2}$, $\kappa_{i,2}$ are positive design constant, $\bar{A}_{i,2} > 0$ is a scalar, which can ensure $|A_{i,2}| < \bar{A}_{i,2}$ just similar with [28].

Substituting (3.24)-(3.27) into (3.23), one has:

$$\begin{aligned}\dot{\mathcal{V}}_{i,2} \leq& \dot{\mathcal{V}}_{i,1} + z_{i,2}(\theta_{i,3}\alpha_{i,2} + \mathcal{J}_{i,2}(\nu_{i,2})) + \frac{1}{2}z_{i,3}^2 + \frac{\kappa_{i,2}^2}{2}\\ &+ (1 - 2\tanh^2(\frac{z_{i,2}}{\lambda_{i,2}}))\mathcal{F}_{i,2}(\bar{x}_{i,2}) - \ell_{i,2}\mathcal{V}_{K_{i,2}} + d_{i,2}\breve{z}_{i,2}^2\\ &- \frac{\mu_{i,2}^2}{\omega_{i,2}} + \frac{\bar{A}_{i,2}^2\mu_{i,2}^2}{2\kappa_{i,2}^2} + \frac{l_{i,2}^2\mu_{i,3}^2}{2} - \frac{3}{2}z_{i,2}^2 - d_{i,2}z_{i,2}^2,\end{aligned} \tag{3.28}$$

where $c_{i,2}, \lambda_{i,2}$ are design parameters, $\nu_{i,2} = [z_{i,2}, x_{i,1}, x_{i,2}, \theta_{i,2}]^T$, $\mathcal{J}_{i,2}(\nu_{i,2}) = g_{i,2} + (\overline{\theta}_{i,2}^3 z_{i,2})/(2l_{i,2}^2) + 5z_{i,2}/2 + d_{i,2}z_{i,2} + (2/z_{i,2})\tanh^2(z_{i,2}/\lambda_{i,2})\mathcal{F}_{i,2} - c_{i,2}z_{i,2}/\theta_{i,2}^2 - \dot{\theta}_{i,2}\beta_{i,2} - \theta_{i,2}\dot{\beta}_{i,2}$, $\mathcal{F}_{i,2}(\bar{x}_{i,2}) = (1/2)\sum_{k=1}^{2} e^{\ell_{i,2}c_{i,k}}/(1-\overline{c}_{i,k})\bar{a}_{i,2,k}^2(x_{i,k})$.

Similar the *Step* $(i, 1)$, one has

$$\begin{aligned}z_{i,2}(W_{i,2}^{*T}S_{i,2}(\breve{\nu}_{i,2}) + \iota_{i,2}) \leq& \frac{z_{i,2}^2}{2\bar{h}_{i,2}^2}||W_{i,2}^{*T}||^2 S_{i,2}^T(\breve{\nu}_{i,2})S_{i,2}(\breve{\nu}_{i,2})\\ &+ \frac{\bar{h}_{i,2}^2}{2} + \frac{1}{2}z_{i,2}^2 + \frac{1}{2}\iota_{i,2}^2.\end{aligned} \tag{3.29}$$

By substituting (3.29) into (3.28) and performing the calculation, we can obtain:

$$\begin{aligned}\dot{\mathcal{V}}_{i,2} \leq& \dot{\mathcal{V}}_{i,1} + z_{i,2}\theta_{i,3}\alpha_{i,2} + \frac{1}{2}z_{i,3}^2 - z_{i,2}^2 + o_{i,2} + d_{i,2}\breve{z}_{i,2}^2\\ &+ (1 - 2\tanh^2(\frac{z_{i,2}}{\lambda_{i,2}}))\mathcal{F}_{i,2}(\bar{x}_{i,2}) - \ell_{i,2}\mathcal{V}_{K_{i,2}} + z_{i,2}\chi_{i,2}\\ &- (\frac{1}{\omega_{i,2}} - \frac{\bar{A}_{i,2}^2}{2\kappa_{i,2}^2})\mu_{i,2}^2 + \frac{l_{i,2}^2}{2}\mu_{i,3}^2 - d_{i,2}z_{i,2}^2\\ &+ \frac{z_{i,2}^2}{2\bar{h}_{i,2}^2}||W_{i,2}^{*T}||^2 S_{i,2}^T(\breve{\nu}_{i,2})S_{i,2}(\breve{\nu}_{i,2}),\end{aligned} \tag{3.30}$$

where $o_{i,2} = \kappa_{i,2}^2/2 + \bar{h}_{i,2}^2/2 + \epsilon_{i,2}^2/2$, $\chi_{i,2} = W_{i,2}^{*T}(S_{i,1}(\nu_{i,2}) - S_{i,1}(\breve{\nu}_{i,2}))$, $\breve{\nu}_{i,2} = [\breve{z}_{i,2}, \breve{x}_{i,1}, \breve{x}_{i,2}]^T$, $\bar{h}_{i,2} > 0$ is a constant, and $|\epsilon_{i,2}| \leq \iota_{i,2}$.

Construct the virtual controller $\alpha_{i,2}$ and the differentiation Nussbaum gain variable $\Phi_{i,2}$ as:

$$\alpha_{i,2} = \mathcal{N}(\Psi_{i,2})(d_{i,2}\breve{z}_{i,2} + \frac{\breve{z}_{i,2}}{2\bar{h}_{i,2}^2}\hat{\ni}_i S_{i,2}^T(\breve{\nu}_{i,2})S_{i,2}(\breve{\nu}_{i,2})), \tag{3.31}$$

$$\dot{\Psi}_{i,2} = d_{i,2}\breve{z}_{i,2}^2 + \frac{\breve{z}_{i,2}^2}{2\bar{h}_{i,2}^2}\hat{\ni}_i S_{i,2}^T(\breve{\nu}_{i,2})S_{i,2}(\breve{\nu}_{i,2}). \tag{3.32}$$

Substituting (3.20), (3.31) and (3.32) into (3.30), we have

$$\begin{aligned}
\dot{\mathcal{V}}_{i,2} \leq & -\sum_{l=1}^{2}((d_{i,l} + \frac{1}{2})z_{i,l}^2 + \ell_{i,l}\mathcal{V}_{K_{i,l}}) + \sum_{l=1}^{2}(z_{i,l}\chi_{i,l} + o_{i,l} \\
& + (\theta_{i,l}^2||W_{i,l}^{*T}||^2 - \hat{\ni}_i)\frac{\breve{z}_{i,l}^2}{2\bar{h}_{i,l}^2}S_{i,l}^T(\breve{\nu}_{i,l})S_{i,l}(\breve{\nu}_{i,l})) \\
& + \sum_{l=1}^{2}(\theta_{i,l}\theta_{i,l+1}\mathcal{N}(\Psi_{i,l}) + 1)\dot{\Psi}_{i,l} + \frac{1}{2}z_{i,3}^2 - \bar{\omega}_{i,2}\mu_{i,2}^2 \\
& + \sum_{l=1}^{2}(1 - 2\tanh^2(\frac{z_{i,l}}{\lambda_{i,l}}))\mathcal{F}_{i,l}(\bar{x}_{i,l}) + \frac{l_{i,2}^2}{2}\mu_{i,3}^2,
\end{aligned} \tag{3.33}$$

where $\bar{\omega}_{i,2} = 1/\omega_{i,2} - \bar{A}_{i,2}^2/(2\kappa_{i,2}^2) - l_{i,1}^2/2$.

Step $(i, j(2 < j < n))$: Choose the jth Lyapunov candidate function: $\mathcal{V}_{i,j} = \mathcal{V}_{i,j-1} + (1/2)z_{i,j}^2 + (1/2)\mu_{i,j}^2 + \mathcal{V}_{K_{i,j}}$, and a Lyapunov-Krasovskii functional

$$\mathcal{V}_{K_{i,j}} = \frac{1}{2}\sum_{l=1}^{j}\frac{e^{\ell_{i,j}c_{i,l}}}{1 - \bar{c}_{i,l}}\int_{t-\tau_{i,l}}^{t} e^{-\ell_{i,j}(t-\xi)}\bar{a}_{i,j,l}^2(x_{i,l}(\xi))d\xi,$$

with $\ell_{i,j}$ is a positive constant.

Similar the above process, the derivative of $\mathcal{V}_{i,j}$ can be rewritten as

$$\begin{aligned}
\dot{\mathcal{V}}_{i,j} \leq & -\sum_{l=1}^{j}((c_{i,l} + \frac{1}{2})z_{i,l}^2 + \ell_{i,l}\mathcal{V}_{K_{i,l}}) + \sum_{l=1}^{j}(z_{i,l}\chi_{i,l} + o_{i,l} \\
& + (\theta_{i,l}^2||W_{i,l}^{*T}||^2 - \hat{\ni}_i)\frac{\breve{z}_{i,l}^2}{2\bar{h}_{i,l}^2}S_{i,l}^T(\breve{\nu}_{i,l})S_{i,l}(\breve{\nu}_{i,l})) + \frac{1}{2}z_{i,j+1}^2 \\
& + \sum_{l=1}^{j}(1 - 2\tanh^2(\frac{z_{i,l}}{\lambda_{i,l}}))\mathcal{F}_{i,l}(\bar{x}_{i,l}) - \sum_{l=2}^{j}\bar{\omega}_{i,l}\mu_{i,l}^2 \\
& + \sum_{l=1}^{j}(\theta_{i,l}\theta_{i,l+1}\mathcal{N}(\Psi_{i,l}) + 1)\dot{\Psi}_{i,l} + \frac{l_{i,j}^2}{2}\mu_{i,j+1}^2,
\end{aligned} \tag{3.34}$$

where $\bar{\omega}_{i,j} = 1/\omega_{i,j} - \bar{A}_{i,j}^2/(2\kappa_{i,j}^2) - l_{i,j-1}^2/2$, $o_{i,j} = \kappa_{i,j}^2/2 + \bar{h}_{i,j}^2/2 + \iota_{i,j}^2/2$.

Step (i, n): The actual control law u_i will be given in this step. From (3.8), the time derivative of nth error surface is given by:

$$\dot{z}_{i,n} = u_i + g_{i,n} + \delta_{i,n}(\bar{x}_{i,n}(\tau_{i,n})) + \Phi_i - \dot{\theta}_{i,n}\beta_{i,n} - \theta_{i,n}\dot{\beta}_{i,n}, \tag{3.35}$$

where $\Phi_i = \varsigma_i^u(t, \bar{x}_{i,n}) + l_i$ can be seen as the compounded disturbance, which will be tackled with constructing a nonlinear disturbance observer in this chapter. Therefore, (3.35) can be rewritten as

$$\begin{aligned} \dot{z}_{i,n} =& f_i^{-1}\mathcal{G}_{i,ni} + \delta_{i,n}(\bar{x}_{i,n}(\tau_{i,n})) + \Phi_i \\ =& f_i^{-1}(W_{i,ni}^{*T}S_{i,ni}(\breve{\nu}_{i,ni}) + \epsilon_{i,ni} + \chi_{i,ni}) \\ &+ \delta_{i,n}(\bar{x}_{i,n}(\tau_{i,n})) + \Phi_i, \end{aligned} \tag{3.36}$$

where $\mathcal{G}_{i,ni} = f_i(u_i + g_{i,n} - \dot{\theta}_{i,n}\beta_{i,n} - \theta_{i,n}\dot{\beta}_{i,n})$, $\chi_{i,ni} = W_{i,ni}^{*T}(S_{i,ni}(\nu_{i,ni}) - S_{i,ni}(\breve{\nu}_{i,ni}))$, $\nu_{i,n_i} = [z_{i,n}, x_{i,1}, \cdots, x_{i,n}, \theta_{i,n}]$, $\breve{\nu}_{i,ni} = [\breve{z}_{i,n}, \breve{x}_{i,n}, \cdots, \breve{x}_{i,n}]^T$, and $f_i > 0$ is a design parameter.

From (3.36), to facilitate stability analysis, an auxiliary variable is introduced:

$$s_i = \breve{z}_{i,n} - \varpi_i = \frac{z_{i,n}}{\theta_{i,n}} - \varpi_i, \tag{3.37}$$

where the intermediate variable ϖ_i is designed as

$$\dot{\varpi}_i = b_i s_i + \frac{\hat{\phi}_i s_i}{2 f_i \bar{h}_{i,ni}^2} S_{i,ni}^T(\breve{\nu}_{i,ni}) S_{i,ni}(\breve{\nu}_{i,ni}), \tag{3.38}$$

with $b_i > 0$ is a design parameter, and $\hat{\phi}_i = \phi_i - \tilde{\phi}_i$ will be given later.

Based on the preceding analysis, the derivative of s_i can be derived

$$\begin{aligned} \dot{s}_i =& \frac{1}{\theta_{i,n}}(f_i^{-1}(W_{i,ni}^{*T}S_{i,ni}(\breve{\nu}_{i,ni}) + \epsilon_{i,ni} + \chi_{i,ni}) + \Phi_i \\ &+ \delta_{i,n}(\bar{x}_{i,n}(\tau_{i,n}))) - \frac{\dot{\theta}_{i,n}}{\theta_{i,n}^2} z_{i,n} - b_i s_i \\ &- \frac{\hat{\phi}_i s_i}{f_i \bar{h}_{i,ni}^2} S_{i,ni}^T(\breve{\nu}_{i,ni}) S_{i,ni}(\breve{\nu}_{i,ni}). \end{aligned} \tag{3.39}$$

Applying Lemma 1.1, the following inequality can be inferred

$$\begin{aligned} &\frac{f_i^{-1}}{\theta_{i,n}}(W_{i,ni}^{*T}S_{i,ni}(\breve{\nu}) + \epsilon_{i,ni} + \chi_{i,ni})s_i \le s_i^2 + \frac{\chi_{i,ni}^2}{2 f_i^2 \underline{\theta}_{i,n}^2} \\ &+ \frac{\bar{h}_{i,ni}^2}{2 f_i \underline{\theta}_{i,n}^2} + \frac{\epsilon_{i,ni}^2}{2 f_i^2} + \frac{\phi_i s_i^2}{2 f_i \bar{h}_{i,ni}^2} S_{i,ni}^T(\breve{\nu}_{i,ni}) S_{i,ni}(\breve{\nu}_{i,ni}), \end{aligned} \tag{3.40}$$

$$\frac{s_i \Phi_i}{f_i \theta_{i,n}} \leq \frac{1}{2} s_i^2 + \frac{\bar{\Phi}_i^2}{2 f_i^2 \underline{\theta}_{i,n}^2}, \tag{3.41}$$

$$-\frac{\dot{\theta}_{i,n}}{\theta_{i,n}^2} z_{i,n} s_i \leq \frac{1}{2} s_i^2 + \frac{\overline{\Theta}_{i,n}^2}{2 \underline{\theta}_{i,n}^4} z_{i,n}^2, \tag{3.42}$$

$$\frac{s_i \delta_{i,n}(\bar{x}_{i,n}(\tau_{i,n}))}{\theta_{i,n}} \leq \frac{f_i^2}{2 \underline{\theta}_{i,n}^2} \sum_{k=1}^{n} \bar{a}_{i,n,k}^2(\bar{x}_{i,k}(t)) + \frac{s_i^2}{2 f_i^2}, \tag{3.43}$$

where $||W_{i,ni}^{*T}||^2/\theta_{i,n}^2 = \phi_i$, $\hat{\phi}_i$ is the estimation of ϕ_i, $\tilde{\phi}_i$ denotes the estimation error of ϕ_i.

From (3.39)–(3.43), it can be inferred that

$$\begin{aligned} s_i \dot{s}_i \leq & -(b_i - 2 - \frac{1}{2 f_i^2}) s_i^2 + \frac{\hat{\phi}_i s^2}{2 f_i \bar{h}_{i,ni}^2} S_{i,ni}^T(\check{\nu}_{i,ni}) S_{i,ni}(\check{\nu}_{i,ni}) \\ & + \frac{\overline{\Theta}_{i,n}^2}{2 \underline{\theta}_{i,n}^4} z_{i,n}^2 + \frac{\chi_{i,ni}^2}{2 f_i^2 \underline{\theta}_{i,n}^2} + \frac{f_i^2}{2 \underline{\theta}_{i,n}^2} \sum_{l=1}^{n} \bar{a}_{i,n,l}^2(\bar{x}_{i,l}(t)) \\ & + \Delta_{i,ni}, \end{aligned} \tag{3.44}$$

where $\Delta_{i,ni} = \frac{\bar{h}_{i,ni}^2}{2 f_i^2} + \frac{\epsilon_{i,ni}^2}{2 f_i^2 \underline{\theta}_{i,n}^2} + \frac{\bar{\Phi}_i^2}{2 f_i^2 \underline{\theta}_{i,n}^2}$.

Building on the previously mentioned intermediate variable s_i, the unknown disturbance observer is formulated as:

$$\hat{\Phi}_i = f_i(s_i - \varepsilon_i), \tag{3.45}$$

where $\dot{\varepsilon}_i = -b_i s_i + \hat{\Phi}_i$, $\hat{\Phi}_i$ is the estimation value of Φ_i, and $\tilde{\Phi}_i = \Phi_i - \hat{\Phi}_i$. Its first-order time derivative along (3.45) is represented by

$$\begin{aligned} \dot{\hat{\Phi}}_i = & \frac{W_{i,ni}^{*T}}{\theta_{i,n}} S_{i,ni}(\check{\nu}_{i,ni}) + \frac{f_i}{\theta_{i,n}} \tilde{\Phi}_i - f_i \hat{\Phi}_i + \frac{g_i}{\theta_{i,n}} \hat{\Phi}_i + \frac{\epsilon_{i,ni}}{\theta_{i,n}} \\ & + \frac{\chi_{i,ni}}{\theta_{i,n}} - \frac{f_i \dot{\theta}_{i,n}}{\theta_{i,n}^2} z_{i,n} - \frac{\hat{\phi}_i s_i}{2 \bar{h}_{i,ni}^2} S_{i,ni}^T(\check{\nu}_{i,ni}) S_{i,ni}(\check{\nu}_{i,ni}) \\ & + \frac{f_i}{\theta_{i,n}} \delta_{i,n}(\bar{x}_{i,n}(\tau_{i,n})), \end{aligned} \tag{3.46}$$

and $\hat{\phi}_i$ is constructed as

$$\dot{\hat{\phi}}_i = \begin{cases} \frac{s_i^2}{2 f_i \bar{h}_{i,ni}^2} S_{i,ni}^T(\check{\nu}_{i,ni}) S_{i,ni}(\check{\nu}_{i,ni}) - \sigma_i \hat{\phi}_i, & \bar{k}_i \leq 0; \\ 0, & \bar{k}_i > 0; \end{cases} \tag{3.47}$$

where $\bar{k}_i = s_i^2 S_{i,ni}^T(\breve{\nu}_{i,ni}) S_{i,ni}(\breve{\nu}_{i,ni})/(2 f_i \bar{h}_{i,ni}^2) - \sigma_i \hat{\phi}_i, \sigma_i$ is a positive design constant. In this chapter, the initial condition $\hat{\phi}_i(0) = \Psi_i$ is chosen as a positive constant, ensuring that $\hat{\phi}$ remains positive and satisfies $\hat{\phi}_i(t) \leq \Psi_i$.

From (3.46), it follows that

$$\begin{aligned}\tilde{\Phi}_i \dot{\tilde{\Phi}}_i =& \tilde{\Phi}_i \dot{\Phi}_i - \frac{W_{i,ni}^{*T}}{\theta_{i,n}} S_{i,ni}(\breve{\nu}_{i,ni}) \tilde{\Phi}_i - \frac{f_i}{\theta_{i,n}} \tilde{\Phi}_i \delta_{i,n}(\bar{x}_{i,n}(\tau_{i,n})) \\ & - \frac{\iota_{i,ni}}{\theta_{i,n}} \tilde{\Phi}_i - \frac{\chi_{i,ni}}{\theta_{i,n}} \tilde{\Phi}_i + \frac{f_i \dot{\theta}_{i,n}}{\theta_{i,n}^2} z_{i,n} \tilde{\Phi}_i - \frac{f_i}{\theta_{i,n}} \tilde{\Phi}_i^2 \\ & + \frac{\hat{\phi}_i s_i}{2\bar{h}_{i,ni}^2} S_{i,ni}^T(\breve{\nu}_{i,ni}) S_{i,ni}(\breve{\nu}_{i,ni}) \tilde{\Phi}_i + f_i^\theta \hat{\Phi}_i \tilde{\Phi}_i,\end{aligned} \tag{3.48}$$

where $f_i^\theta = f_i - f_i/\theta_{i,n}$.

By utilizing Lemma 1.1, we obtain

$$\tilde{\Phi}_i \dot{\Phi}_i \leq \frac{\varrho_i^1}{2} \tilde{\Phi}_i^2 + \frac{\bar{\dot{\Phi}}_i^2}{2\varrho_i^1}, \tag{3.49}$$

$$-\frac{W_{i,ni}^{*T}}{\theta_{i,n}} \varphi_{i,ni} \tilde{\Phi}_i \leq \frac{\varrho_i^2}{2} \tilde{\Phi}_i^2 + \frac{\phi_i h_{i,n_i}}{2\varrho_i^2}, \tag{3.50}$$

$$f_i^\theta \hat{\Phi}_i \tilde{\Phi}_i \leq \frac{\varrho_i^3}{2} \tilde{\Phi}_i^2 + \frac{(f_i^\theta)^2}{2\varrho_i^3} \bar{\Phi}_i^2 - f_i^\theta \tilde{\Phi}_i^2, \tag{3.51}$$

$$\frac{(f_i^\theta)^2}{2\varrho_i^3} \bar{\Phi}_i^2 \leq \frac{(\theta_{i,n}^2 + 1 + 2\underline{\theta}_{i,n})}{2\varrho_i^3 \underline{\theta}_{i,n}^2} f_i^2 \bar{\Phi}_i^2, \tag{3.52}$$

$$-\frac{\iota_{i,ni}}{\theta_{i,n}} \tilde{\Phi}_i \leq \frac{\varrho_i^4}{2} \tilde{\Phi}_i^2 + \frac{\epsilon_{i,ni}^2}{2\varrho_i^4 \underline{\theta}_{i,n}^2}, \tag{3.53}$$

$$-\frac{\chi_{i,ni}}{\theta_{i,n}} \tilde{\Phi}_i \leq \frac{\varrho_i^5}{2} \tilde{\Phi}_i^2 + \frac{\chi_i^2}{2\varrho_i^5 \underline{\theta}_{i,n}^2}, \tag{3.54}$$

$$\frac{f_i \dot{\theta}_{i,n}}{\theta_{i,n}^2} z_{i,n} \tilde{\Phi}_i \leq \frac{\varrho_i^6}{2} \tilde{\Phi}_i^2 + \frac{f_i^2 \overline{\Theta}_{i,n}^2}{2\varrho_i^6 \underline{\theta}_{i,n}^4} z_{i,n}^2, \tag{3.55}$$

$$\begin{aligned}& \frac{\hat{\phi}_i s_i}{2\bar{h}_{i,ni}^2} S_{i,ni}^T(\breve{\nu}_{i,n}) S_{i,ni}(\breve{\nu}_{i,ni}) \tilde{\Phi}_i - \frac{f_i}{\theta_{i,n}} \tilde{\Phi}_i \delta_{i,n}(\bar{x}_{i,n}(\tau_{i,n})) \\ \leq & \frac{\varrho_i^7}{2} \tilde{\Phi}_i^2 + \frac{\Psi_i^2 h_{i,n_i}^2}{8\varrho_i^7 \bar{h}_{i,ni}^4} s_i^2 + \frac{f_i^2}{2\underline{\theta}_{i,n}^2} \sum_{k=1}^{n} \bar{a}_{i,n,k}^2(x_{i,k}) + \frac{1}{2} \tilde{\Phi}_i^2,\end{aligned} \tag{3.56}$$

where $\varrho_i^1, \varrho_i^2, \varrho_i^3, \varrho_i^4, \varrho_i^5, \varrho_i^6, \varrho_i^7$ are positive constants.

Substituting (3.49)-(3.56) into (3.48), (3.48) can be rewritten as

$$
\begin{aligned}
\tilde{\Phi}_i\dot{\tilde{\Phi}}_i \leq & -(f_i - \frac{1}{2} - \frac{\varrho_i}{2})\tilde{\Phi}_i^2 + \frac{\chi_{i,ni}^2}{2\varrho_i^5\underline{\theta}_{i,n}^2} + \frac{f_i^2\overline{\Theta}_{i,n}^2}{2\varrho_i^6\underline{\theta}_{i,n}^4}z_{i,n}^2 \\
& + \frac{\Psi_i^2 h_{i,n_i}}{8\varrho_i^7\bar{h}_{i,ni}^4}s_i^2 + \frac{f_i^2}{2\underline{\theta}_{i,n}^2}\sum_{k=1}^{n}\bar{a}_{i,n,k}^2(x_{i,k}) + \Omega_i,
\end{aligned} \tag{3.57}
$$

where $\varrho_i = \sum_{l=1}^{7}\varrho_i^l$, $\Omega_i = \bar{\bar{\Phi}}_i^2/(2\varrho_i^1) + \phi_i h_{i,n_i}/(2\varrho_i^2) + (\underline{\theta}_{i,n}^2 + 1 + 2\underline{\theta}_{i,n})/(2\varrho_i^3\underline{\theta}_{i,n}^2)f_i^2\bar{\Phi}_i^2 + \epsilon_{i,ni}^2/(2\varrho_i^4\underline{\theta}_{i,n}^2)$.

The nth Lyapunov candidate function $\mathcal{V}_{i,n}$ is selected as:

$$
\begin{aligned}
\mathcal{V}_{i,n} = & \mathcal{V}_{i,n-1} + \frac{1}{2}z_{i,n}^2 + \frac{1}{2}\mu_{i,n}^2 + \mathcal{V}_{K_{i,n}} + \frac{1}{2}s_i^2 + \frac{1}{2}\tilde{\phi}_i^2 \\
& + \frac{1}{2}\tilde{\Phi}_i^2 + \frac{1}{2}\tilde{\ni}_i^2,
\end{aligned} \tag{3.58}
$$

where $\tilde{\ni}_i = \ni_i - \hat{\ni}_i$ denotes the estimation error of $\ni_i$, and a Lyapunov-Krasovskii function

$$
\mathcal{V}_{K_{i,n}} = \frac{1}{2}\sum_{k=1}^{n}\frac{e^{\ell_{i,n}c_{i,k}}}{1-\bar{c}_{i,k}}\int_{t-\tau_{i,k}}^{t} e^{-\ell_{i,n}(t-\xi)}\bar{a}_{i,n,k}^2(x_{i,k}(\xi))d\xi,
$$

with $\ell_{i,n}$ is a positive constant.

From (3.8), the expression of $\dot{\mathcal{V}}_{i,n}$ is expressed as:

$$
\begin{aligned}
\dot{\mathcal{V}}_{i,n} = & \dot{\mathcal{V}}_{i,n-1} + z_{i,n}(u_i + g_{i,n} + \delta_{i,n}(\bar{x}_{i,n}(\tau_{i,n})) + \Phi_i \\
& - \dot{\theta}_{i,n}\beta_{i,n} - \theta_{i,n}\dot{\beta}_{i,n}) + \mu_{i,n}\dot{\mu}_{i,n} + \dot{\mathcal{V}}_{K_{i,n}} + s_i\dot{s}_i \\
& + \tilde{\phi}_i\dot{\tilde{\phi}}_i + \tilde{\Phi}_i\dot{\tilde{\Phi}}_i + \tilde{\ni}_i\dot{\tilde{\ni}}_i.
\end{aligned} \tag{3.59}
$$

Similar to the above steps, the following inequality is obtained

$$
\begin{aligned}
z_{i,n}\delta_{i,n}(\bar{x}_{i,n}(\tau_{i,n})) + \dot{\mathcal{V}}_{K_{i,n}} \leq & -\ell_{i,n}\mathcal{V}_{K_{i,n}} + \frac{1}{2}z_{i,n}^2 \\
& + \frac{1}{2}\sum_{k=1}^{n}\frac{e^{-\ell_{i,n}c_{i,k}}}{1-\bar{c}_{i,k}}\bar{a}_{i,n,k}^2(\bar{x}_{i,l}(t)),
\end{aligned} \tag{3.60}
$$

$$
\mu_{i,n}\dot{\mu}_{i,n} \leq -\frac{\mu_{i,n}^2}{\omega_{i,n}} + \frac{\bar{A}_{i,n}^2}{2\kappa_{i,n}^2}\mu_{i,n}^2 + \frac{\kappa_{i,n}^2}{2}, \tag{3.61}
$$

where $\kappa_{i,n}$ is a positive constant.

Substituting (3.44), (3.47), (3.57), (3.60) and (3.61) into (3.59), one has

$$\begin{aligned}
\dot{\mathcal{V}}_{i,n} \leq & \dot{\mathcal{V}}_{i,n-1} + z_{i,n}(u_i + \Phi_i) - \underline{b}_i s_i^2 - \underline{f}_i \tilde{\Phi}_i^2 - d_{i,n} z_{i,n}^2 \\
& - \left(\frac{1}{\omega_{i,n}} - \frac{\bar{A}_{i,n}^2}{2\kappa_{i,n}^2}\right)\mu_{i,n}^2 - \frac{\sigma_i}{2}\tilde{\phi}_i^2 - \ell_{i,n}\mathcal{V}_{K_{i,n}} - \frac{3}{2} z_{i,n}^2 \\
& + \left(1 - 2\tanh^2\left(\frac{z_{i,n}}{\lambda_{i,n}}\right)\right)\mathcal{F}_{i,n}(\bar{x}_{i,n}) + d_{i,n}\breve{z}_{i,n}^2 + \Omega_i \\
& + \left(\frac{1}{2 f_i^2 \underline{\theta}_{i,ni}^2} + \frac{1}{2\varrho_i^5 \underline{\theta}_{i,n}^2}\right)\chi_{i,ni}^2 + \frac{\kappa_{i,n}^2}{2} + \frac{\sigma_i}{2}\phi_i^2 \\
& + \Delta_{i,ni} + z_{i,n}\mathcal{J}_{i,n}(\nu_{i,n}) + \tilde{\ni}_i \dot{\hat{\ni}}_i,
\end{aligned} \tag{3.62}$$

where $\mathcal{J}_{i,n}(\nu_{i,n}) = g_{i,n} - \dot{\theta}_{i,n}\beta_{i,n} - \theta_{i,n}\dot{\beta}_{i,n} - d_{i,n}z_{i,n}/\theta_{i,n}^2 + f_i^2\overline{\Theta}_{i,n}^2/(2\varrho_i^6\underline{\theta}_{i,n}^4)z_{i,n} + \overline{\Theta}_{i,n}^2/(2\underline{\theta}_{i,n}^4)z_{i,n} + d_{i,n}z_{i,n} + 5z_{i,n}/2 + (1/z_{i,n})\tanh^2(z_{i,n}/\lambda_{i,n})\sum_{l=1}^{n}(e^{\ell_{i,n}c_{i,l}})/(1 - \overline{c}_{i,l})\bar{a}_{i,2,l}^2(x_{i,l}(t)) + (f_i^2/\underline{\theta}_{i,n}^2)(2/z_{i,n})\tanh^2(z_{i,n}/\lambda_{i,n})\sum_{l=1}^{n}\bar{a}_{i,2,l}^2(x_{i,l}(t))$, $\mathcal{F}_{i,n}(\bar{x}_{i,n}) = f_i^2/\underline{\theta}_{i,n}^2\sum_{k=1}^{n}\bar{a}_{i,n,k}^2(\bar{x}_{i,k}(t)) + \sum_{k=1}^{n}e^{-\ell_{i,n}c_{i,k}}/(2(1 - \overline{c}_{i,k}))\bar{a}_{i,n,k}^2(\bar{x}_{i,k}(t))$,
$\underline{b}_i = b_i - 2 - 1/(2f_i^2) - \Psi_i^2 h_{i,n_i}^2/(8\varrho_i^7\bar{h}_{i,ni}^4)$, $\underline{f}_i = f_i - \varrho_i/2 - 1/2$, $\nu_{i,n} = [z_{i,n}, x_{i,1}, \cdots, x_{i,n}, \theta_{i,n}]^T$, $\breve{\nu}_{i,n} = [\breve{z}_{i,n}, \breve{x}_{i,1}, \cdots, \breve{x}_{i,n}]^T$, $c_{i,n}$ is a positive constant.

We approach the term $z_{i,n}\mathcal{S}_{i,n}(\nu_{i,n})$, in the same manner, and $\dot{\mathcal{V}}_{i,n-1}$ has been derived in *Step* $(i, n-1)$. Hence, (3.62) can be rewritten as

$$\begin{aligned}
\dot{\mathcal{V}}_{i,n} \leq & -\sum_{l=1}^{n}\left(\left(d_{i,l} + \frac{1}{2}\right)z_{i,l}^2 + \ell_{i,l}\mathcal{V}_{K_{i,l}}\right) + \sum_{l=1}^{n}(z_{i,l}\chi_{i,l} + o_{i,l}) \\
& + \sum_{l=1}^{n-1}\left((\theta_{i,l}^2\|W_{i,l}^{*T}\|^2 - \hat{\ni}_i)\frac{\breve{z}_{i,l}^2}{2\bar{h}_{i,l}^2}S_{i,l}^T(\breve{\nu}_{i,l})S_{i,l}(\breve{\nu}_{i,l})\right) \\
& + \sum_{l=1}^{n-1}(\theta_{i,l}\theta_{i,l+1}\mathcal{N}(\Psi_{i,l}) + 1)\dot{\Psi}_{i,l} - \sum_{l=2}^{n}\bar{\omega}_{i,l}\mu_{i,l}^2 - \frac{1}{2}z_{i,n}^2 \\
& + z_{i,n}(u_i + \Phi_i) - \underline{b}_i s_i^2 - \underline{f}_i\tilde{\Phi}_i^2 - \frac{\sigma_i}{2}\tilde{\phi}_i^2 + d_{i,n}\breve{z}_{i,n}^2 \\
& + \left(\frac{1}{2f_i^2\underline{\theta}_{i,ni}^2} + \frac{1}{2\varrho_i^5\underline{\theta}_{i,n}^2}\right)\chi_{i,ni}^2 + \frac{\sigma_i}{2}\phi_i^2 + \Delta_{i,ni} + \Omega_i \\
& + \theta_{i,n}^2\|W_{i,n}^{*T}\|^2\frac{\breve{z}_{i,n}^2}{2\bar{h}_{i,n}^2}S_{i,n}^T(\breve{\nu}_{i,n})S_{i,n}(\breve{\nu}_{i,n}) + \tilde{\ni}_i\dot{\hat{\ni}}_i \\
& + \sum_{l=1}^{n}\left(1 - 2\tanh^2\left(\frac{z_{i,l}}{\lambda_{i,l}}\right)\right)\mathcal{F}_{i,l}(\bar{x}_{i,l}),
\end{aligned} \tag{3.63}$$

where $\bar{\omega}_{i,n} = 1/\omega_{i,n} - \bar{A}_{i,n}^2/(2\kappa_{i,n}^2) - l_{i,n-1}^2/2$, $o_{i,n} = \kappa_{i,n}^2/2 + \bar{h}_{i,n}^2/2 + \epsilon_{i,n}^2/2$.

Given the vulnerability of communication networks about the CPSs, each data transmission carries the risk of being hijacked. Developing a strategy to minimize the frequency of data transmissions while enhancing the defense capabilities of communication networks is of utmost importance. The event-triggered strategy is recognized for its effectiveness in conserving communication resources, where transmission occurs only when a specific condition is met. Consequently, this chapter focuses on designing an event-triggered adaptive control scheme in the presence of deception attacks. In this section, the relative threshold-based event-triggered approach is employed. Specifically, when the measurement error surpasses the predetermined threshold, the controller is updated. In other words, the event-triggered control signal can be expressed as

$$u_i(t) = v_i(t_k^i), \forall t \in [t_k^i, t_{k+1}^i). \tag{3.64}$$

The event-triggered mechanism is designed as

$$t_{k+1}^i = \inf\{t > t_k^i \,|\, |e_i(t)| \geq a_i |u_i(t)| + a_i^{'}\}, \tag{3.65}$$

where $v_i(t_k^i)$ is successfully transition control signal, $e_i(t) = v_i(t) - u_i(t)$ stands the measurement error, $0 < a_i < 1$ and $a_i^{'}$ denote the positive design parameters.

From (3.65), one has

$$v_i(t) = (1 + \zeta_i(t) a_i) u_i(t) + \zeta_i^{'}(t) a_i^{'}, \tag{3.66}$$

where $\zeta_i(t), \zeta_i^{'}(t) \in [-1, 1]$ are time-varying variables.

Thus, u_i can be represented as

$$u_i = \frac{v_i(t)}{1 + \zeta_i(t) a_i} - \frac{\zeta_i^{'}(t) a_i^{'}}{1 + \mu_i(t) a_i}. \tag{3.67}$$

The transition control input $v_i(t)$ and adaptive law $\hat{\ni}_i$ are designed as

$$v_i(t) = -(1 + a_i)(\alpha_{i,n} \tanh(\frac{\breve{z}_{i,n}\alpha_{i,n}}{\wp_i}) + \bar{a}_i \tanh(\frac{\breve{z}_{i,n}\bar{a}_i}{\wp_i})), \tag{3.68}$$

$$\alpha_{i,n} = \mathcal{N}(\Psi_{i,n})(d_{i,n}\breve{z}_{i,n} + \frac{\breve{z}_{i,n}}{2\bar{h}_{i,n}^2} \hat{\ni}_i S_{i,n}^T(\breve{v}_{i,n}) S_{i,n}(\breve{v}_{i,n})) - \hat{\Phi}_i, \tag{3.69}$$

$$\dot{\Psi}_{i,n} = d_{i,n}\breve{z}_{i,n}^2 + \frac{\breve{z}_{i,n}^2}{2\bar{h}_{i,n}^2} \hat{\ni}_i S_{i,n}^T(\breve{v}_{i,n}) S_{i,n}(\breve{v}_{i,n}), \tag{3.70}$$

$$\dot{\hat{\ni}}_i = \sum_{l=1}^{n} \frac{\breve{z}_{i,l}^2}{2\bar{h}_{i,l}^2} S_{i,l}^T(\breve{v}_{i,l}) S_{i,l}(\breve{v}_{i,l}) - \sigma_i^{'} \hat{\ni}_i, \tag{3.71}$$

where $\sigma_i^{'}$, $\wp_i$ are design positive parameters, $\bar{a}_i > a_i^{'}/(1 - a_i)$.

Based on the aforementioned design and Lemma 1.2, we can derive

$$
\begin{aligned}
z_{i,n}u_i =& \frac{v_i(t)z_{i,n}}{1+\zeta_i(t)a_i} - \frac{\zeta_i^{'}(t)a_i^{'}z_{i,n}}{1+\zeta_i(t)a_i} \\
\leq& \theta_{i,n}\left(\frac{\breve{z}_{i,n}v_i(t)}{1+a_i} + \left|\frac{a_i^{'}\breve{z}_{i,n}}{1-a_i}\right|\right) \\
\leq& \theta_{i,n}\left(-\breve{z}_{i,n}\alpha_{i,n}\tanh\left(\frac{\breve{z}_{i,n}\alpha_{i,n}}{\wp_i}\right) + |\breve{z}_{i,n}\alpha_{i,n}|\right. \\
&\left. + \breve{z}_{i,n}\alpha_{i,n} - \breve{z}_{i,n}\bar{a}_i\tanh\left(\frac{\breve{z}_{i,n}\bar{a}_i}{\wp_i}\right) + |\breve{z}_{i,n}\bar{a}_i|\right) \\
\leq& 0.557\,\wp_i\theta_{i,n} + \theta_{i,n}\breve{z}_{i,n}\alpha_{i,n} \\
\leq& 0.557\,\wp_i\overline{\theta}_{i,n} + \theta_{i,n}\mathcal{N}(\Psi_{i,n})\dot{\Psi}_{i,n} - z_{i,n}\hat{\Phi}_i.
\end{aligned} \tag{3.72}
$$

Substituting (3.71) and (3.75) into (3.63), one has

$$
\begin{aligned}
\dot{\mathcal{V}}_{i,n} \leq& -\sum_{l=1}^{n}\left((d_{i,l}+\frac{1}{2})z_{i,l}^2 + \ell_{i,l}\mathcal{V}_{K_{i,l}}\right) + \sum_{l=1}^{n}(z_{i,l}\chi_{i,l} + o_{i,l}) \\
&+ \sum_{l=1}^{n}\left((\theta_{i,l}^2\|W_{i,l}^{*T}\|^2 - \eth_i)\frac{\breve{z}_{i,l}^2}{2\bar{h}_{i,l}^2}S_{i,l}^T(\breve{v}_{i,l})S_{i,l}(\breve{v}_{i,l})\right) \\
&+ \sum_{l=1}^{n-1}(\theta_{i,l}\theta_{i,l+1}\mathcal{N}(\Psi_{i,l})+1)\dot{\Psi}_{i,l} - \sum_{l=2}^{n}\bar{\omega}_{i,l}\mu_{i,l}^2 \\
&+ z_{i,n}\tilde{\Phi}_i - \underline{b}_i s_i^2 - \underline{f}_i\tilde{\Phi}_i^2 - \frac{\sigma_i}{2}\tilde{\phi}_i^2 + \frac{\sigma_i}{2}\phi_i^2 - \frac{1}{2}z_{i,n}^2 \\
&+ \left(\frac{1}{2f_i^2\underline{\theta}_{i,ni}^2} + \frac{1}{2\varrho_i^5\theta_{i,n}^2}\right)\chi_{i,ni}^2 + \Delta_{i,ni} + \Omega_i \\
&+ (\theta_{i,n}\mathcal{N}(\Phi_{i,n})+1)\dot{\Psi}_{i,n} + 0.557\wp_i\overline{\theta}_{i,n} + \sigma_i^{'}\tilde{\eth}_i\hat{\eth}_i \\
&+ \sum_{l=1}^{n}\left(1-2\tanh^2\left(\frac{z_{i,l}}{\lambda_{i,l}}\right)\right)\mathcal{F}_{i,l}(\bar{x}_{i,l}) \\
&- \sum_{l=1}^{n}\frac{\breve{z}_{i,l}^2}{2\bar{h}_{i,l}^2}\tilde{\eth}_i S_{i,l}^T(\breve{v}_{i,l})S_{i,l}(\breve{v}_{i,l}).
\end{aligned} \tag{3.73}
$$

With the prior theoretical groundwork established, the key result regarding the consensus control mechanism for the investigated cyber-physical systems is summarized in the following theorem.

Theorem 3.1 *For the nonlinear CPSs experiencing unknown time-delays and deception attacks, If the Assumptions 3.1–3.3 are hold and the virtual control laws* (3.18), (3.31) *and the actual control law* (3.68), *auxiliary variables* (3.19), (3.32), (3.70), *and the adaptive law* (3.71) *are both well defined, then all states belonging* *to the closed-loop systems* (3.1) *are bounded. Furthermore, the output consensus errors will be converged to an arbitrary small region around the origin by adjusting these design parameters appropriately.*

Proof. It is easy to obtain the following inequalities:

$$\sigma_i^{'}\tilde{\ni}_i\hat{\ni}_i \leq \frac{\sigma_i^{'}}{2}\ni_i^2 - \frac{\sigma_i^{'}}{2}\tilde{\ni}_i^2, \tag{3.74}$$

$$z_{i,n}\tilde{\Phi}_i \leq \frac{1}{2}z_{i,n}^2 + \frac{1}{2}\tilde{\Phi}_i^2, \tag{3.75}$$

$$z_{i,l}\chi_{i,l} \leq \frac{1}{2}z_{i,l}^2 + \frac{1}{2}\bar{\chi}_{i,l}^2, \tag{3.76}$$

where $\chi_{i,l}$ is bounded as $|\chi_{i,l}| \leq \bar{\chi}_{i,l}$, and $\bar{\chi}_{i,l}$ is an unknown constant.

Substituting (3.74)-(3.75) into (3.73), it yields

$$\begin{aligned}\dot{\mathcal{V}}_{i,n} \leq & -\sum_{l=1}^{n} d_{i,l}z_{i,l}^2 - \sum_{l=1}^{n}\ell_{i,l}\mathcal{V}_{K_{i,l}} - \sum_{l=2}^{n}\bar{\omega}_{i,l}\mu_{i,l}^2 - \frac{\sigma_i^{'}}{2}\tilde{\ni}_i^2 \\ & + \sum_{l=1}^{n}(\theta_{i,l}^*\mathcal{N}(\Psi_{i,l}) + 1)\dot{\Psi}_{i,l} - \underline{b}_i s_i^2 - (\underline{f}_i - \frac{1}{2})\tilde{\Phi}_i^2 \\ & + \sum_{l=1}^{n}(1 - 2\tanh^2(\frac{z_{i,l}}{\lambda_{i,l}}))\mathcal{F}_{i,l}(\bar{x}_{i,l}) + \Delta_i - \frac{\sigma_i}{2}\tilde{\phi}_i^2,\end{aligned}$$

where $\Delta_i = (1/(2f_i^2\underline{\theta}_{i,ni}^2) + 1/(2\varrho_i^5\underline{\theta}_{i,n}^2))\bar{\chi}_{i,ni}^2 + 0.557\wp_i\overline{\theta}_{i,n} + (\sigma_i^{'}/2)\ni_i^2 + \sum_{l=1}^{n}(1/2)\bar{\chi}_{i,l}^2 + \Delta_{i,ni} + \Omega_i + (\sigma_i/2)\phi_i^2 + \sum_{l=1}^{n} o_{i,l}$, $\theta_i^* = \max\{\theta_{i,l}\theta_{i,l+1}, \theta_{i,n}; l = 1, 2, \cdots, n-1\}$, $\ni_i = \max\{\theta_{i,l}^2\|\varepsilon_{i,l}^{*T}\|^2, l = 1, 2, \ldots, n\}$, $\hat{\iota}_i$ is the estimation value of $\ni_i$.

The overall Lyapunov function is constucted as: $\mathcal{V}(t) = \sum_{i=1}^{N}\mathcal{V}_{i,n}(t)$, and the derivative of $V(t)$ implies

$$\begin{aligned}\dot{\mathcal{V}} \leq & -C\mathcal{V} + \sum_{i=1}^{N}\Delta_i + \sum_{i=N}^{N}\sum_{l=1}^{n}(\theta_{i,l}^*\mathcal{N}(\Psi_{i,l}) + 1)\dot{\Psi}_{i,l} \\ & + \sum_{i=n}^{N}\sum_{l=1}^{n}(1 - 2\tanh^2(\frac{z_{i,l}}{\lambda_{i,l}}))\mathcal{F}_{i,l}(\bar{x}_{i,l}),\end{aligned} \tag{3.77}$$

where $C = \min\{2d_{i,l}, \ell_{i,l}, 2\underline{b}_i, 2\underline{f}_i - 1, 2\bar{\lambda}_{i,j}, \sigma_i, \sigma_i^{'}, i = 1, 2, \ldots, N; l = 1, 2, \ldots, n; j = 2, \ldots, n\}$. In order to guarantee the stability of system (3.1), relevant parameters should satisfy the following condition: $\underline{b}_i > 0$; $2\underline{f}_i - 1 > 0$; $\bar{\omega}_{i,j} > 0$. It is important to note that $\sum_{i=1}^{N}\Delta_i$ is the constant term, which is determined by the choice of parameters. Subsequently, we will thoroughly examine the fourth term of (3.77) by applying Lemma 3.2 and 3.3, considering the following three cases. We define a compact set $\Gamma_{z_{i,l}} := \{(i,l) | |z_{i,l}| < 0.8814\lambda_{i,l}, i = 1, 2, \ldots, N; l = 1, 2, \ldots, n\}$.

Case 1: $z_{i,l} \in \Gamma_{z_{i,l}}$ with $i = 1, 2, \ldots, N; l = 1, 2, \ldots, n$. Then, we guarantee a prescribed convergence domain of the output consensus error $|\breve{z}_{i,1}| \leq 0.8814\lambda_{i,1}/\underline{\theta}_{i,1}$. The boundary of $\breve{z}_{i,l}$ will converge to an arbitrarily small domain around the origin by selecting the appropriate parameter $\lambda i, l$. In the next step, we need to verify the boundness of $y_i - y_j$.

Define $\breve{x}_1 = [\breve{x}_{1,1}, \ldots, \breve{x}_{N,1}]^T$, $\omega_1 = [\omega_{1,1}, \ldots, \omega_{N,1}]^T$, $\breve{z}_1 = [\breve{z}_{1,1}, \ldots, \breve{z}_{N,1}]^T$, According to the definition (3.8), one has

$$\dot{\omega}_1 = -\mathcal{L}_c\breve{x}_1 = -\mathcal{L}_c\breve{z}_1 - \mathcal{L}_c\omega_1.$$

We know that $\mathcal{G}$ is strongly connected with the help of Assumption 3.2, so the matrix $\mathcal{L}_c$ has a zero eigenvalue whose the eigenvector is 1_N, and the other eigenvalues are in the open right half plane. $\mathcal{L}_c = \mathcal{P}\mathcal{J}\mathcal{P}^{-1}$. $\mathcal{J} = diag\{0, \varsigma\}$ is the Jordan canonical form of $\mathcal{L}_c$, and $\mathcal{P}$ is positive definite matrix consisting of columns integrated by the right eigenvectors of $\mathcal{L}_c$. Defining $\Lambda = h^{-1}\beta_1$, we have

$$\dot{\Lambda} = -\mathcal{J}\Lambda - \mathcal{J}\mathcal{P}^{-1}\breve{z}_1.$$

It is obviously that $\dot{\Lambda}_1 = 0$ and $\Lambda_1(t) = \dot{\Lambda}_1(0)$ where Λ_1 is the first row of Λ and

$$\mathcal{J}\mathcal{P}^{-1} = \begin{bmatrix} 0_{1\times N} \\ \Pi_{(N-1)\times N} \end{bmatrix}.$$

Let $\hat{\Lambda} = [\Lambda_2, \ldots, \Lambda_N]^T$, $\hat{J} = diag\{\varrho\}$, $\breve{z}_2 = [\breve{z}_{2,1}, \ldots, \breve{z}_{N,1}]^T$ then

$$\dot{\hat{\Lambda}} = -\hat{\mathcal{J}}\hat{\Lambda} - \Pi\breve{z}_2.$$

Because $\hat{\mathcal{J}} > 0$ and $||\breve{z}_2||$ is bounded, thus $||\Lambda||$ is bounded. Consider the Lyapunov function $\mathcal{V}_\Lambda = \hat{\Lambda}^T\mathcal{Q}\hat{\Lambda}$, where $\mathcal{Q}$ is the solution of $\hat{\mathcal{J}}\mathcal{Q} + \mathcal{Q}\hat{\mathcal{J}} = -2\mathcal{I}$. Thus, the relational expression of $\dot{\mathcal{V}}_\Lambda$ can be inferred:

$$\dot{\mathcal{V}}_\Lambda \leq -\hat{\Lambda}^T\hat{\Lambda} + 2||\mathcal{Q}\Pi||\,||\breve{z}_2||^2.$$

Since $\int_0^\infty ||\breve{z}_2||^2 d\tau$ is also bounded. Therefore

$$\int_0^\infty \hat{\Lambda}^T\hat{\Lambda}d\tau \leq -\mathcal{V}_\Lambda(t) + \mathcal{V}_\Lambda(0) + 2||\mathcal{Q}\Pi|| \int_0^\infty ||\breve{z}_2||^2 d\tau,$$

which represents $\int_0^\infty \hat{\Lambda}^T\hat{\Lambda}d\tau$ is bounded. According to the *Barbalat*'s lemma, $\lim_{t\to\infty} ||\hat{\Lambda}|| = 0$. Since the first column vector of $\mathcal{P}$ is $1_\mathcal{N}$, the following formula holds

$$\lim_{t\to\infty} \breve{x}_1 = 1_{\mathcal{N}} h^T \breve{x}_1(0), \tag{3.78}$$

where h^{-1} is the first row of h^{-1}. On the basis of (3.78), we can further obtain that

$$\lim_{t\to\infty} (\breve{x}_{i,1} - \breve{x}_{j,1}) = 0, \forall i, j \in N.$$

Further, one has

$$\lim_{t\to\infty} (y_i - y_j) = 0, \forall i, j \in N.$$

Therefore, the output of the leaderless-following MASs will be reach consensus asymptotically.

Case 2: $z_{i,l} \notin \Gamma_{z_{i,l}}$ with $i = 1, 2, \ldots, N; l = 1, 2, \ldots, n$. Then, with the help of the inequality $1 - 2\tanh(z_{i,l}/\lambda_{i,l}) \le 0$, (3.77) becomes

$$\dot{\mathcal{V}} \le -C\mathcal{V} + \sum_{i=1}^{N} \Delta_i + \sum_{i=N}^{N} \sum_{l=1}^{n} (\theta_{i,l}^{*} \mathcal{N}(\Psi_{i,l}) + 1)\dot{\Psi}_{i,l}. \tag{3.79}$$

Multiplying both sides of (3.79) by e^{at} and integrating it over $[0, t]$, we have

$$\begin{aligned} V \le & e^{bt} \sum_{i=N}^{N} \sum_{l=1}^{n} \int_0^t (\theta_{i,l}^{*} \mathcal{N}(\Psi_{i,l}) + 1)\dot{\Psi}_{i,l} e^{bt} d\tau + \mathcal{V}(0) \\ & + \frac{\sum_{i=1}^{n} \Delta_i}{C}. \end{aligned} \tag{3.80}$$

From [24] and Lemma 3.4, it can be seen that $z_{i,l}$, $\hat{\vartheta}_i$, $\tilde{\Phi}_i$, s_i, ϕ_i are uniformly ultimately bounded with $i = 1, 2, \ldots, N; l = 1, 2, \ldots, n$. Thus, $\breve{z}_{i,l}$ is also bounded that can be made arbitrarily small by adjusting C. The boundness of the convergence consensus error can be demonstrated with the way of *Case* 1.

Case 3: In this case, we divide the set $z_{i,l}$ into two subsets $z_{i_1,l_1} \in \Gamma_{z_{i_1,l_1}}$ and $z_{i_2,l_2} \notin \Gamma_{z_{i_2,l_2}}$ with $i_1 \in S_{i_1}, l_1 \in S_{l_1}, i_2 \in S_{i_2}, l_2 \in S_{l_2}$. With the help of *Case* 1, if $z_{i_1,l_1} \in \Gamma_{z_{i_1,l_1}}$ holds, it is obvious that $|z_{i_1,l_1}| < 0.8814\delta_{i_1,l_1}$. Therefore, $\breve{z}_{i_1,1}$ is also bounded with $\forall i_l \in S_{i_l}$. For $z_{i_2,l_2} \notin \Gamma_{z_{i_2,l_2}}$, we can deduce that $1 - 2\tanh(z_{i_2,l_2}/\lambda_{i_2,l_2}) \le 0$. By using the way of *Case* 2, $z_{i_2,1}$, $\forall i_2 \in S_{i_2}$ is also bounded. According to the demonstration in *Case* 1, the consensus error $y_i - y_j$ will converge around the origin with a arbitrarily small neighborhood by adjusting the design parameter appropriately.

After taking the above three cases into consideration, the proof is completed.

Remark 3.5 To prevent the Zeno behavior, we suppose that time constant $T^* > 0$ can be found, satisfying $\forall k \in z^+, t_{k+1} - t_k \le T^*$. From the event-triggered mechanism (3.65), we have

$$\frac{d|e_i(t)|}{dt} = \frac{d}{dt}\sqrt{e_i^2} = \text{sign}(e_i)\dot{e}_i \le |\dot{v}_i(t)|.$$

According to (3.68), the functions v_i and $\dot{v}_i$ are composed of a set of bounded signals in the close-loop systems. Therefore, there exists a positive constant $\bar{v}_i$ satisfying $|\dot{v}_i| \leq \bar{v}_i$. For the initial condition $e_i(t_k) = 0$, $\lim_{t \to t_{k+1}} e_i(t) = a_i|u_i(t)| + a_i^{'} = \breve{\varpi}_i > 0$, the time of trigger interval $T^* \geq t_{k+1} - t_k \geq \frac{\breve{\varpi}_i}{\bar{v}_i}$ can be obtained. It indicates that Zeno behavior is impossible.

Remark 3.6 According to the above analysis, we can achieve a good control performance by adjusting the parameter C. For instance, we can design large $d_{i,j}$, b_i, f_i, σ_i and $\sigma_i^{'}$, small $\omega_{i,j}$ and $\bar{h}_{i,j}$. Increasing $d_{i,j}$, b_i, f_i, σ_i, $\sigma_i^{'}$ will accelerate the convergence speed of the system state. In the meanwhile, the control cost will augment correspondingly. The designers can make decision according to the practical engineering requirement.

3.3 Simulation Results

To demonstrate the effectiveness of the proposed control strategy, we provide the following examples in this section. We examine the nonlinear cyber physical MASs composed of four agents 1–4, with the adjacency matrix representing their communication relationship given by $\mathcal{A} = [0, 0, 1, 1; 1, 0, 0, 0; 0, 0, 0, 1; 0, 1, 0, 0]$.

Example 3.1 The dynamics equation of the ith agent is described as follows:

$$\begin{cases} \dot{x}_{i,1} = x_{i,2} + g_{i,1}(x_{i,1}) + \delta_{i,1}(x_{i,1}(\tau_{i,1})) \\ \dot{x}_{i,2} = u_i + \varsigma_i^u(t, \bar{x}_{i,n}) + g_{i,2}(\bar{x}_{i,2}) + \delta_{i,2}(\bar{x}_{i,2}(\tau_{i,2})) + l_i \\ x_{i,j} = x_{i,j} + \varsigma_{i,j}(t, x_{i,j}), \end{cases}$$

where $i = 1, 2, 3, 4$, the unknown nonlinear functions are chosen as $g_{i,1} = x_{i,1}^2 \sin(x_{i,1}) - 2x_{i,1}$, $g_{i,2} = x_{i,2}/(1 + x_{i,1}^2)$, $\delta_{i,1} = x_{i,1}(t - \tau_{i,1}(t))\sin(x_{i,1}(t - \tau_{i,1}))$, $\delta_{i,2} = x_{i,1}(t - \tau_{i,1}(t))x_{i,2}(t - \tau_{i,2}(t))$, the external disturbances are modeled by $l_i = 0.01\sin(t) + \cos(t)x_{i,1}\sin(x_{i,2})$. Moreover, the time-varying delay functions are chosen as $\tau_{i,1} = 0.1 + 0.05\sin(t)$, $\tau_{i,2} = 0.2 + 0.15\cos(t)$. The deception attacks are given by $\varsigma_{i,1}(t, x_{i,1}) = (0.5 + 1.4\sin(t))x_{i,1}$, $\varsigma_{i,2}(t, x_{i,2}) = (0.5 + 1.4\cos(t))x_{i,2}$, and $\varsigma_i^u(t, \bar{x}_{i,2}) = x_{i,1}x_{i,2}^2\sin(2t)$, which are assumed unknown. Select the Nussbaum gain as $\mathcal{N}(\Psi_{i,j}) = \Psi_{i,j}^2\cos(\Psi_{i,j})$ to tackle the issue of the unknown parameters.

The initial condition of state variables and adaptive parameters are defined as: $x(0) = [x_{11}(0), x_{12}(0); x_{21}(0), x_{22}(0); x_{31}(0), x_{32}(0); x_{41}(0), x_{42}(0)]^T = [0.1, 0.2; -0.1, -0.1; 0.2, 0.1; -0.2, -0.2]^T$, $\varsigma = [0.2, 0.15, 0.18, 0.1]^T$, $\theta = [0.15, 0.2, 0.25, 0.3]^T$, and the initial values of the other variables are defined as zero. Besides, the design parameters are selected as $d_{i,1} = 0.2$, $d_{i,2} = 100$, $\bar{h}_{1,1} = 10$, $\bar{h}_{1,2} = 5$, $\bar{h}_{2,1} = 3$, $\bar{h}_{2,2} = 3$, $\bar{h}_{3,1} = 5$, $\bar{h}_{3,2} = 5$, $\bar{h}_{4,1} = 4$, $\bar{h}_{4,2} = 4$, $f_1 = 65$, $f_2 = 65$, $f_3 = 45$, $f_4 = 65$, $b_i = 50$, $\sigma_i = 15$, $\sigma_1^{'} = 10$, $\sigma_2^{'} = 10$, $\sigma_3^{'} = 10$,

$\sigma_4' = 40$, $\bar{h}_{1,n_i} = 5$, $\bar{h}_{2,n_i} = 3$, $\bar{h}_{3,n_i} = 5$, $\bar{h}_{4,n_i} = 2$, $a_i = 0.25$, $a_i' = 0.15$, $\bar{a}_i = 0.5$, $\wp_i = 0.11$, $\omega_i = 0.02$. The simulation results are demonstrated in Figs. 3.1, 3.2, 3.3 and 3.4.

Figure 3.1 illustrates the responses of $x_{i,j}$ and $\breve{x}_{i,j}$ for four agents with $i = 1, \cdots, 4$; $j = 1, 2$. It is evident that the proposed control strategy successfully achieves leaderless-following consensus of the agents' outputs in a short time. From Fig. 3.2, it is clear that the adaptive laws remain bounded. The trajectories of the actual control inputs under the actuator deception attacks are shown in Fig. 3.3. Figure 3.4 presents the trajectories of the compounded disturbance observers $\hat{\Phi}_i$ and the auxiliary variables s_i. It can be perfectly verified that the Zeno behavior is excluded for all

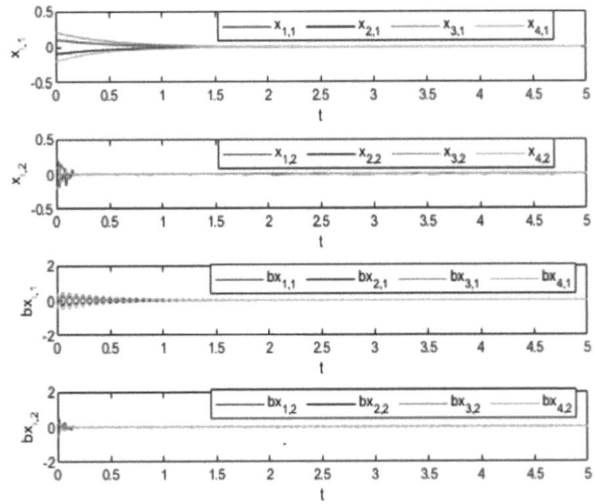

Fig. 3.1 Trajectories of $x_{i,j}$ and compromised states $\breve{x}_{i,j}$ with $i = 1, 2, 3, 4$; $j = 1, 2$

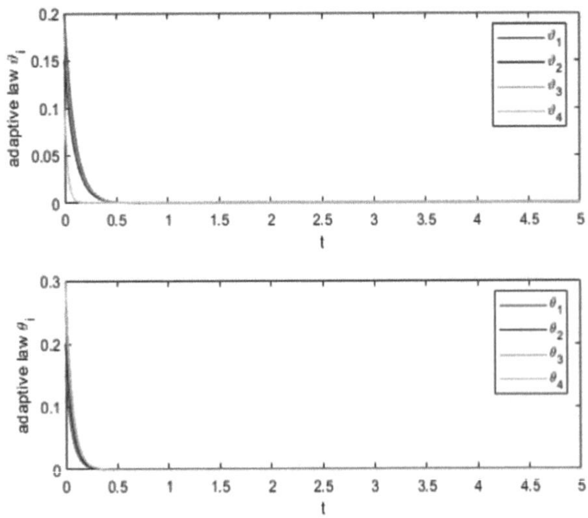

Fig. 3.2 Adaptive parameters W_i and θ_i with $i = 1, 2, 3, 4$

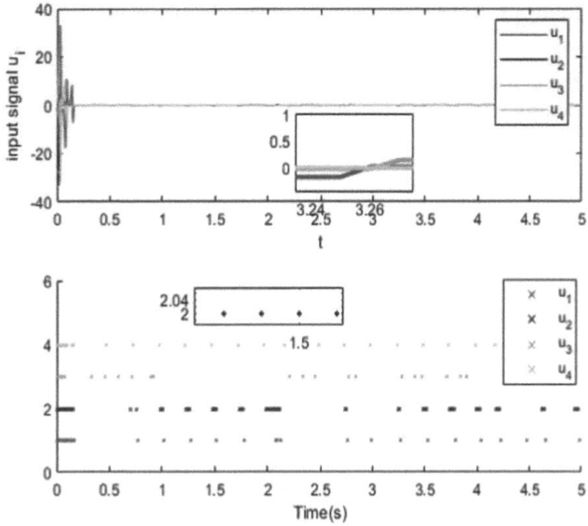

Fig. 3.3 Evolution of control signals under unknown deception attacks and event triggered instants with $i = 1, 2, 3, 4$

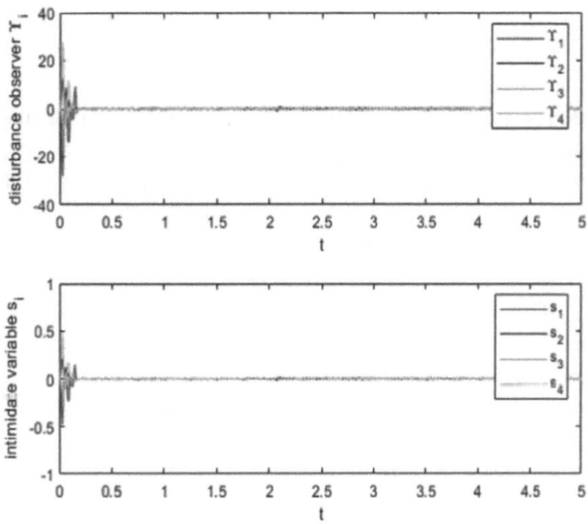

Fig. 3.4 Trajectories of compounded disturbance Υ_i and s_i with $i = 1, 2, 3, 4$

agents from Fig. 3.3. Finally, table 3.1 highlights that the proposed event-triggered mechanism significantly alleviates the communication load, confirming its efficiency in reducing data transmission frequency.

Example 3.2 A nonlinear system consisting of two chemical reactors with delayed recycle streams is simulated in this part. Inspired by [29], its dynamics can be represented as

Table 3.1 Udate times of control laws by two ways

Agents	Agent 1	Agent 2	Agent 3	Agent 4
Event triggered control	33	78	23	32
Time triggered control	500	500	500	500

$$
\begin{cases}
\dot{x}_{i,1} = -\mathcal{C}_{i,1}x_{i,1} - \dfrac{1}{\mathcal{M}_{i,1}}x_{i,1} + \dfrac{1-\mathcal{Q}_{i,2}}{\mathcal{N}_{i,1}}x_{i,2} + \delta_{i,1} \\
\dot{x}_{i,2} = -\mathcal{C}_{i,2}x_{i,2} - \dfrac{1}{\mathcal{M}_{i,2}}x_{i,2} - \dfrac{1}{\mathcal{M}_{i,2}}x_{i,1}x_{i,2} \\
\qquad + \dfrac{\mathcal{R}_{i,2}}{\mathcal{N}_{i,2}}u_i + \delta_{i,2} \\
y = x_{i,1}, i = 1, \cdots, 4
\end{cases}
$$

where $x_{i,1}$ and $x_{i,2}$ denote the compositions, $\mathcal{C}_{i,1}$ and $\mathcal{C}_{i,2}$ represent the reaction constants, $\mathcal{N}_{i,1}$ and $\mathcal{N}_{i,2}$ are the reactor volumes, $\mathcal{R}_{i,2}$ and $\mathcal{Q}_{i,2}$ denote the feed rate and recycle flow rate, $\mathcal{M}_{i,1}$ and $\mathcal{M}_{i,2}$ are the reactor residence times, $k_{i,1}$ and $k_{i,2}$ are time delay functions. The systems parameters are defined as: $\mathcal{C}_{1,1} = \mathcal{C}_{i,2} = 0.3$, $\mathcal{M}_{i,1} = \mathcal{M}_{i,2} = 2$, $\mathcal{N}_{i,1} = \mathcal{N}_{i,2} = \mathcal{Q}_{i,2} = \mathcal{R}_{i,2} = 0.5$, $\delta_{i,1} = -0.5x_{i,1}(t - \tau_{i,1}(t))$, $\delta_{i,2} = 0.5x_{i,1}(t - \tau_{i,1}) + 0.5x_{i,2}(t - \tau_{i,2}(t))$, $\tau_{i,1}(t) = 0.3 + 0.15\sin(t)$, $\tau_{i,2}(t) = 0.2 + 0.15\cos(t)$. In addition, the deception attacks are chosen as: $\varsigma_{i,1}(t, x_{i,1}) = (0.5 + \cos(t))x_{i,1}$, $\varsigma_{i,2}(t, x_{i,2}) = (0.8 + \sin(t))x_{i,2}$, and $\varsigma_i^u(t, \bar{x}_{i,2}) = 0.1\cos(t)\sin(x_{i,1}x_{i,2})$.

The initial condition of state variables and adaptive parameters are defined as: $x(0) = [x_{11}(0), x_{12}(0); x_{21}(0), x_{22}(0); x_{31}(0), x_{32}(0); x_{41}(0), x_{42}(0)]^T = [0.5, 0.4; -0.5, -0.2; 0.3, 0.2; -0.3, -0.4]^T$, $\varpi = [1.4, 1.3, 1.2, 1.1]^T$, $\theta = [2, 1.9, 1.8, 1.7]^T$, and the initial values of the other variables are defined as zero. Besides, the design parameters are selected as $d_{i,1} = 0.1$, $d_{i,2} = 23$, $\bar{h}_{i,1} = 5$, $\bar{h}_{i,2} = 5$, $f_i = 3$, $b_i = 10$, $\sigma_i = 4$, $\sigma_i^{'} = 4$, $\bar{h}_{i,n_i} = 5$, $a_i = 0.15$, $a_i^{'} = 0.15$, $\bar{a}_i = 1.5$, $\wp_i = 0.15$, $\omega_i = 0.02$. The stabilization performance is illustrated in Fig. 3.5. The curves of adaptive laws are shown in Fig. 3.6. Moreover, the Fig. 3.7 demonstrates the Zeno behavior is avoided effectively. These results collectively verify that the proposed control approach is capable of successfully addressing the cooperative control problem of cyber-physical systems subject to unknown state delays and deception attacks.

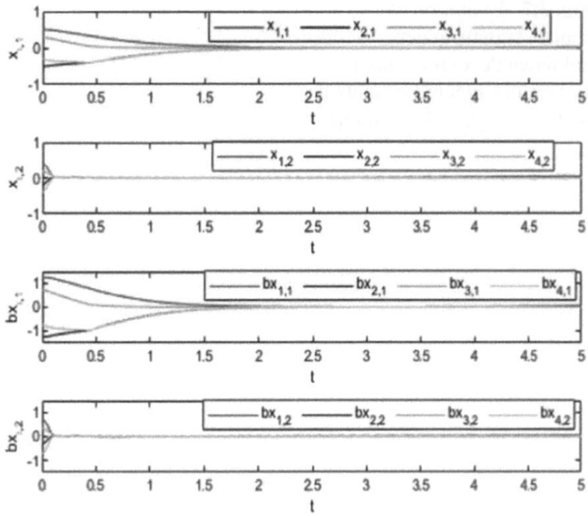

Fig. 3.5 Trajectories of actual states $x_{i,j}$ and compromised states $\breve{x}_{i,j}$ with $i = 1, 2, 3, 4$; $j = 1, 2$ (Example 3.2)

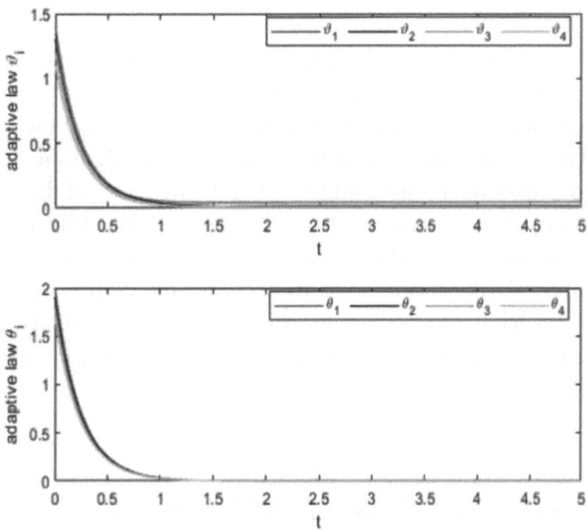

Fig. 3.6 Trajectories of adaptive laws W_i and θ_i with $i = 1, 2, 3, 4$ (Example 3.2)

3.4 Conclusion

3.4.1 Contribution

The event-triggered cooperative adaptive neural dynamic surface control method has been presented against the leaderless-following nonlinear cyber physical systems with time delay and deception attacks. According to the approximation property of neural network, it is used to deal with the unknown nonlinear uncertainties and construct a disturbance observer directly based on compromised state variables. Under

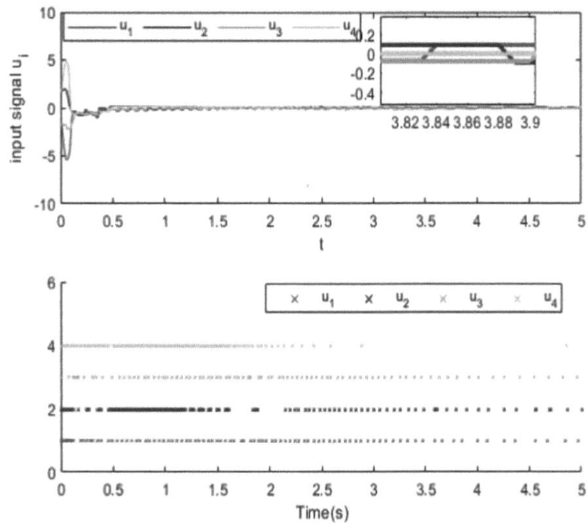

Fig. 3.7 Evolution of control signals under unknown deception attacks and event triggered instants with $i = 1, 2, 3, 4$ (Example 3.2)

the unknown deception attacks, there is the main difficulty that the true system state variables can't be applied directly in the process of controller design. Therefore, compared with existing works, the Nussbaum gain technique is combined with the dynamic surface control method to address the issue of unknown control gains caused by sensor deception and to eliminate the effect of the "explosion of complexity". This novel strategy significantly simplifies the controller design process. The effect of state time-delays are eliminated by choosing the appropriate Lyapunov-Krasovskii functions. Furthermore, so as to decrease the computation burden, we have adopted the event-triggered control strategy and less adaptive laws under the framework of adaptive mechanism. Finally, the simulation example is shown to verify the availability of the proposed control strategy.

3.4.2 Limitations of the Current Work

Despite the promising results, several limitations remain:

Computational Complexity: The integration of radial basis function (RBF) neural networks for approximating unknown nonlinearities and the design of compound disturbance observers introduce considerable computational overhead. This may limit real-time implementation, particularly in large-scale or resource-constrained systems.

Sensitivity to Deception Severity: Although compromised state variables are used in the controller design, the current approach may suffer from performance degradation under severe or highly dynamic deception attacks, especially when sensor falsification is frequent or coordinated.

Simplified Triggering Strategy: While the relative threshold-based event-triggered mechanism reduces communication frequency, its static nature may not optimally adapt to varying system dynamics, potentially leading to delays in critical responses or suboptimal performance.

3.4.3 Future Directions

To address the aforementioned limitations and further improve the practicality and robustness of the proposed strategy, future research may focus on the following directions:

Reduced-Order Observers: Developing lightweight observer designs to lower computational overhead while maintaining estimation accuracy.

Deception-Resilient State Estimation: Incorporate resilient estimation techniques, such as robust observers or fault detection filters, to enhance system tolerance against high-magnitude or frequent sensor deception attacks, particularly in dynamic or adversarial environments.

Advanced Event-Triggering Schemes: Design more flexible and adaptive event-triggering strategies, such as dynamic thresholds or hybrid time-event triggers, to balance communication savings with rapid control response in time-sensitive tasks.

By pursuing these directions, the proposed approach can be further extended to meet the challenges of complex, real-world distributed systems that require robust, scalable, and adaptive control under uncertainty and limited communication.

References

1. Zhang Y, Li H, Sun J, He W (2019) Cooperative adaptive event-triggered control for multiagent systems with actuator failures. IEEE Trans Syst Man Cybern Syst 49(9):1759–1768
2. Chen K, Wang J, Zhang Y, Liu Z (2020) Leader-following consensus for a class of nonlinear strick-feedback multiagent systems with state time-delays. IEEE Trans Syst Man Cybern Syst 50(7):2351–2361
3. Mao J, Karimi HR, Xiang Z (2019) Observer-based adaptive consensus for a class of nonlinear multiagent systems. IEEE Trans Syst Man Cybern Syst 49(9):1893–1900
4. Wang F, Chen B, Lin C, Li X (2017) Distributed adaptive neural control for stochastic nonlinear multiagent systems. IEEE Trans Cybern 47(7):1795–1803
5. Zhang Y, Sun J, Liang H, Li H (2020) Event-triggered adaptive tracking control for multiagent systems with unknown disturbances. IEEE Trans Cybern 50(3):890–901
6. Cao L, Li H, Dong G, Lu R (2021) Event-triggered control for multiagent systems with sensor faults and input saturation. IEEE Trans Syst Man Cybern Syst 51(6):3855–3866
7. Liu L, Sun H, Ma L, Zhang J, Bo Y (2021) Quasi-consensus control for a class of time-varying stochastic nonlinear time-delay multiagent systems subject to deception attacks. IEEE Trans Syst Man Cybern Syst 51(11):6863–6873

8. Wang A, Liu L, Qiu J, Feng G (2019) Event-triggered robust adaptive fuzzy control for a class of nonlinear systems. IEEE Trans Fuzzy Syst 27(8):1648–1658
9. Li H, Zhang C, Liu S, Zhang X (2022) A dynamic gain approach to consensus control of nonlinear multiagent systems with time delays. IEEE Trans Cybern 52(7):6993–7001
10. Su Y, Shi Y, Sun C (2021) Distributed model predictive control for tracking consensus of linear multiagent systems with additive disturbances and time-varying communication delays. IEEE Trans Cybern 51(7):3813–3823
11. Zhou X, Liang W, Shimizu S, Ma J, Jin Q (2021) Siamese neural network based few-shot learning for anomaly detection in industrial cyber-physical systems. IEEE Trans Ind Inf 17(8):5790–5798
12. Wang W, Han Z, Liu K, Li J (2021) Distributed adaptive resilient formation control of uncertain nonholonomic mobile robots under deception attacks. IEEE Trans Circuits Syst I Regular 68(9):3822–3835
13. He W, Yan Z, Sun C, Chen Y (2017) Adaptive neural network control of a flapping wing micro aerial vehicle with disturbance observer. IEEE Trans Cybern 47(10):3452–3465
14. Li D, Li D (2015) Adaptive controller design-based neural networks for output constraint continuous stirred tank reactor. Neurocomputing 153:159–163
15. Jin X, Haddad WM, Yucelen T (2017) An adaptive control architecture for mitigating sensor and actuator attacks in cyber-physical systems. IEEE Trans Autom Control 62(11):6058–6064
16. Zhang Y, Li H, Sun J, He W (2019) Cooperative adaptive event-triggered control for multiagent systems with actuator failures. IEEE Trans Syst Man Cybern Syst 49(9):1759–1768
17. Wang C, Hill DJ, Ge S, Chen G (2006) An ISS-modular approach for adaptive neural control of pure-feedback systems. Automatica 42(5):723–731
18. Ge SS, Hong F, Lee TH, Hang CC (2002) Adaptive neural control of nonlinear time-delay systems with unknown virtual control coefficients. In: Proceedings of the 41st IEEE conference on decision and control, vol 1. IEEE, pp 961–966
19. Ye X, Jiang J (1998) Adaptive nonlinear design without a priori knowledge of control directions. IEEE Trans Autom Control 43(11):1617–1621
20. Zhang T, Ge SS (2009) Adaptive neural network tracking control of MIMO nonlinear systems with unknown dead zones and control directions. IEEE Trans Neural Networks 20(3):483–497
21. Li D, Li D (2017) Adaptive neural tracking control for nonlinear time-delay systems with full state constraints. IEEE Trans Syst Man Cybern Syst 47(7):1590–1601
22. Chen M, Ge SS (2015) Adaptive neural output feedback control of uncertain nonlinear systems with unknown hysteresis using disturbance observer. IEEE Trans Ind Electron 62(12):7706–7716
23. Liu W, Li P (2019) Disturbance observer-based fault-tolerant adaptive control for nonlinearly parameterized systems. IEEE Trans Ind Electron 66(11):8681–8691
24. Yu J, Shi P, Lin C, Yu H (2020) Adaptive neural command filtering control for nonlinear MIMO systems with saturation input and unknown control direction. IEEE Trans Cybern 50(6):2536–2545
25. Shojaei F, Arefi MM, Khayatian A, Karimi HR (2019) Observer-based fuzzy adaptive dynamic surface control of uncertain nonstrict feedback systems with unknown control direction and unknown dead-zone. IEEE Trans Syst Man Cybern Syst 49(11):2340–2351
26. Wu Y, Xie X (2021) Robust adaptive control for state-constrained nonlinear systems with input saturation and unknown control direction. IEEE Trans Syst Man Cybern Syst 51(2):1192–1202
27. Ma J, Zheng Z, Li P (2015) Adaptive dynamic surface control of a class of nonlinear systems with unknown direction control gains and input saturation. IEEE Trans Cybern 45(4):728–741
28. Wu Y, Pan Y, Chen M, Li H (2021) Quantized adaptive finite-time bipartite NN tracking control for stochastic multiagent systems. IEEE Trans Cybern 51(6):2870–2881
29. Hua C, Guan X, Shi P (2005) Robust backstepping control for a class of time-delayed systems. IEEE Trans Autom Control 50(6):894–899

Chapter 4
Event-Triggered Adaptive Fault-Tolerant Control for a Class of Nonlinear Multiagent Systems with Sensor and Actuator Faults

Most of the research work related to fault-tolerant control problems is carried out under the classical time-driven control scheme, which leads to excessive waste of communication resources. Event-triggered control method has attracted a lot of researchers ' attention in this frontier field because of its advantages in saving communication resources and maintaining good control performance. Unlike periodic time-driven control, event-triggered control only updates the control signal at some discrete moments that violate the predefined triggering conditions, thereby effectively reducing the data transmission frequency. Therefore, event-triggered control methods have emerged for various control objectives, such as linear systems [1], pure feedback systems [2], multi-agent systems [3], stochastic systems [4], etc. The problems of the above models cover unmeasurable states, prescribed performance, unknown external disturbances and arbitrary switching. The author of [5] proposed an adaptive tracking control method based on event-triggered and observer for nonlinear systems with asymmetric output constraints and non-strict feedback form. In addition, for a class of nonlinear non-strict feedback systems in [6], an event-triggered adaptive finite-time fuzzy control method is introduced to obtain faster convergence speed. In addition, the event-triggered control strategy for full-state constrained nonlinear systems is also considered in [7, 8].

In recent years, with the complexity and networking of industrial intelligent systems, the problem of control stability of nonlinear multi-agent systems in the face of hardware faults such as sensor aging and actuator failure has become increasingly prominent. Especially in the circuit system, the sensor signal distortion and actuator output deviation caused by component damage and aging have become the key factors affecting the system performance. When the sensor fault occurs, the decrease of the accuracy of the measurement signal directly leads to the deviation of the control instruction, while the actuator fault will cause the nonlinear deviation between the actual output and the control input. Under the dual influence, the traditional control strategy is difficult to maintain the expected performance. Current research shows that event-triggered control shows unique advantages in solving the problem

X. Wang et al., *Optimal Adaptive Cooperative Control for Nonlinear Multiagent Systems*,
https://doi.org/10.1007/978-981-95-2964-3_4

of communication resource optimization in fault scenarios because it can dynamically adjust the timing of signal update according to the system state. However, the existing schemes still have technical bottlenecks in dealing with time-varying fault parameter coupling and multi-type fault cooperative compensation.

4.1 Problem Formulation and Preliminaries

4.1.1 System Description

Consider the following nonlinear strict-feedback MASs both subject to sensor and actuator faults:

$$\begin{cases} \dot{x}_{i,k} = x_{i,k+1} + f_{i,k}(x_{i,k}) + d_{i,k}(t, x_{i,k}) \\ \dot{x}_{i,n} = u_i^F + f_{i,n}(x_i) + d_{i,n}(t, x_i) \\ y_i = x_{i,1} \\ y_i^F = \kappa_i y_i, \end{cases} \tag{4.1}$$

where $x_{i,k} = [x_{i,1}, x_{i,2}, \ldots, x_{i,k}]^T \in \mathbb{R}^k$, $x_{i,n} = [x_{i,1}, x_{i,2}, \ldots, x_{i,n}]^T \in \mathbb{R}^n$ with $x_{i,j} \in \mathbb{R}$ being the jth state of ith agent, u_i^F represents the actuator fault input, y_i and y_i^F denote the output and the actual measurement value of the MASs respectively, with $i = 1, \ldots, N; k = 1, \ldots, n-1; j = 1, \ldots, n. 0 < \kappa_i < 1$ indicates that each sensor may suffer an unknown constant loss scaling failure, and let $\beta_i = 1/\kappa_i$. $f_{i,k}(\cdot)$ stands for a smooth but unknown nonlinear function, which satisfies the local Lipschitz condition. It can be denoted as $|f_{i,j}(x_{i,j}) - f_{i,j}(x_{i,j}^*)| \le m_{i,j}|x_{i,j} - x_{i,j}^*|$, in which $m_{i,j} > 0$ is an unknown constant. $d_{i,k}(\cdot)$ denotes the unknown external disturbance satisfying $|d_i(x_{i,j})| \le d_{i,j}$ where $d_{i,j}$ is an unknown positive constant. Define the $(N+1)$th agent as the leader, and its output signal as the reference signal $y_r = y_{N+1} \in \mathbb{R}$. The actuator faults can be represented as

$$u_i^F(t) = (1 - \theta_i)u_i(\cdot) + \bar{u}_i(t), 0 < \theta_i < 1, \tag{4.2}$$

where $u_i(\cdot)$ denotes the saturated controller input, $\bar{u}_i$ represents unknown function satisfying $|\bar{u}_i| \le \chi_i'$. The saturated function expression can be represented as

$$u_i(\cdot) = \text{sat}(v_i) = \begin{cases} \text{sign}(v_i)u_{Mi} & |v_i| \ge u_{Mi}, \\ v_i & |v_i| < u_{Mi}. \end{cases} \tag{4.3}$$

The saturated function can be approximately transformed as:

$$g_i(v_i) = u_{Mi} \times \tanh(\frac{v_i}{u_{Mi}}) = u_{Mi}\frac{e^{\frac{v_i}{u_{Mi}}} - e^{\frac{-v_i}{u_{Mi}}}}{e^{\frac{v_i}{u_{Mi}}} + e^{\frac{-v_i}{u_{Mi}}}}, \tag{4.4}$$

$$u_i(\cdot) - \text{sat}(v_l) = g_i(v_i) + \varrho_i, \tag{4.5}$$

where $|\varrho_i| = |\text{sat}(v_i) - g_i(v_i)| < \bar{\varrho}_i \leq u_{Mi}(1 - \tanh(1))$.

According to the mean-value theorem, one has

$$\begin{aligned} g_i(v_i) &= g_i(\hat{v}_i) + (v_i - \hat{v}_i)\frac{\partial g_i(\cdot)}{\partial v_i}|_{v_i=\tilde{v}_i} \\ &= g_i(\hat{v}_i) + (v_i - \hat{v}_i)g_i^*(\tilde{v}_i), \end{aligned} \tag{4.6}$$

with $\tilde{v}_i \in (\hat{v}_i, v_i)$, $g_i^*(\tilde{v}_i) = \frac{\partial g_i(\cdot)}{\partial v_i}|_{v_i=\tilde{v}_i}$. Let $\hat{v}_i = 0$, (4.2) becomes

$$\begin{aligned} u_i^F(t) &= (1 - \theta_i)g_i^*(\tilde{v}_i)v_i + (1 - \theta_i)\varrho_i + \bar{u}_i(t) \\ &= l_i v_i + \bar{u}_i^{'}, \end{aligned} \tag{4.7}$$

where $l_i = (1 - \theta_i)g_i^*(\tilde{v}_i)$, $\bar{u}_i^{'} = (1 - \theta_i)\varrho_i + \bar{u}_i(t)$. Based on the property of the saturation function (4.4), one obtains $g_i^*(\tilde{v}_i) \in (0, 1]$ and $l_i \in (0, 1)$. Specially, $\bar{u}^{'}$ can be viewed as the compound bounded function satisfying $|\bar{u}_i^{'}| \leq \bar{\varrho}_i + \bar{\chi}_i^{'} \leq \chi_i$.

Remark 4.1 The proposed system dynamics model (4.1) has a wide range of applications in many practical engineering fields, covering typical scenarios such as chemical reactors [10], robotic systems [11], and flight control systems [12]. Different from the existing frontier research [13, 14], this chapter for the first time incorporates sensor faults and actuator faults into a unified analysis framework, breaking through the limitations of single fault scenario in traditional research. It is worth noting that although sensor faults are common in industrial practice, most of the existing studies are still limited to the assumption that the fault parameters are known or have clear upper and lower bounds. By constructing an integrated fault-tolerant control architecture, this chapter realizes the collaborative processing of two types of faults, which significantly expands the applicability of theoretical results in complex engineering environments.

Control Objective: The primary purpose about this chapter is to develop the event-triggered adaptive output feedback fault-tolerant control strategy for the nonlinear multiagent systems such that: (1) in the CLS, all signals are semiglobally uniformly ultimately bounded (SGUUB); (2) the output signals of all followers can synchronously track the leader to the desired accuracy; (3) the sensor faults and actuator faults can be compensated appropriately.

In this chapter, the radial basis function neural networks (RBFNNs) will be utilized to approximate the unknown smooth nonlinear function $h(z) : \mathbb{R}^n \rightarrow \mathbb{R}$ as $h_{NN}(z) = W^{*T} S(z)$, where $z \in \Omega$ with $\Omega \subseteq \mathbb{R}^q$ is a compact set, $W^* = [W_1^*, W_2^*, \ldots, W_l^*]^T \in$

$\mathbb{R}^l$ denotes the desired weight vector, $S(z) = [S_1(z), S_2(z), \ldots, S_l(z)]^T \in \mathbb{R}^l$ is the basis function vector with the sufficiently large node number l. The basic function $S_i(z)$ is chosen as Gaussian function $S_i(z) = \exp\left[-\frac{(z-\upsilon_i)^T(z-\upsilon_i)}{\eta_i^2}\right], i = 1, \ldots, l$ where $\upsilon_i = [\upsilon_{i1}, \upsilon_{i2}, \ldots, \upsilon_{iq}]^T$ is the center of receptive filed, η_i represents the width of this function, and the inequality $0 < S_i^T(Z)S_i(Z) \leq 1$ holds, obviously. According to the property of $RBFNNs$, the nonlinear uncertainties in MASs (4.1) can be represented as follows: $f_{i,j}(X) = W_{i,j}^T S_{i,j}(X) + \varepsilon_{i,j}(X), \forall X \in \Omega$, where $W_{i,j}^T$ is the optimal weight vector and $\varepsilon_{i,j}(X)$ is the bounded approximation error, there exists a positive constant ϵ satisfying $|\varepsilon_{i,j}| \leq \epsilon_{i,j}$. More knowledge about RBF NNs can be seen in [15].

4.1.2 Command Filtering

Motivated by the reference [16], we introduced the command filter to avoid the explosion of differentiation brought by the traditional backstepping method. It can be described as

$$\dot{x}_{1,c} = \omega_n x_{2,c}, \tag{4.8}$$

$$\dot{x}_{2,c} = -2\zeta\omega_n x_{2,c} - \omega_n(x_{1,c} - \alpha_1), \tag{4.9}$$

where α_1 represents the input signal satisfying $|\dot{\alpha}_1| \leq \rho_1$ and $|\ddot{\alpha}_1| \leq \rho_2$ with $\rho_i \geq 0$, $i = 1, 2$, and $x_{1,c}$ denotes the output signal of the filter. Accordingly, the initial condition of the command filter satisfies $x_{1,c}(0) = \alpha_1(0)$, and $x_{2,c}(0) = 0$. For any $\iota > 0$, there exist constants $\zeta \in (0, 1]$ and $\omega_n \in (0, +\infty)$ such that $|x_{1,c} - \alpha_1| \leq \iota$ and $|\dot{x}_{1,c}|$, $|\ddot{x}_{1,c}|$, $|\dddot{x}_{1,c}|$ are all bounded.

To this end, some necessary assumptions and lemmas are elaborated.

Assumption 4.1 [17] The graph $\mathcal{G}$ contains a directed spanning tree.

Assumption 4.2 The output of the leader y_r and its first-order time derivative are continuous, known, and bounded.

Remark 4.2 The *Assumption* 4.1 is widely used in the consensus control for MASs, such as [18–20]. It means that all follower agents can obtain the information from the leader agent, directly or indirectly. The *Assumption* 4.2 is standard to demonstrate the boundedness of tracking signal (can see in [18, 19] et al.), which plays a vital role in the design procedure of control strategy.

Lemma 4.1 *For $\tilde{\varphi} \in \mathbb{R}$, the function $V(\tilde{\varphi}(t)) = \frac{1}{3}|\tilde{\varphi}(t)|^3$ is differentiable, and the first-order time differentiation of $V(\tilde{\varphi}(t))$ is $\dot{V}(\tilde{\varphi}(t)) = \tilde{\varphi}^2\dot{\tilde{\varphi}}sgn(\tilde{\varphi})$.*

Proof. Demonstrate the differentiability of $V(\tilde{\varphi}(t))$ at $\tilde{\varphi}(t) = 0$. According to the sign of $\tilde{\varphi}$, we can obtain the following piecewise function:

$$\dot{V}(\tilde{\varphi}(t)) = \begin{cases} \tilde{\varphi}^2\dot{\tilde{\varphi}}, & \tilde{\varphi} \geq 0; \\ -\tilde{\varphi}^2\dot{\tilde{\varphi}}, & \tilde{\varphi} < 0; \end{cases} \tag{4.10}$$

where $\dot{V}(0^+) = \dot{V}(0^-)$ such that $\dot{V}(\tilde{\varphi})$ is continuous at $\tilde{\varphi}(t) = 0$. Thus, we can call that $V(\tilde{\varphi}(t))$ is differentiable at $\tilde{\varphi}(t) = 0$.

4.2 NN State Observer Design

In consideration of the unmeasurable states, we will construct the state observers to estimate the state variables. From the above analysis,

$$\begin{cases} \dot{\hat{x}}_{i,1} = \hat{x}_{i,2} + \hat{W}_{i,1}^T S_{i,1}(\hat{\bar{x}}_{i,1}) + r_{i,1}(\hat{\beta}_i y_i^F - \hat{x}_{i,1}) \\ \dot{\hat{x}}_{i,k} = \hat{x}_{i,k+1} + \hat{W}_{i,k}^T S_{i,k}(\hat{\bar{x}}_{i,k}) + r_{i,k}(\hat{\beta}_i y_i^F - \hat{x}_{i,1}) \\ \dot{\hat{x}}_{i,n} = \hat{l}_i v_i + \hat{W}_{i,k}^T S_{i,k}(\hat{\bar{x}}_{i,k}) + r_{i,n}(\hat{\beta}_i y_i^F - \hat{x}_{i,1}), \end{cases} \tag{4.11}$$

where $\hat{\beta}_i$, $\hat{l}_i$, $\hat{W}_{i,j}^T$ are the estimation of β_i, l_i and $W_{i,j}^T$ respectively; $r_{i,1}$, $r_{i,k}$, $r_{i,n}$ are design positive constants which can be well defined such that the matrix $A_i = \begin{pmatrix} -r_{i,1} & \\ \vdots & I_{n-1} \\ -r_{i,n} & \cdots \; 0 \end{pmatrix}$ is Hurwitz stable. Define $e_{i,k} = x_{i,k} - \hat{x}_{i,k}$, we can obtain the following equation:

$$\dot{e}_i = A_i e_i + \Delta f_i + \tilde{W}_i^T S_i + \varepsilon_i + B_i \tilde{l}_i v_i + R_i \tilde{\beta}_i y_i^F + d_i, \tag{4.12}$$

where $e_i = [e_{i,1}, \ldots, e_{i,n}]^T$, $\Delta f_i = [\Delta f_{i,1}, \ldots, \Delta f_{i,n}]^T$, $\tilde{W}_i^T S_i = [\tilde{W}_{i,1}^T S_{i,1}(\hat{\bar{x}}_{i,1}), \ldots, \tilde{W}_{i,n}^T S_{i,n}(\hat{\bar{x}}_{i,n})]^T$, $\varepsilon_i = [\varepsilon_{i,1}, \ldots, \varepsilon_{i,n}]^T$, $B_i = [0, \ldots, 1]$, $R_i = [r_{i,1}, \ldots, r_{i,n}]^T$, $d_i = [d_{i,1}, \ldots, d_{i,n}]^T$.

Define the Lyapunov function candidate function:

$$V_{i,0} = \frac{1}{2} e_i^T P_i e_i,$$

where $P_i^T = P_i$ is a positive definite matrix such that $A_i^T P_i + P_i A_i = -Q_i$, Q_i is any given symmetric positive definite matrix. $\dot{V}_{i,0}$ can be calculated as

$$\begin{aligned} \dot{V}_{i,0} =& \frac{1}{2} e_i^T (A_i^T P_i + P_i A_i) e_i + e_i^T P_i (\Delta f_i + \tilde{W}_i^T S_i \\ &+ \varepsilon_i + B_i \tilde{l}_i v_i + R_i \tilde{\beta}_i y_i^F + d_i). \end{aligned} \tag{4.13}$$

By utilizing the Young's inequality, one has

$$\dot{V}_{i,0} \leq -\Pi_{i,0}||e_i||^2 + \frac{1}{2}\sum_{l=1}^{n} \tilde{W}_{i,l}^T \tilde{W}_{i,l} + \frac{1}{2}||P_i||^2 \tilde{l}_i^2 v_i^2 + \frac{1}{2}(\tilde{\beta}_i y_i^F)^2 + D_{i,0}, \tag{4.14}$$

where $m_i^2 = \sum_{j=1}^{n} m_{i,j}^2$, and $\tilde{\beta}_i, \tilde{l}_i$ are the estimation error of β_i and l_i. $\tilde{W}_{i,j}^T = W_{i,j}^{*T} - \hat{W}_{i,j}^T$ is the estimation error of the optimal weight $W_{i,j}^{*T}$, $\Pi_{i,0} = \lambda_{\min}(Q_i) - 2||P_i||^2 - \frac{1}{2}||P_i||^2||R_i||^2 - \frac{1}{2}m_i^2 - \frac{1}{2}$, $D_{i,0} = \frac{1}{2}||\bar{\varepsilon}_i||^2 + \frac{1}{2}||\bar{d}_i||^2$, $\lambda_{\min}(Q_i)$ represents the minimum eigenvalue of matrix Q_i.

The leader r could only communicate with a part of the followers such that the tracking error $y_i - y_r$ can not be directly used to the design process of the controller in this chapter. To address the knotty problem, we introduce a distributed sliding-mode estimator for each follower to estimate the output signal of the leader:

$$\begin{cases} \dot{\hat{y}}_{ri} = -b_i \sum_{j=0}^{N} a_{i,j}(\hat{y}_{ri} - \hat{y}_{rj}) - q_i \operatorname{sgn}\left[\sum_{j=0}^{N} a_{i,j}(\hat{y}_{ri} - \hat{y}_{rj})\right] \\ \ddot{\hat{y}}_{ri} = -b_i \sum_{j=0}^{N} a_{i,j}(\dot{\hat{y}}_{ri} - \dot{\hat{y}}_{rj}) - q_i \operatorname{sgn}\left[\sum_{j=0}^{N} a_{i,j}(\dot{\hat{y}}_{ri} - \dot{\hat{y}}_{rj})\right], \end{cases} \tag{4.15}$$

where $\hat{y}_{ri}$ denotes the estimation value of the follower i for the leader y_r, b_i, q_i are positive design constants, $\hat{y}_{r0} = y_r$. If $q_i > |\dot{y}_r|$, then $\hat{y}_{ri}$ converges to y_r in finite time. Let $\tilde{y}_{ri} = \hat{y}_{ri} - y_r$. The derivative of $\tilde{y}_{ri}$ yields

$$\dot{\tilde{y}}_{ri} = -b_i \sum_{j=0}^{N} a_{i,j}(\dot{\hat{y}}_{ri} - \dot{\hat{y}}_{rj}) - q_i \operatorname{sgn}[\sum_{j=0}^{N} a_{i,j}(\dot{\hat{y}}_{ri} - \dot{\hat{y}}_{rj}] - \dot{y}_r, \tag{4.16}$$

with $\tilde{y}_{r0} = 0$. From the proof of [[21], Theorem 3.1], we can obtain that when $q_i > |\dot{y}_r|$, $\tilde{y}_{ri}$ is bounded and $\tilde{y}_{ri} \to 0, \forall i \in \mathbb{N} := \{1, \ldots, N\}$ in finite time with the settling time $T_{ri} = \frac{\max_{i \in \mathbb{N}}\{|\hat{y}_{ri} - y_r(0)|\}}{q_i - \sup_{t \geq 0}||\dot{y}_r||}$.

4.3 Event-Triggered Leader-Following Adaptive Consensus Control Protocol

In this section, the command filter inversion method is used to derive the observer-based event-triggered adaptive fault-tolerant control strategy. The overall design process follows strict recursive logic and is divided into n steps. This method smoothes the virtual control law by introducing a first-order low-pass filter, which effectively avoids the explosion of computational complexity caused by repeated derivation in the traditional inversion method. At the same time, a compensation system is constructed to dynamically correct the filtering error. In the recursive design process, combined with the real-time estimation of the unmeasurable state by the state observer, the fault compensation mechanism is embedded into each design link, and finally a control architecture with both computational efficiency and fault tolerance performance is formed. First of all, we defined the following coordinate transformation:

$$\begin{cases} \nu_{i,1} = y_i - \hat{y}_{ri} \\ \nu_{i,k} = \hat{x}_{i,k} - x_{i,kc}, \end{cases} \tag{4.17}$$

where $x_{i,kc}$ is the output signal with $\alpha_{i,k-1}$ being the input signal of the command filter, $i = 1, \ldots, N; k = 2, \ldots, n$.

To offset the impact of the error brought by the command filter, the compensating system is defined as follows:

$$\begin{cases} \dot{\xi}_{i,1} = -c_{i,1}\xi_{i,1} + \xi_{i,2} + (x_{i,2_c} - \alpha_{i,1}) \\ \dot{\xi}_{i,k} = -c_{i,k}\xi_{i,k} - \xi_{i,k-1} + \xi_{i,k+1} + (x_{i,(k+1)_c} - \alpha_{i,k}) \\ \dot{\xi}_{i,n} = -c_{i,n}\xi_{i,n} - \xi_{i,n-1}, i = 1, \ldots, N; k = 1, \ldots, n, \end{cases} \tag{4.18}$$

where $c_{i,j}$ is the design constant, the initial condition of the system is set as $\xi(0) = 0$. The filtering error $x_{i,(k+1)_c} - \alpha_{i,k}$ meets $|x_{i,(k+1)c} - \alpha_{i,k}| \leq \iota_i$. Based on the Lemma 3 in [16], the boundedbess of the compensating system (4.18) is demonstrated, i.e., $||\xi(t)|| \leq \frac{\iota}{2\kappa_0}$, where $\kappa_0 = 0.5 \min\{c_{i,1}, c_{i,k}, c_{i,n}\}$.

Then, we design the compensating tracking error signals as

$$\begin{cases} z_{i,1} = \nu_{i,1} - \xi_{i,1} = \breve{z}_{i,1} + \tilde{\beta}_i y_i^F \\ z_{i,k} = \nu_{i,k} - \xi_{i,k}, i = 1, \ldots, N; k = 1, \ldots, n, \end{cases} \tag{4.19}$$

where $z_{i,1}$ is unavailable duo to the existence of the sensor faults, but $\breve{z}_{i,1} = \hat{\beta}_i y_i^F - \xi_{i,1} - \hat{y}_{ri}$ is available by employing the adaptive parameter law.

According to the above coordinate transformation, the detailed design procedure of controller will be presented.

Step$(i, 1)$: Choose the Lyapunov function candidate:

$$V_{i,1} = V_{i,0} + \frac{1}{2\zeta_{i,1}}\tilde{W}_{i,1}^T\tilde{W}_{i,1} + \frac{1}{2}z_{i,1}^2 + \frac{1}{3\eta_i}|\tilde{\beta}_i|^3, \tag{4.20}$$

where $\zeta_{i,1}$, η_i are positive design parameters.

From (4.20), we can obtain the information about $\dot{V}_{i,1}$

$$\begin{aligned}\dot{V}_{i,1} =&\dot{V}_{i,0} + \frac{1}{\zeta_{i,1}}\tilde{W}_{i,1}^T\dot{\hat{W}}_{i,1} + z_{i,1}(\xi_{i,2} + x_{i,2c} + W_{i,1}^T S_{i,1}(\hat{\bar{x}}_{i,1})\\ &+ \Delta f_{i,1} + \varepsilon_{i,1} + d_{i,1} + z_{i,2} + e_{i,2} - \dot{\hat{y}}_{ri} - \dot{\xi}_{i,1})\\ &- \frac{1}{\eta_i}\tilde{\beta}_i^2\dot{\hat{\beta}}_i\mathrm{sgn}(\tilde{\beta}_i).\end{aligned} \tag{4.21}$$

According to the Young's inequality, one has

$$\begin{aligned}z_{i,1}(z_{i,2} + e_{i,2} + \Delta f_{i,1} + \varepsilon_{i,1} + d_{i,1}) \le& \frac{5}{2}z_{i,1}^2 + \frac{1}{2}m_{i,1}^2 e_{i,1}^2\\ &+\frac{1}{2}\bar{d}_{i,1}^2 + \frac{1}{2}\bar{\varepsilon}_{i,1}^2 + \frac{1}{2}e_{i,2}^2 + \frac{1}{2}z_{i,2}^2.\end{aligned} \tag{4.22}$$

Substituting (4.22) and (4.18) into (4.21), we have

$$\begin{aligned}\dot{V}_{i,1} \le&\dot{V}_{i,0} + z_{i,1}(\hat{W}_{i,1}^T S_{i,1}(\hat{\bar{x}}_{i,1}) + \tilde{W}_{i,1}^T S_{i,1}(\hat{\bar{x}}_{i,1}) + c_{i,1}\xi_{i,1}\\ &+ \alpha_{i,1} - \dot{\hat{y}}_{ri}) + \frac{5}{2}z_{i,1}^2 + \frac{1}{2}m_{i,1}^2 e_{i,1}^2 + \frac{1}{2}\bar{d}_{i,1}^2 + \frac{1}{2}\bar{\varepsilon}_{i,1}^2\\ &+ \frac{1}{2}e_{i,2}^2 + \frac{1}{2}z_{i,2}^2 - \frac{1}{\zeta_{i,1}}\tilde{W}_{i,1}^T\dot{\hat{W}}_{i,1} - \frac{1}{\eta_i}\tilde{\beta}_i^2\dot{\hat{\beta}}_i\mathrm{sgn}(\tilde{\beta}_i).\end{aligned} \tag{4.23}$$

We design the 1st virtual control law as follows:

$$\alpha_{i,1} = -c_{i,1}\breve{z}_{i,1} - c_{i,1}\xi_{i,1} - \hat{W}_{i,1}^T S_{i,1}(\hat{\bar{x}}_{i,1}) + \dot{\hat{y}}_{ri}. \tag{4.24}$$

Substituting (4.24) into (4.23), one has

$$\begin{aligned}\dot{V}_{i,1} \le&\dot{V}_{i,0} + z_{i,1}(-c_{i,1}z_{i,1} + c_{i,1}\tilde{\beta}_i y_i^F + \tilde{W}_{i,1}^T S_{i,1}(\hat{\bar{x}}_{i,1}))\\ &+ \frac{5}{2}z_{i,1}^2 + \frac{1}{2}m_{i,1}^2 e_{i,1}^2 + \frac{1}{2}\bar{d}_{i,1}^2 + \frac{1}{2}\bar{\varepsilon}_{i,1}^2 + \frac{1}{2}e_{i,2}^2 + \frac{1}{2}z_{i,2}^2\\ &- \frac{1}{\zeta_{i,1}}\tilde{W}_{i,1}^T\dot{\hat{W}}_{i,1} - \frac{1}{\eta_i}\tilde{\beta}_i^2\dot{\hat{\beta}}_i\mathrm{sgn}(\tilde{\beta}_i),\\ \le&\dot{V}_{i,0} - (c_{i,1} - \frac{5}{2})z_{i,1}^2 + (\breve{z}_{i,1} + \tilde{\beta}_i y_i^F)(c_{i,1}\tilde{\beta}_i y_i^F\\ &+ \tilde{W}_{i,1}^T S_{i,1}(\hat{\bar{x}}_{i,1})) + \frac{1}{2}m_{i,1}^2 e_{i,1}^2 + \frac{1}{2}\bar{d}_{i,1}^2 + \frac{1}{2}\bar{\varepsilon}_{i,1}^2\\ &+ \frac{1}{2}e_{i,2}^2 + \frac{1}{2}z_{i,2}^2 - \frac{1}{\zeta_{i,1}}\tilde{W}_{i,1}^T\dot{\hat{W}}_{i,1} - \frac{1}{\eta_i}\tilde{\beta}_i^2\dot{\hat{\beta}}_i\mathrm{sgn}(\tilde{\beta}_i).\end{aligned} \tag{4.25}$$

Using the Young's inequality, we have

$$(\breve{z}_{i,1}+\tilde{\beta}_i y_i^F)\tilde{W}_{i,1}^T S_{i,1}(\hat{\breve{x}}_{i,1})) \leq \breve{z}_{i,1}\tilde{W}_{i,1}^T S_{i,1}(\hat{\breve{x}}_{i,1})) + \frac{1}{2}(\tilde{\beta}_i y_i^F)^2 + \frac{1}{2}\tilde{W}_{i,1}^T \tilde{W}_{i,1}, \tag{4.26}$$

$$(\breve{z}_{i,1}+\tilde{\beta}_i y_i^F)c_{i,1}\tilde{\beta}_i y_i^F \leq c_{i,1}(\tilde{\beta}_i y_i^F)^2 + \frac{c_{i,1}^2}{4h_i} + h_i \breve{z}_{i,1}^2(\tilde{\beta}_i y_i^F)^2. \tag{4.27}$$

Substituting (4.14), (4.26) and (4.27) into (4.25), we obtain

$$\begin{aligned}\dot{V}_{i,1} \leq &-\Pi_i||e_i||^2-(c_{i,1}-\frac{5}{2})z_{i,1}^2+\frac{1}{2}\sum_{l=1}^n \tilde{W}_{i,l}^T\tilde{W}_{i,l}+\frac{1}{2}z_{i,2}^2\\ &+\frac{1}{2}||P_i||^2\bar{l}_i^2 v_i^2+\frac{1}{2}\tilde{W}_{i,1}^T\tilde{W}_{i,1}+(c_{i,1}+1)(\tilde{\beta}_i y_i^F)^2+\frac{c_{i,1}^2}{4h_i}\\ &+h_i\breve{z}_{i,1}^2(\tilde{\beta}_i y_i^F)^2+\frac{1}{\zeta_{i,1}}\tilde{W}_{i,1}^T(\zeta_{i,1}\breve{z}_{i,1}S_{i,1}(\hat{\breve{x}}_{i,1})-\dot{\hat{W}}_{i,1})\\ &-\frac{1}{\eta_i}\tilde{\beta}_i^2\dot{\hat{\beta}}_i\operatorname{sgn}(\tilde{\beta}_i)+D_{i,1},\end{aligned} \tag{4.28}$$

where $\Pi_i = \Pi_{i,0} - \frac{1}{2}m_{i,1}^2 - \frac{1}{2}$, $D_{i,1} = D_{i,0} + \frac{1}{2}\bar{d}_{i,1}^2 + \frac{1}{2}\bar{\varepsilon}_{i,1}^2 + \frac{c_{i,1}^2}{4h_i}$.

Design the parameter adaptive laws $\hat{W}_{i,1}$ and $\hat{\beta}_i$ as

$$\dot{\hat{W}}_{i,1} = \zeta_{i,1}\breve{z}_{i,1}S_{i,1}(\hat{\breve{x}}_{i,1}) - \sigma_{i,1}\hat{W}_{i,1}, \tag{4.29}$$

$$\dot{\hat{\beta}}_i = \begin{cases}\eta_i(1+c_{i,1}+h_i\breve{z}_{i,1}^2)(y_i^F)^2-\mu_i\hat{\beta}_i, & \bar{\eta}_i > 0;\\ 0, & \bar{\eta}_i \leq 0;\end{cases} \tag{4.30}$$

where $\bar{\eta}_i = \eta_i(1+c_{i,1}+h_i\breve{z}_{i,1}^2)(y_i^F)^2 - \mu_i\hat{\beta}_i$, $\sigma_{i,1} > 0$, $\mu_i > 0$ are design constants.

Substituting (4.29) and (4.30) into (4.28), (4.28) becomes

$$\begin{aligned}\dot{V}_{i,1} \leq &-\Pi_i||e_i||^2-(c_{i,1}-\frac{5}{2})z_{i,1}^2+\frac{1}{2}\sum_{l=1}^n \tilde{W}_{i,l}^T\tilde{W}_{i,l}+\frac{1}{2}z_{i,2}^2\\ &+\frac{1}{2}||P_i||^2\bar{l}_i^2 v_i^2+\frac{1}{2}\tilde{W}_{i,1}^T\tilde{W}_{i,1}+\frac{\sigma_{i,1}}{\zeta_{i,1}}\tilde{W}_{i,1}^T\hat{W}_{i,1}\\ &+\frac{\mu_i}{\eta_i}\tilde{\beta}_i^2\hat{\beta}_i\operatorname{sgn}(\tilde{\beta}_i)+D_{i,1}.\end{aligned} \tag{4.31}$$

Step$(i, 2)$: The 2nd Lyapunov candidate function is constructed as

$$V_{i,2} = V_{i,1} + \frac{1}{2}z_{i,2}^2 + \frac{1}{2\zeta_{i,2}}\tilde{W}_{i,2}^T\tilde{W}_{i,2}, \tag{4.32}$$

where $\zeta_{i,2}$ is a positive design parameter.

Differentiating $z_{i,2}$ results in

$$\begin{aligned}\dot{z}_{i,2} =& \dot{\hat{x}}_{i,2} - \dot{x}_{i,2c} - \dot{\xi}_{i,2}\\ =& \hat{W}_{i,2}^T S_{i,2}(\hat{\bar{x}}_{i,2}) + r_{i,2}(\hat{\beta}_i y_i^F - \hat{x}_{i,1})\\ &+ z_{i,3} + x_{i,3c} + \xi_{i,3} - \dot{x}_{i,2c} - \dot{\xi}_{i,2}.\end{aligned} \tag{4.33}$$

According to equation (4.33), the derivative of $V_{i,2}$ can be obtained

$$\begin{aligned}\dot{V}_{i,2} =& \dot{V}_{i,1} + z_{i,2}(\tilde{W}_{i,2}^T S_{i,2}(\hat{\bar{x}}_{i,2}) + \hat{W}_{i,2}^T S_{i,2}(\hat{\bar{x}}_{i,2}) + r_{i,2}(\hat{\beta}_i y_i^F\\ &- \hat{x}_{i,1}) - \tilde{W}_{i,2}^T S_{i,2}(\hat{\bar{x}}_{i,2}) + z_{i,3} + x_{i,3c} + \xi_{i,3} - \dot{x}_{i,2c}\\ &- \dot{\xi}_{i,2}) - \frac{1}{\zeta_{i,2}}\tilde{W}_{i,2}^T \dot{\hat{W}}_{i,2}.\end{aligned} \tag{4.34}$$

By utilizing the Young's inequality, one has

$$z_{i,2} z_{i,3} \le \frac{1}{2} z_{i,2}^2 + \frac{1}{2} z_{i,3}^2, \tag{4.35}$$

$$- z_{i,2}\tilde{W}_{i,2}^T S_{i,2} \le \frac{1}{2} z_{i,2}^2 + \frac{1}{2}\tilde{W}_{i,2}^T \tilde{W}_{i,2}. \tag{4.36}$$

Substituting (4.18), (4.35) and (4.36) into (4.34), one has

$$\begin{aligned}\dot{V}_{i,2} \le& \dot{V}_{i,1} + z_{i,2}(\alpha_{i,2} + c_{i,2}\xi_{i,2} + \xi_{i,1} + \hat{W}_{i,2}^T S_{i,2}(\hat{\bar{x}}_{i,2})\\ &+ r_{i,2}(\hat{\beta}_i y_i^F - \hat{x}_{i,1}) - \dot{\xi}_{i,2}) z_{i,2}^2 + \frac{1}{2} z_{i,3}^2 + \frac{1}{2}\tilde{W}_{i,2}^T \tilde{W}_{i,2}\\ &+ \frac{1}{\zeta_{i,2}}\tilde{W}_{i,2}^T(\zeta_{i,2} z_{i,2} S_{i,2}(\hat{\bar{x}}_{i,2}) - \dot{\hat{W}}_{i,2}).\end{aligned} \tag{4.37}$$

Construct the following adaptive law and virtual control law:

$$\dot{\hat{W}}_{i,2} = \zeta_{i,2} z_{i,2} S_{i,2}(\hat{\bar{x}}_{i,2}) - \sigma_{i,2}\hat{W}_{i,2}, \tag{4.38}$$

$$\begin{aligned}\alpha_{i,2} =& - c_{i,2} v_{i,2} - \xi_{i,1} - \frac{3}{2} z_{i,2} - \hat{W}_{i,2}^T S_{i,2}(\hat{\bar{x}}_{i,2})\\ &- r_{i,2}(\hat{\beta}_i y_i^F - \hat{x}_{i,1}) + \dot{x}_{i,2c}.\end{aligned} \tag{4.39}$$

Substituting (4.31), (4.38) and (4.39) into (4.37) leads to

$$\begin{aligned}\dot{V}_{i,2} \le& - \varPi_i ||e_i||^2 - (c_{i,1} - \frac{5}{2}) z_{i,1}^2 - c_{i,2} z_{i,2}^2 + \sum_{j=1}^{2}(\frac{1}{2}\tilde{W}_{i,j}^T \tilde{W}_{i,j}\\ &+ \frac{\sigma_{i,j}}{\zeta_{i,j}}\tilde{W}_{i,j}^T \hat{W}_{i,j}) + \frac{1}{2}||P_i||^2 \bar{l}_i^2 v_i^2 + \frac{\mu_i}{\eta_i}\tilde{\beta}_i^2 \hat{\beta}_i \operatorname{sgn}(\tilde{\beta}_i)\\ &+ \sum_{j=1}^{n}\tilde{W}_{i,j}^T \tilde{W}_{i,j} + \frac{1}{2} z_{i,3}^2 + D_{i,1}.\end{aligned} \tag{4.40}$$

$Step(i, k\,(2 < k < n))$: We define the kth Lyapunov candidate function as

$$V_{i,k} = V_{i,k-1} + \frac{1}{2}z_{i,k}^2 + \frac{1}{2\eta_i}\tilde{W}_{i,k}^T\tilde{W}_{i,k}, \tag{4.41}$$

where $\eta_i > 0$ is a design constant.

Similar the above $Step(i, 2)$, we construct the following adaptive law and virtual control law:

$$\dot{\hat{W}}_{i,k} = \zeta_{i,k}z_{i,k}S_{i,k}(\hat{\bar{x}}_{i,k}) - \sigma_{i,k}\hat{W}_{i,k}, \tag{4.42}$$

$$\begin{aligned}\alpha_{i,k} =& - c_{i,k}v_{i,k} - \xi_{i,k-1} - \frac{3}{2}z_{i,k} - \hat{W}_{i,k}^T S_{i,k}(\hat{\bar{x}}_{i,k}) \\ & - r_{i,k}(\hat{\beta}_i y_i^F - \hat{x}_{i,1}) + \dot{x}_{i,kc}.\end{aligned} \tag{4.43}$$

Substituting the above equation into the derivative of $V_{i,k}$, we have

$$\begin{aligned}\dot{V}_{i,k} \leq& - \Pi_i||e_i||^2 - (c_{i,1} - \frac{5}{2})z_{i,1}^2 - \sum_{j=2}^{k} c_{i,k}z_{i,k}^2 \\ & + \sum_{j=1}^{n} \tilde{W}_{i,j}^T\tilde{W}_{i,j} + \sum_{j=1}^{k}(\frac{\sigma_{i,j}}{\zeta_{i,j}}\tilde{W}_{i,j}^T\hat{W}_{i,j} + \frac{1}{2}\tilde{W}_{i,j}^T\tilde{W}_{i,j}) \\ & + \frac{1}{2}||P_i||^2\tilde{l}_i^2 v_i^2 + \frac{\mu_i}{\eta_i}\tilde{\beta}_i^2\hat{\beta}_i \mathrm{sgn}(\tilde{\beta}_i) + \frac{1}{2}z_{i,k+1}^2 + D_{i,1}.\end{aligned} \tag{4.44}$$

$Step(i, n)$: Construct the nth Lyapunov candidate function

$$V_{i,n} = V_{i,n-1} + \frac{1}{2}z_{i,n}^2 + \frac{1}{2\zeta_{i,n}}\tilde{W}_{i,n}^T\tilde{W}_{i,n} + \frac{1}{2\tau_i}\tilde{l}_i^2 + \frac{l_i}{2\pi_i}\tilde{\phi}_i^2, \tag{4.45}$$

where $\zeta_{i,n}$, τ_i, π_i are positive design constants. Let $\phi_i = 1/l_i$, $\tilde{\phi}_i = \phi_i - \hat{\phi}_i$, $\hat{\phi}_i$ and $\tilde{\phi}_i$ are the estimation and estimation error of ϕ_i, respectively.

Differentiating $z_{i,n}$ results in

$$\begin{aligned}z_{i,n}\dot{z}_{i,n} =& z_{i,n}(\hat{l}_i v_i + \hat{W}_{i,k}^T S_{i,k}(\hat{\bar{x}}_{i,k}) + r_{i,n}(\hat{\beta}_i y_i^F - \hat{x}_{i,1}) \\ & + \tilde{W}_{i,k}^T S_{i,k}(\hat{\bar{x}}_{i,k}) - \tilde{W}_{i,k}^T S_{i,k}(\hat{\bar{x}}_{i,k}) - \dot{x}_{i,nc} - \dot{\xi}_{i,n}).\end{aligned} \tag{4.46}$$

Similar with the aforementioned steps, one has

$$- z_{i,n}\tilde{W}_{i,n}^T S_{i,n} \leq \frac{1}{2}z_{i,n}^2 + \frac{1}{2}\tilde{W}_{i,n}^T\tilde{W}_{i,n}. \tag{4.47}$$

According the above results, the first-order time derivative of $V_{i,n}$ can be obtained

$$\begin{aligned}\dot{V}_{i,n} \leq & \dot{V}_{i,n-1} + z_{i,n}(\hat{l}_i v_i + \hat{W}_{i,k}^T S_{i,k}(\hat{\bar{x}}_{i,k}) + r_{i,n}(\hat{\beta}_i y_i^F - \hat{x}_{i,1}) \\ & - \dot{x}_{i,nc} - \dot{\xi}_{i,n}) + \frac{1}{\zeta_{i,n}} \tilde{W}_{i,n}^T (\zeta_{i,n} z_{i,n} S_{i,n}(\hat{\bar{x}}_{i,n}) \\ & - \sigma_{i,n} \dot{\hat{W}}_{i,n}^T) - \frac{1}{\tau_i} \tilde{l}_i \dot{\hat{l}}_i - \frac{l_i}{\pi_i} \tilde{\phi}_i \dot{\hat{\phi}}_i + \frac{1}{2} z_{i,n}^2. \end{aligned} \tag{4.48}$$

Design the switching event-triggered mechanism as follows:

$$v_i(t) = \vartheta_i(t_k^i), \quad \forall t \in [t_k^i, t_{k+1}^i), \tag{4.49}$$

$$t_{k+1}^i = \begin{cases} \inf\{t > t_k^i | \, |\rho_i(t)| \geq \delta_i |v_i(t)| + \varpi_{i,1}\}, \\ \text{if} \ \ |v_i(t)| \leq E_i; \\ \inf\{t > t_k^i | \, |\rho_i(t)| \geq \varpi_{i,2}\}, \\ \text{if} \ \ |v_i(t)| > E_i; \end{cases} \tag{4.50}$$

where $\rho_i(t) = \vartheta_i(t) - v_i(t)$ is the measurement error, t_k^i, t_{k+1}^i denote the current update instant and the next update time of the control input signal about the ith agent, respectively. When the measurement error $|\rho_i(t)|$ exceeds the predefined threshold, the control input updates, i.e., $v_i(t) = \vartheta_i(t_{k+1}^i)$. Thus, the communication resource that the controllers cost will be decreased effectively by the proposed event-triggered adaptive control strategy.

Next, the following parameters are defined as:

$$\delta_i = \begin{cases} \delta_{i,1}, & |v_i(t)| \leq E_i; \\ 0, & |v_i(t)| > E_i, \end{cases} \quad \varpi_i = \begin{cases} \varpi_{i,1}, & |v_i(t)| \leq E_i; \\ \varpi_{i,2}, & |v_i(t)| > E_i, \end{cases} \tag{4.51}$$

where $0 < \delta_{i,1} < 1$, E_i, $\varpi_{i,1}$ and $\varpi_{i,2}$ are positive design parameters. Thus, there exists $\varrho_{i,1}(t)$ and $\varrho_{i,2}(t)$ satisfying $\varrho_{i,1}(t) \in [0, 1]$, $\varrho_{i,2}(t) \in [0, 1]$ such that the following equation holds:

$$\vartheta_i(t) = (1 + \delta_i \varrho_{i,1}(t)) v_i(t) + \varpi_i \varrho_{i,2}(t), \tag{4.52}$$

$$v_i(t) = \frac{\vartheta_i(t)}{1 + \varrho_{i,1}(t)\delta_i} - \frac{\varpi_i \varrho_{i,2}(t)}{1 + \varrho_{i,1}(t)\delta_i}. \tag{4.53}$$

Furthermore, we construct the following parameter adaptive laws and actual control law:

$$\dot{\hat{\phi}}_i = - \frac{\pi_i \bar{\vartheta}_i(t) z_{i,n}}{1 + \delta_i} - \pi_i' \hat{\phi}_i, \tag{4.54}$$

$$\dot{\hat{l}}_i = - \frac{\tau_i \vartheta_i(t) z_{i,n}}{1 + \delta_i} - \tau_i' \hat{l}_i, \tag{4.55}$$

$$\begin{aligned}\vartheta_i(t) =&\hat{\phi}_i\vartheta_i(t), &(4.56)\\ \dot{\hat{W}}_{i,n} =&\zeta_{i,n}z_{i,n}S_{i,n}(\hat{\bar{x}}_{i,n}) - \sigma_{i,n}\hat{W}_{i,n}, &(4.57)\\ \bar{\vartheta}_i(t) =& -(1+\delta_i)(\alpha_{i,n}\tanh(\frac{\alpha_{i,n}z_{i,n}}{\gamma_i}) + \varpi_i\tanh(\frac{\varpi_i z_{i,n}}{\gamma_i})), &(4.58)\\ \alpha_{i,n} =&c_{i,n}v_{i,n} + \xi_{i,n-1} + \hat{W}_{i,n}^T S_{i,n}(\hat{x}_{i,n})\\ &+ r_{i,n}(\hat{\beta}_i y_i^F - \hat{x}_{i,1}) + z_{i,n} + z_{i,n}\hat{l}_i^2 - \dot{x}_{i,nc}, &(4.59)\end{aligned}$$

where $\tau_i^{'}$, $\pi_i^{'}$, $\sigma_{i,n}$, $\gamma_{i,n}$ are positive design parameters, and the initial conditions of adaptive laws are defined as $\hat{\phi}_i(0) > 0$, $\hat{l}_i(0) > 0$. Based on the property about hyperbolic tangent function in [22], $-\alpha\tanh(\alpha/\gamma) \leq 0$ ($\forall\alpha \in \mathbb{R}, \gamma > 0$) such that $z_{i,n}\bar{\vartheta}_i(t) < 0$ and $z_{i,n}\vartheta_i(t) < 0$. Therefore, with the help of the Lemma 1.2, the following formulas hold:

$$\begin{aligned} z_{i,n}\hat{l}_i v_i =&\frac{z_{i,n}\hat{l}_i\vartheta_i(t)}{1+\varrho_{i,1}(t)\delta_i} - \frac{\varrho_{i,2}(t)z_{i,n}\hat{l}_i\varpi_i}{1+\varrho_{i,1}(t)\delta_i}\\ \leq&(1+\tilde{l}_i\tilde{\phi}_i - \tilde{l}_i\phi_i - l_i\tilde{\phi}_i)\frac{z_{i,n}\bar{\vartheta}_i(t)}{1+\delta_i}\\ &- \frac{\varrho_{i,2}(t)z_{i,n}\hat{l}_i\varpi_i}{1+\varrho_{i,1}(t)\delta_i}\\ \leq&(\tilde{l}_i\tilde{\phi}_i - \tilde{l}_i\phi_i - l_i\tilde{\phi}_i)\frac{z_{i,n}\bar{\vartheta}_i(t)}{1+\delta_i} - |z_{i,n}\alpha_{i,n}|\\ &- |z_{i,n}\varpi_i| - \frac{\varrho_{i,2}(t)z_{i,n}\hat{l}_i\varpi_i}{1+\varrho_{i,1}(t)\delta_i} + 0.557\gamma_i, \end{aligned} \quad (4.60)$$

$$\begin{aligned} -\frac{1}{\tau_i}\tilde{l}_i\dot{\hat{l}}_i - \frac{l_i}{\pi_i}\tilde{\phi}_i\dot{\hat{\phi}}_i =&(\tilde{l}_i\phi_i + l_i\tilde{\phi}_i - \tilde{l}_i\tilde{\phi}_i)\frac{z_{i,n}\bar{\vartheta}_i(t)}{1+\delta_i}\\ &+ \frac{\tau_i^{'}}{\tau_i}\tilde{l}_i\hat{l}_i + \frac{\pi_i^{'}}{\pi_i}\tilde{\phi}_i\hat{\phi}_i. \end{aligned} \quad (4.61)$$

Substituting the above formulas into (4.48), one has

$$\begin{aligned} \dot{V}_{i,n} \leq&\dot{V}_{i,n-1} + z_{i,n}(\alpha_{i,n} - c_{i,n}z_{i,n} - \frac{1}{2}z_{i,n} - z_{i,n}\hat{l}_i^2)\\ &+ \frac{\sigma_{i,n}}{b_{i,n}}\tilde{W}_{i,n}^T\hat{W}_{i,n} + \frac{\tau_i^{'}}{\tau_i}\tilde{l}_i\hat{l}_i + \frac{\pi_i^{'}}{\pi_i}\tilde{\phi}_i\hat{\phi}_i - |z_{i,n}\alpha_{i,n}|\\ &- |z_{i,n}\varpi_i| + \frac{1}{2}\tilde{W}_{i,n}^T\tilde{W}_{i,n} - \frac{\varrho_{i,2}(t)z_{i,n}\hat{l}_i\varpi_i}{1+\varrho_{i,1}(t)\delta_i} + 0.557\gamma_i. \end{aligned} \quad (4.62)$$

By utilizing the Young's inequality, we can obtain

$$-\frac{\varrho_{i,2}(t)z_{i,n}\hat{l}_i\varpi_i}{1+\varrho_{i,1}(t)\delta_i} \le z_{i,n}^2\tilde{l}_i^2 + \frac{1}{4}(\frac{\varpi_i}{1-\delta_i})^2. \quad (4.63)$$

The information of $V_{i,n-1}$ has been obtained in the $Step(i, n-1)$. Substituting (4.63) into (4.62), we obtain

$$\begin{aligned}\dot{V}_{i,n} \le &-\Pi_i||e_i||^2 - (c_{i,1}-\frac{5}{2})z_{i,1}^2 - \sum_{j=2}^{n} c_{i,j}z_{i,j}^2 \\ &+\sum_{j=1}^{n}(\frac{\sigma_{i,j}}{\zeta_{i,j}}\tilde{W}_{i,j}^T\hat{W}_{i,j} + \tilde{W}_{i,j}^T\tilde{W}_{i,j}) + \frac{1}{2}||P_i||^2G_i\tilde{l}_i^2 \\ &+\frac{\tau_i^{'}}{\tau_i}\tilde{l}_i\hat{l}_i + \frac{\pi_i^{'}l_i}{\pi_i}\tilde{\phi}_i\hat{\phi}_i + \frac{\mu_i}{\eta_i}\tilde{\beta}_i^2\hat{\beta}_i\text{sgn}(\tilde{\beta}_i) + D_i. \end{aligned} \quad (4.64)$$

where $D_i^{'} = D_{i,1} + 0.557\gamma_i + \frac{1}{4}(\frac{\varpi_i}{1-\delta_i})^2$. In there, we assume that there exists the positive constant G_i satisfying $v_i^2 \le G_i$. This assumption is reasonable in a lot of actual engineering, such as the flight systems [9], the multi-agent systems [23], the mobile robot systems [24], the chemical systems [25] etc..

By applying Young's inequality, one has

$$\frac{\sigma_{i,j}}{b_{i,j}}\tilde{W}_{i,j}^T\hat{W}_{i,j} \le \frac{\sigma_{i,j}}{2\zeta_{i,j}}W_{i,j}^{*T}W_{i,j}^* - \frac{\sigma_{i,j}}{2\zeta_{i,j}}\tilde{W}_{i,j}, \quad (4.65)$$

$$\frac{\tau_i^{'}}{\tau_i}\tilde{l}_i\hat{l}_i \le \frac{\tau_i^{'}}{2\tau_i}l_i^2 - \frac{\tau_i^{'}}{2\tau_i}\tilde{l}_i^2, \quad (4.66)$$

$$\frac{\pi_i^{'}l_i}{\pi_i}\tilde{\phi}_i\hat{\phi}_i \le \frac{\pi_i^{'}l_i}{2\pi_i}\phi_i^2 - \frac{\pi_i^{'}l_i}{2\pi_i}\tilde{\phi}_i^2, \quad (4.67)$$

$$\frac{\mu_i}{\eta_i}\tilde{\beta}_i^2\hat{\beta}_i\text{sgn}(\tilde{\beta}_i) \le \frac{\mu_i}{3\eta_i}\beta_i^3 - \frac{\mu_i}{3\eta_i}|\tilde{\beta}_i|^3. \quad (4.68)$$

Substituting (4.65)–(4.68) into (4.64) yields

$$\begin{aligned}\dot{V}_{i,n} \le &-\Pi_i||e_i||^2 - (c_{i,1}-\frac{5}{2})z_{i,1}^2 - \sum_{j=2}^{n} c_{i,j}z_{i,j}^2 \\ &-\sum_{j=1}^{n}(\frac{\sigma_{i,j}}{2\zeta_{i,j}}-1)\tilde{W}_{i,j}^T\tilde{W}_{i,j} - (\frac{\tau_i^{'}}{2\tau_i} - \frac{||P_i||^2G_i}{2})\tilde{l}_i^2 \\ &-\frac{\pi_i^{'}l_i}{2\pi_i}\tilde{\phi}_i^2 - \frac{\mu_i}{3\eta_i}|\tilde{\beta}_i|^3 + D_i, \end{aligned} \quad (4.69)$$

where $D_i = D_i^{'} + \sum_{j=1}^{n}\frac{\sigma_{i,j}}{2\zeta_{i,j}}W_{i,j}^{*T}W_{i,j}^* + \frac{\tau_i^{'}}{2\tau_i}l_i^2 + \frac{\pi_i^{'}l_i}{2\pi_i}\phi_i^2 + \frac{\mu_i}{3\eta_i}\beta_i^3$.

According to the above discussion, the main results of this chapter are concluded as follows.

Theorem 4.1 *Consider the MASs* (4.1) *with sensor and actuator faults under Assumptions 4.1–4.2. Then, the proposed switched event-triggered adaptive fault-tolerant control strategy [i.e.,*(4.24), (4.29), (4.38), (4.39), (4.42), (4.43), (4.56), (4.57)*] ensures that all signals in the CLS are SGUUB, and the consensus tracking errors satisfy* $||y_i - y_r|| \leq \sqrt{2(1-e^{Ct})D/C + 2V(0)e^{-Ct} + \iota/2\kappa_0}$. *In the meanwhile, the Zeno phenomenon can be avoided effectively.*

Proof. Construct the overall Lyapunov function

$$V = \sum_{i=1}^{N} V_{i,n} = \sum_{i=1}^{N}(\frac{1}{2}e_i^T P_i e_i + \frac{1}{3\eta_i}|\tilde{\beta}_i|^3 + \frac{1}{2\tau_i}\tilde{l}_i^2 + \frac{l_i}{2\pi_i}\tilde{\phi}_i^2)$$
$$+ \frac{1}{2}\sum_{i=1}^{N}\sum_{k=1}^{n}(z_{i,k}^2 + \frac{1}{2\zeta_{i,k}}\tilde{W}_{i,k}^T\tilde{W}_{i,k}). \tag{4.70}$$

Calculating the derivation of (4.70), one has:

$$\dot{V} = \sum_{i=1}^{N}\dot{V}_{i,n} \leq -\sum_{i=1}^{N}\Pi_i||e_i||^2 - \sum_{i=1}^{N}(c_{i,1} - \frac{5}{2})z_{i,1}^2$$
$$-\sum_{i=1}^{N}\frac{\pi_i' l_i}{2\pi_i}\tilde{\phi}_i^2 - \sum_{i=1}^{N}\sum_{j=1}^{n}(\frac{\sigma_{i,j}}{2\zeta_{i,j}} - 1)\tilde{W}_{i,j}^T\tilde{W}_{i,j}$$
$$-\sum_{i=1}^{N}(\frac{\tau_i'}{2\tau_i} - \frac{||P_i||^2 G_i}{2})\tilde{l}_i^2 - \sum_{i=1}^{N}\sum_{j=2}^{n}c_{i,j}z_{i,j}^2$$
$$-\sum_{i=1}^{N}\frac{\mu_i}{3\eta_i}|\tilde{\beta}_i|^3 + \sum_{i=1}^{N}D_i. \tag{4.71}$$

The above inequality further becomes

$$\dot{V} \leq -CV + D, \tag{4.72}$$

where $C = \min\{\frac{2\Pi_i}{\lambda_{\min}(P_i)}, 2c_{i,1} - 5, 2c_{i,j}, \sigma_{i,k} - 2\zeta_{i,k}, l_i\pi_i', \tau_i' - ||P_i||^2 G_i\tau_i, \mu_i; k = 1, \ldots, n; j = 2, \ldots, n; i = 1, \ldots, N\}$, $D = \sum_{i=1}^{N} D_i$, $\lambda_{\min}(P_i)$ represents the minimum eigenvalue of matrix P_i. To guarantee the stability of the system, the design parameters should satisfy the following condition: $\frac{2\Pi_i}{\lambda_{\min}(P_i)} > 0$; $2c_{i,1} - 5 > 0$; $\sigma_{i,k} - 2\zeta_{i,k} > 0$; $\tau_i' - ||P_i||^2 G_i\tau_i > 0$.

Multiplying (4.72) by e^{Ct} on both sides, one has

$$e^{Ct}\dot{V} + Ce^{Ct}V \leq De^{Ct},$$
$$V(t) \leq (1 - e^{Ct})\frac{D}{C} + V(0)e^{-Ct}. \tag{4.73}$$

Therefore, it is easy to get the following inequality:

$$||z_{i,j}|| \leq \sqrt{2(1 - e^{Ct})\frac{D}{C} + 2V(0)e^{-Ct}}, \tag{4.74}$$
$$||\hat{W}_{i,j}|| \leq \sqrt{2\zeta_{i,j}((1 - e^{Ct})\frac{D}{C} + V(0)e^{-Ct})}, \tag{4.75}$$

with $j = 1, \ldots, n; i = 1, \ldots, N$. By the same way, the boundedness of e_i, $\hat{\beta}_i$, $\hat{l}_i$ and $\hat{\phi}_i$ can also be verified. Therefore, in the closed-loop system, all signals are bounded. Further, the output tracking error formula can be obtained: $y_i - y_r = z_{i,1} + \xi_{i,1} + (\hat{y}_{ri} - y_r)$. According to (4.16), the tracking error is ultimately bounded because of the result that $\hat{y}_{ri}$ can converge to y_r in finite time T_{ri}. Then, we can obtain the expression about the bound of the tracking error as follows:

$$||y_i - y_r|| \leq \sqrt{2(1 - e^{Ct})\frac{D}{C} + 2V(0)e^{-Ct}} + \frac{\iota}{2\kappa_0}, \tag{4.76}$$

where the tracking error can converge a bounded domain by adjusting the design parameters properly. For instance, according to the above analysis, we can choose large $c_{i,1}, \ldots, c_{i,n}$, μ_i, small ϖ_i, τ_i and π_i to obtain the desire tracking performance. Moreover, increasing $c_{i,1}, \ldots, c_{i,n}$ and μ_i can expedite the system states' convergent speed, the control cost will be higher correspondingly. As a consequence, we should make decisions in combination with specific situations in actual engineering systems.

Next, we will exclude the Zeno phenomenon under the event-triggered mechanism. According to equation $\rho_i(t) = \vartheta_i(t) - v_i(t)$, we have

$$\frac{d|\rho_i(t)|}{dt} = \dot{\rho}_i \text{sign}(\dot{\rho}_i) \leq |\dot{\vartheta}_i|. \tag{4.77}$$

Since $\dot{\vartheta}_i$ is a function consisting of bounded variable y_i^F ,$x_{i,1}, \ldots, x_{i,n}, \xi_{i,1}, \ldots, \xi_{i,n}, \hat{W}_{i,1}, \ldots, \hat{W}_{i,n}, \hat{\phi}_i, \hat{l}_i, \hat{\beta}_i, \dot{x}_{i,2c}, \ldots, \dot{x}_{i,nc}, \hat{y}_{ri}, \dot{\hat{y}}_{ri}$, we can find the upper bound of $|\dot{\vartheta}_i|$ such that $|\dot{\vartheta}_i| \leq \vartheta_i^*$. Thus, we can have

$$|\rho_i(t)| = \int_{t_k^i}^{t_{k+1}^i} |\dot{\rho}_i| dt \leq \int_{t_k^i}^{t_{k+1}^i} \vartheta_i^* dt$$
$$\leq \vartheta_i^*(t_{k+1}^i - t_k^i). \tag{4.78}$$

From (4.50), the lower bound of $|\rho_i(t)|$ can be obtained, i.e., $|\rho_i(t)| \geq \varpi_i$. Therefore, the following inequality holds:

$$t_{k+1}^i - t_k^i \geq \frac{\varpi_i}{\vartheta_i^*} > 0. \tag{4.79}$$

As the lower bound of the inter-event interval is positive, the Zeno behavior can be excluded. Then, we finished the proof of $Theorem$ 4.1.

4.4 Simulation Study

The presented event-triggered adaptive neural consensus control strategy will be applied in the following numerical example for demonstrating its effectiveness in this section.

We study the nonlinear multi-agent systems consisting of three follower agents 1–3 and the leader agent r. In Fig:4.1, the topology showing the relationship of agents is provided. In addition, the dynamics of the follower in multiagent system (4.1) can be represented as the following differentiation equation:

$$\begin{aligned}
\dot{x}_{i,1} &= x_{i,2} + x_{i,1}\sin(x_{i,1}) + 0.1\cos(t)\sin(x_{i,1}) \\
\dot{x}_{i,2} &= u_i^F + x_{i,2}^2\sin(x_{i,2}) + 0.1\sin(t)\cos(x_{i,1}x_{i,2}) \\
y_i^F &= \kappa_i y_i \\
y_{i,1} &= x_{i,1},
\end{aligned} \tag{4.80}$$

where $\kappa_i = 0.8$, $u_i^F = (1-\theta_i)u_i + 0.5\cos(x_{i,1})\sin(x_{i,2})$ with $\theta_i = 0.3$, and the bounded values of the saturated control inputs are defined as $u_{Mi} = 20 (i = 1, 2, 3)$. The dynamics of leader r is described as: $y_r = 0.2 + 0.5\sin(t)$. According to the directed graph topology, the adjacency matrix $\mathcal{A}$ can be obtained:

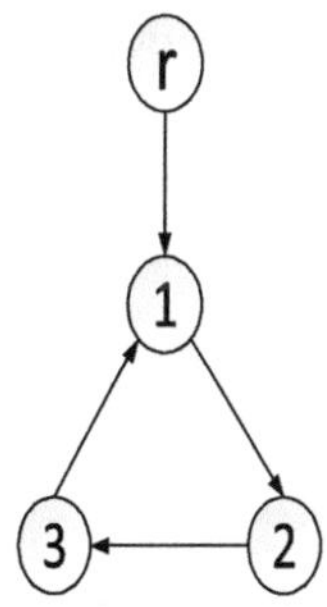

Fig. 4.1 Directed topology graph for leader-follow agents

$$\mathcal{A} = \begin{bmatrix} 0 & 0 & 1 \\ 1 & 0 & 0 \\ 0 & 1 & 0 \end{bmatrix},$$

and the weight matrix between the followers and the leader is $\mathcal{A}_0 = \text{diag}\{1, 0, 0\}$. In this chapter, the RBF NN is utilized to construct the state observer, in which the neural network $W_{i,1}^T S_{i,1}$ contains 12 nodes with centers spaced evenly in the interval $[-6, 6]$ and the neural network $W_{i,2}^T S_{i,2}$ contains 24 nodes with centers spaced evenly in the interval $[-6, 6]$. The Gaussian functions are chosen as $S_{i,1}(\hat{x}_{i,1}) = e^{-\frac{(\hat{x}_{i,1}-6+i)^2}{4}}$, $S_{i,2}(\hat{\bar{x}}_{i,2}) = e^{-\frac{(\hat{x}_{i,1}-6+0.5i)^2+(\hat{x}_{i,2}-6+0.5i)^2}{8}}$, and the initial states of followers, the neural-network-based observers, the sliding mode estimators are set as:

$$\begin{cases} x_1(0) = [x_{1,1}(0), x_{2,1}(0), x_{3,1}(0)]^T = [0.1, -0.1, 0.2]^T; \\ x_2(0) = [x_{1,2}(0), x_{2,2}(0), x_{3,2}(0)]^T = [0.2, -0.1, 0.1]^T; \\ \hat{x}_1(0) = [\hat{x}_{1,1}(0), \hat{x}_{2,1}(0), \hat{x}_{3,1}(0)]^T = [0.2, -0.2, 0.3]^T; \\ \hat{x}_2(0) = [\hat{x}_{1,2}(0), \hat{x}_{2,2}(0), \hat{x}_{3,2}(0)]^T = [0.2, -0.2, 0.2]^T; \\ \hat{y}_{ri}(0) = [\hat{y}_{r1}(0), \hat{y}_{r2}(0), \hat{y}_{r3}(0)]^T = [0.2, 0.3, 0.4]^T. \end{cases}$$

The design parameters are chosen as: $\xi = 0.8$, $\omega_n = 150$; $r_{1,1} = 20$, $r_{1,2}$=20, $r_{2,1}$=20, $r_{2,2} = 20, r_{3,1} = 20, r_{3,2} = 20$; $c_{1,1} = 8, c_{1,2} = 25, c_{2,1} = 5, c_{2,2} = 6, c_{3,1} = 8, c_{3,2} = 5$; $\sigma_{1,1} = 1, \sigma_{1,2} = 1, \sigma_{2,1} = 10, \sigma_{2,2} = 5, \sigma_{3,1} = 1, \sigma_{3,2} = 1$; $\zeta_{1,1} = \zeta_{1,2} = 0.4$, $\zeta_{2,1} = \zeta_{2,2} = 1$, $\zeta_{3,1} = \zeta_{3,2} = 0.4$; $h_1 = h_2 = h_3 = 1$, $\mu_1 = 7$, $\mu_2 = 8.8$, $\mu_3 = 6.6$, $\eta_1 = \eta_2 = \eta_3 = 1$; $\pi_1 = 0.5$, $\pi_1' = 0.05$, $\pi_2 = 0.5$, $\pi_2' = 0.05$, $\pi_3 = 0.5$, $\pi_3' = 0.05$; $\tau_1' = 0.1$, $\tau_2' = 0.5$, $\tau_3' = 0.1$, $\tau_1 = 1$, $\tau_2 = 1$, $\tau_3 = 0.8$. In addition, the sliding mode estimator parameters are chosen as $b_i = 0.5, q_i = 0.6 > |\dot{y}_r|$. The initial values of the adaptive laws are defined as:

$$\begin{cases} \hat{\beta}(0) = [\hat{\beta}_1(0), \hat{\beta}_2(0), \hat{\beta}_3(0)]^T = [1, 1, 1]^T; \\ \hat{l}(0) = [\hat{l}_1(0), \hat{l}_2(0), \hat{l}_3(0)]^T = [1, 1, 1.1]^T; \\ \hat{\phi}(0) = [\hat{\phi}_1(0), \hat{\phi}_2(0), \hat{\phi}_3(0)]^T = [0.7, 0.8, 0.9]^T. \end{cases}$$

The event-triggered mechanism are designed as:

$$u_i(t) = \vartheta_i(t_k^i), \quad \forall t \in [t_k^i, t_{k+1}^i) \tag{4.81}$$

$$t_{k+1}^i = \begin{cases} \inf\{t > t_k^i \mid |\rho_i(t)| \geq 0.15|v_i(t)| + 1.5\}, \\ \text{if } \ |u_i(t)| \leq 10; \\ \inf\{t > t_k^i \mid |\rho_i(t)| \geq 2.5\}, \\ \text{if } \ |u_i(t)| > 10; \end{cases} \tag{4.82}$$

where u_i denotes the actual control, ρ_i represents the measurement error with $i = 1, 2, 3$.

Table 4.1 Update times of control laws by two ways

Agents	Agent 1	Agent 2	Agent 3
The time-triggered control	50000	50000	50000
The method presented here	407	452	390

From Figs. 4.2, 4.3, 4.4, 4.5, 4.6, 4.7, 4.8, 4.9, 4.10, 4.11 and 4.12, we can obtain the simulation results. The curves of the output of 3 agents and the reference trajectory from leader r are depicted in Fig. 4.2. As we can see, under the proposed control strategy in this chapter, all followers will track the leader accurately. In Fig. 4.3, the curves of state variable $x_{i,2}$ are shown. Figures 4.4, 4.5 and 4.6 depict the states of the the followers and the corresponding neural network observers. The estimation trajectories about reference signal by using sliding mode estimator are shown in Fig. 4.7. Figures 4.8 and 4.9 depict the trajectories of the parameter adaptive laws. Figure 4.10 shows the continuous and unsaturated control input signals. Further, the trajectories of the event-triggered and saturated controller inputs are shown in Fig. 4.11. Figure 4.12 depicts the Event triggered instants. It can be verified explicitly that the Zeno behavior is excluded for all follower agents. According to the Table 4.1, there is an apparent contrast between two different control methods about the update times of control signal, and the proposed control strategy in this chapter can release the pressure of data transmission effectively.

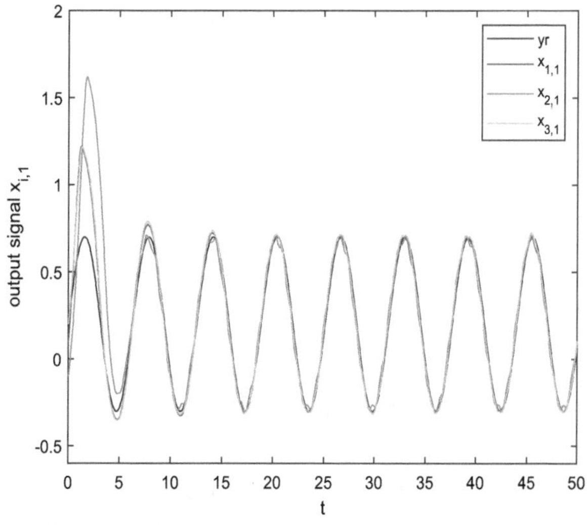

Fig. 4.2 Trajectories of $y_i(i = 1, 2, 3)$, the desired tracking signal y_r

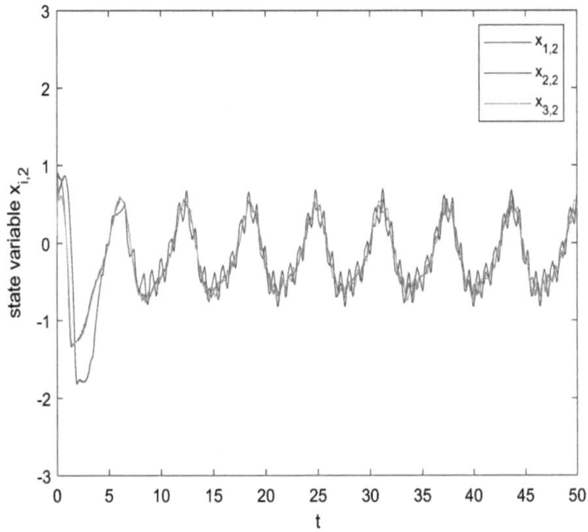

Fig. 4.3 Trajectories of state $x_{i,2}(i = 1, 2, 3)$

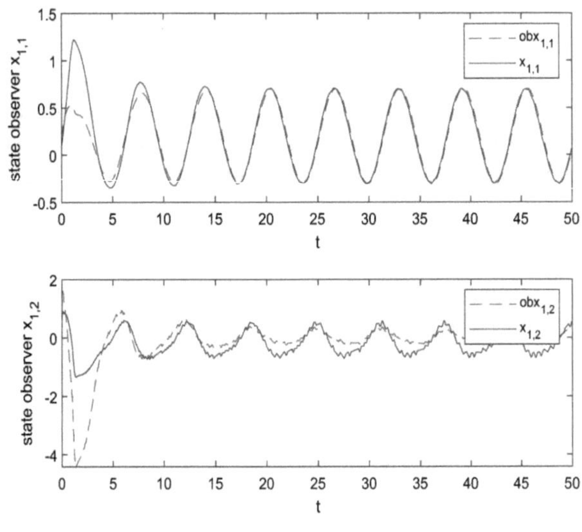

Fig. 4.4 Trajectories of the agent 1's states and its estimates

4.5 Conclusion

4.5.1 Contribution

The main contribution of this chapter lies in the first-time research on the leader-follower consensus control problem of nonlinear multi-agent systems with sensor faults, actuator faults and input saturation. Unlike the traditional coordinate transformation method, a sliding mode estimator is used to dynamically approximate the

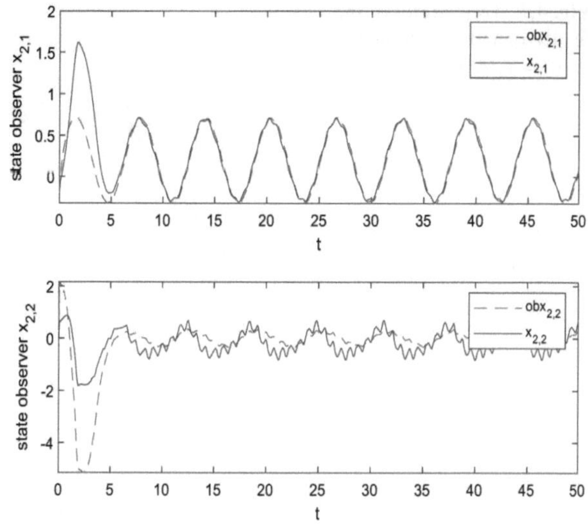

Fig. 4.5 Trajectories of the agent 2's states and its estimates

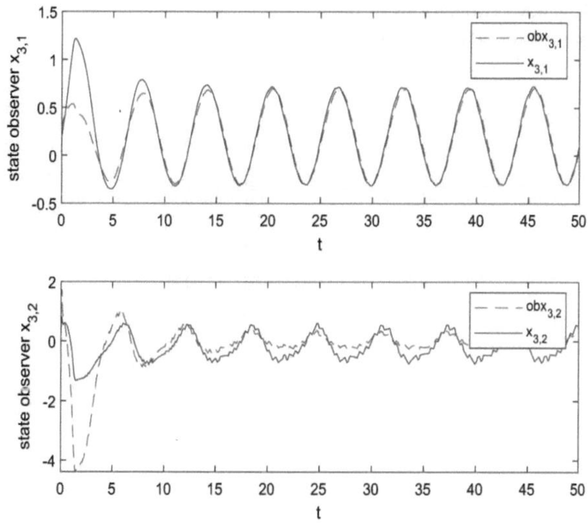

Fig. 4.6 Trajectories of the agent 3's states and its estimates

leader's trajectory, and a control framework under fault and saturation constraints is constructed. The command filter backstepping technique is proposed, which uses a low-pass filter to smooth the virtual control law and corrects the filtering error through a compensation term, effectively solving the differential explosion problem in traditional backstepping control and reducing the computational burden of high-dimensional systems. A dynamic threshold switching event-triggered mechanism is designed, which adaptively adjusts the triggering threshold according to the

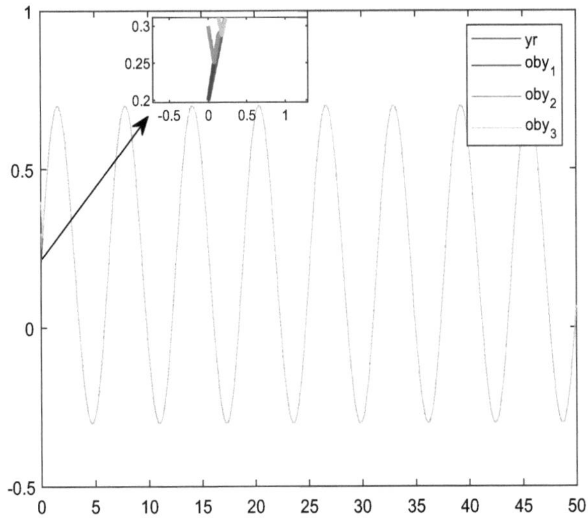

Fig. 4.7 Trajectories of the leader y_r and its estimates $\hat{y}_{ri}(i = 1, 2, 3)$

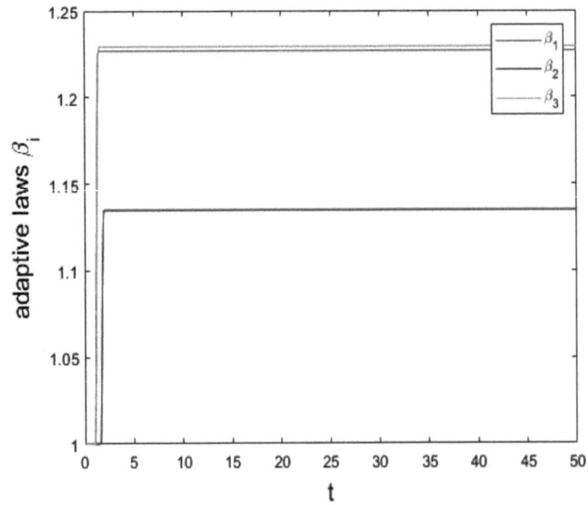

Fig. 4.8 Trajectories of the parameter adaptive laws $\beta_i(i = 1, 2, 3)$

control input amplitude to balance the communication overhead and control accuracy in real time, providing a flexible solution for multi-agent cooperative control in bandwidth-limited scenarios of industrial Internet of Things.

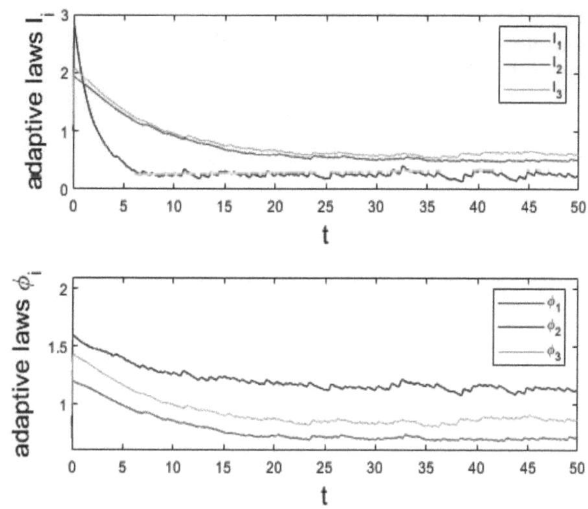

Fig. 4.9 Trajectories of the parameter adaptive laws l_i and $\phi_i (i = 1, 2, 3)$

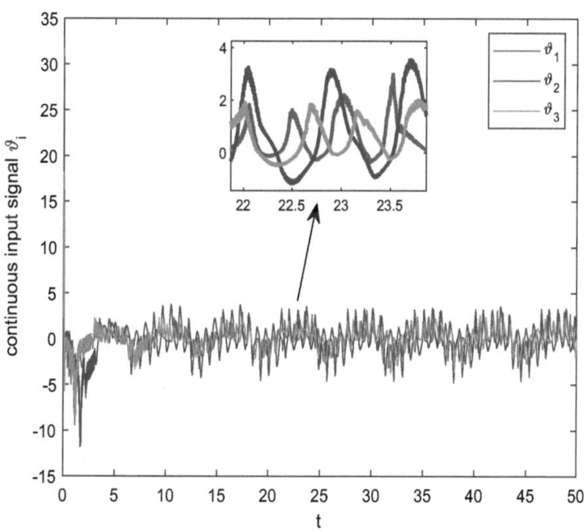

Fig. 4.10 Trajectories of the actual controller inputs $\vartheta_i (i = 1, 2, 3)$

4.5.2 Limitations of the Current Work

Despite the promising results, several limitations remain:

High computational complexity: The radial basis function neural network is used for state observation and function approximation. Combined with the design of sliding mode estimator and compensation system, it brings significant computational overhead, which may hinder real-time implementation in large-scale multiagent systems or resource-constrained platforms.

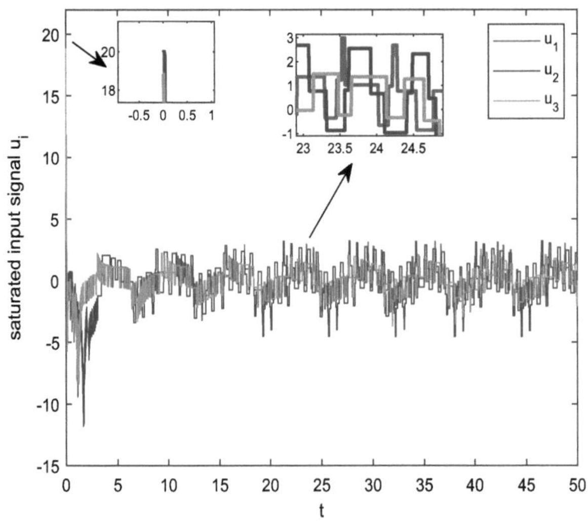

Fig. 4.11 Trajectories of the saturated controller inputs $u_i (i = 1, 2, 3)$

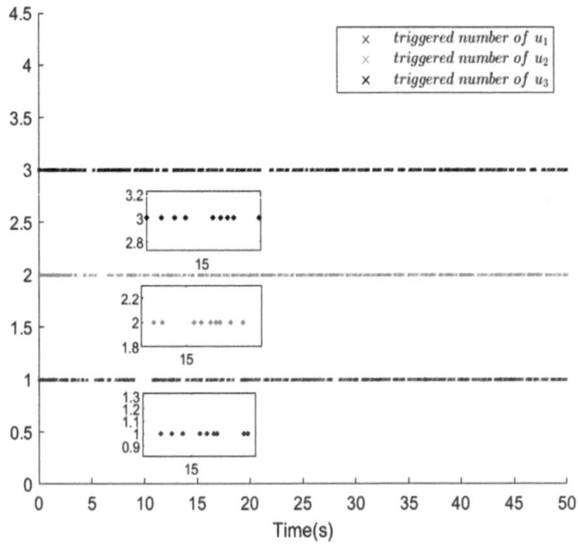

Fig. 4.12 Event triggered instants

High sensitivity of parameters: The control performance is greatly influenced by design parameters such as observer gain, filter time constant, and adaptive law coefficient. Improper parameter tuning can lead to decreased tracking accuracy, intensified transient oscillations, and even instability. A large amount of manual calibration is required.

Limitations of the event-triggered mechanism: The static threshold design of the switching event-triggered mechanism cannot optimally adapt to the dynamic changes of the system. In a highly dynamic environment, it may lead to excessive triggering, wasting resources or causing control accuracy degradation due to update delays.

4.5.3 Future Directions

Based on the research in this chapter, future exploration can focus on the following directions. In terms of technology integration, integrate deep learning with edge computing to enhance fault diagnosis and prediction capabilities, as well as the real-time performance of local control. In terms of scenario expansion, apply the strategy to fields such as intelligent transportation and unmanned aerial vehicle clusters to verify the adaptability of heterogeneous systems. At the theoretical level, construct a multi-agent model with time delay and parameter perturbation to improve the theory of cooperative control for compound faults. In the field of communication optimization, use game theory to quantify the balance between communication and control, dynamically adjust the transmission strategy based on information entropy to improve resource utilization efficiency.

References

1. Deng C, Che W-W, Wu Z-G (2021) A dynamic periodic event-triggered approach to consensus of heterogeneous linear multiagent systems with time-varying communication delays. IEEE Trans Cybern 51(4):1812–1821
2. Qiu J, Sun K, Wang T, Gao H (2019) Observer-based fuzzy adaptive event-triggered control for pure-feedback nonlinear systems with prescribed performance. IEEE Trans Fuzzy Syst 27(9):2152–2162
3. Wang X, Wang H, Huang T, Kurths J (2023) Neural-network-based adaptive tracking control for nonlinear multiagent systems: the observer case. IEEE Trans Cybern 53(2):138–150
4. Zou W, Shi P, Xiang Z, Shi Y (2020) Consensus tracking control of switched stochastic nonlinear multiagent systems via event-triggered strategy. IEEE Trans Neural Netw Learn Syst 31(5):1036–1045
5. Wang A, Liu L, Qiu J, Feng G (2019) Event-triggered robust adaptive fuzzy control for a class of nonlinear systems. IEEE Trans Fuzzy Syst 27(9):1648–1658
6. Wang A, Liu L, Qiu J, Feng G (2020) Finite-time adaptive fuzzy control for nonstrict-feedback nonlinear systems via an event-triggered strategy. IEEE Trans Fuzzy Syst 28(9):2164–2174
7. Huang Y, Wang J, Wang F, He B (2021) Event-triggered adaptive finite-time tracking control for full state constraints nonlinear systems with parameter uncertainties and given transient performance. ISA Trans 108(1):131–143
8. Liu Y, Zhu Q, Zhao N (2021) Event-triggered adaptive fuzzy control for switched nonlinear systems with state constraints. Inf Sci 562(1):28–43
9. Shen Q, Jiang B, Cocquempot V (2014) Adaptive fuzzy observer-based active fault-tolerant dynamic surface control for a class of nonlinear systems with actuator faults. IEEE Trans Fuzzy Syst 22(9):338–349

10. Li D-J, Li D-P (2015) Adaptive controller design-based neural networks for output constraint continuous stirred tank reactor. Neurocomputing 153(1):159–163
11. Liu Z, Lai G, Zhang Y, Chen CLP (2015) Adaptive fuzzy tracking control of nonlinear time-delay systems with dead-zone output mechanism based on a novel smooth model. IEEE Trans Fuzzy Syst 23(9):1998–2011
12. He W, Yan Z, Sun C, Chen Y (2017) Adaptive neural network control of a flapping wing micro aerial vehicle with disturbance observer. IEEE Trans Cybern 47(10):3452–3465
13. Zhang L, Yang G-H (2018) Observer-based fuzzy adaptive sensor fault compensation for uncertain nonlinear strict-feedback systems. IEEE Trans Fuzzy Syst 26(9):2301–2310
14. Zhai D, An L, Li X, Zhang Q (2018) Adaptive fault-tolerant control for nonlinear systems with multiple sensor faults and unknown control directions. IEEE Trans Neural Netw Learn Syst 29(10):4436–4446
15. Chen CLP, Wen G-X, Liu Y-J, Wang F-Y (2014) Adaptive consensus control for a class of nonlinear multiagent time-delay systems using neural networks. IEEE Trans Neural Netw Learn Syst 25(7):1217–1226
16. Dong W, Farrell JA, Polycarpou MM, Djapic V, Sharma M (2012) Command filtered adaptive backstepping. IEEE Trans Control Syst Technol 20(3):566–580
17. Chen K, Wang J, Zhang Y, Liu Z (2020) Leader-following consensus for a class of nonlinear strict-feedback multiagent systems with state time-delays. IEEE Trans Syst Man Cybern Syst 50(10):2351–2361
18. Zhang L, Chen B, Lin C, Shang Y (2022) Fuzzy adaptive fixed-time consensus tracking control of high-order multiagent systems. IEEE Trans Fuzzy Syst 30(3):567–578
19. Wu Y, Liang H, Zhang Y, Ahn CK (2021) Cooperative adaptive dynamic surface control for a class of high-order stochastic nonlinear multiagent systems. IEEE Trans Cybern 51(10):5214–5224
20. Zhang F, Chen Y-Y (2021) Fuzzy adaptive containment control for nonlinear nonaffine pure-feedback multiagent systems. IEEE Trans Fuzzy Syst 29(10):2878–2889
21. Cao Y, Ren W, Meng Z (2010) Decentralized finite-time sliding mode estimators and their applications in decentralized finite-time formation tracking. Syst Control Lett 59(10):522–529
22. Ren B, San P P, Ge S S, Lee T H (2009) Adaptive dynamic surface control for a class of strict-feedback nonlinear systems with unknown backlash-like hysteresis. In: American control conference, pp 4482–4487
23. Li Z, Liu X, Ren W, Xie L (2013) Distributed tracking control for linear multiagent systems with a leader of bounded unknown input. IEEE Trans Autom Control 58(4):518–523
24. Fierro R, Lewis F (1997) Control of a nonholonomic mobile robot: backstepping kinematics into dynamics. In: Proceedings of 1995 34th IEEE conference on decision and control, vol 14, pp 149–163
25. Kabore R, Wang H (2001) Design of fault diagnosis filters and fault-tolerant control for a class of nonlinear systems. IEEE Trans Autom Control 46(10):1805–1810

Chapter 5
Event-Triggered Adaptive Finite-Time Control for Switched Cyberphysical Systems with Uncertain Deception Attacks

This study explores finite-time event-triggered adaptive control for switched cyber-physical systems (CPSs) subject to unknown deception attacks. To address system complexity, a command filter with finite-time convergence properties is employed, while a disturbance observer incorporating fuzzy logic is designed to approximate external perturbations. The proposed scheme integrates Nussbaum gain techniques to counteract performance deterioration caused by deceptive attacks. Through the common Lyapunov function framework, rigorous stability analysis establishes finite-time uniform boundedness guarantees for all closed-loop signals, with numerical simulations validating the theoretical results.

Cyber-physical systems (CPSs) merge computing, networking, and physical processes, enabling critical applications in smart infrastructure and defense systems. A major security concern in these systems is deception attacks, which compromise integrity by altering sensor measurements and control commands. Existing studies have proposed adaptive control methods for nonlinear CPSs, but many rely on idealized communication models that do not account for real-world bandwidth constraints [1, 2].

To overcome these challenges, this work investigates resilient control strategies for switched CPSs by employing fuzzy systems and neural networks as function approximators [3–6]. Prior research has primarily emphasized asymptotic stability [7, 8], whereas finite-time control approaches—crucial for achieving rapid response and precise tracking performance—have recently attracted growing attention [9–12].Techniques such as dynamic surface control (DSC) and command filtering have been adopted to reduce design complexity in finite-time control schemes [10, 11].

In contrast to time-triggered approaches, event-triggered schemes effectively minimize redundant communication while enhancing resource efficiency [13–15]. This chapter presents an innovative finite-time control framework incorporating adaptive fuzzy logic and event-triggered mechanisms for switched cyber-physical systems, with the key advancement being the guarantee of uniform boundedness within finite time.

X. Wang et al., *Optimal Adaptive Cooperative Control for Nonlinear Multiagent Systems*,
https://doi.org/10.1007/978-981-95-2964-3_5

5.1 Problem Formulation and Preliminaries

5.1.1 System Formulation

Consider the following nonlinear cyberphysical system subject to uncertain deception attacks and backlash-like hysteresis:

$$\begin{cases} \dot{x}_1 = f_{1,\varsigma(t)}(x_1) + x_2 \\ \dot{x}_i = f_{i,\varsigma(t)}(\bar{x}_i) + x_{i+1} \\ \dot{x}_n = f_{n,\varsigma(t)}(\bar{x}_n) + u_s(v) + \rho_u(\bar{x}_n, t) + \delta(\bar{x}_n, t) \\ \hat{x}_k = x_k + \rho_x(x_k, t) \\ y = x_1 \end{cases} \tag{5.1}$$

where $\bar{x}_i = [x_1, \cdots, x_i]^{\mathrm{T}}$ denotes the state variable with $i = 2, \ldots, n-1$, u_s, y, ρ_u are the control input, system output and actuator attack respectively. $\delta(\bar{x}_n, t)$ represents the unknown but bounded external disturbance. $\hat{x}_k$ refers to as the state variable compromised by the sensor attacks, $\rho_x(x_k, t)$ represents the uncertain sensor deception attack with $k = 1, \ldots, n$. $f_{k,\varsigma(t)}$ denotes the unknown but smooth nonlinear function, and $\varsigma(t) : [0, +\infty) \to \mathcal{M}_i = 1, \ldots, M_i$ is the switching signal. We claim that the sth switched subsystem is activated when $\varsigma(t) = s$. $u_s(v)$ indicates the backlash-like hysteresis type of nonlinear impacts on actuator, which can be depicted as $\frac{du_s}{dt} = \varpi|\frac{dv}{dt}|(\phi v(t) - u_s) + \varrho\frac{dv}{dt}$ with v being the input of backlash-like hysteresis, and ϖ, ϕ, ϱ being unknown positive constants, ϕ satisfies $\phi \geq \varrho$. Based on the results in [16], it yields

$$u_s(v) = \phi v(t) + h(v), \tag{5.2}$$

where $h(v) = (u_s(0) - \phi v(0))e^{-\varpi(v-v(0))\mathrm{sign}\dot{v}} + e^{-\varpi v \mathrm{sign}\dot{v}} \int_{v(0)}^{v} (\varrho - \phi)e^{-\varpi\tau \mathrm{sign}\dot{v}} d\tau$ is bounded, such that $h(v) \leq \bar{h}$, $\bar{h}$ is unknown positive constant.

Therefore, model (5.1) can be recast as

$$\begin{cases} \dot{x}_i = f_{i,\varsigma(t)}(\bar{x}_i) + x_{i+1} \\ \dot{x}_n = f_{n,\varsigma(t)}(\bar{x}_n) + \phi_0 v(t) + g(v) + \delta(\bar{x}_n, t) \\ y = x_1 \end{cases} \tag{5.3}$$

where $g(v) = (\phi - \phi_0)v(t) + \rho_u(\bar{x}_n, t) + h(v)$, ϕ_0 is the known constant.

Control Objective: Our goal is to design a finite-time adaptive fuzzy control scheme with event-triggered mechanisms for nonlinear switched CPSs under deception attacks and backlash-like hysteresis, ensuring all system signals reach and remain near equilibrium within finite time.

To achieve this objective, some necessary lemmas and assumptions are listed.

Assumption 5.1 : Suppose that $\rho_x(x_i, t)$ and $\rho_u(\bar{x}_n, t)$ satisfy $\rho_x(x_i, t) = a_i(t)x_i(t)$, and $\rho_u(t, \bar{x}_n(t)) = \beta(t)\varrho(\bar{x}_n(t))$ with $i = 1, 2, \ldots, n$, respectively, where $a_i(t)$, $\beta(t)$ are the unknown time-varying signals, and $\varrho(\bar{x}_n)$ is the unknown continuous nonlinear function. In addition, the variable a_i should satisfy $1 + a_i(t) \neq 0$. Without loss of generality, we assume that the sign of $1 + a_i(t)$ is positive. Thus, there exists two unknown positive constants $\bar{a}_i$, $\underline{a}_i$ and $\bar{\beta}_i$ satisfying $\underline{a}_i \leq a_i(t) \leq a_i$ and $\beta(t) \leq \bar{\beta}$.

Remark 5.1 : In this chapter, we make the following coordinate transformation: $\ell_i = 1/(1 + a_i(t))$. Then, it is obvious that the sign of ℓ_i is always positive and there exists unknown positive constants $\underline{\ell}_i$, $\bar{\ell}_i$ and $\bar{\ell}_{i,d}$ such that $\underline{\ell}_i \leq \ell_i \leq \bar{\ell}_i$, and $|\dot{\ell}_i| \leq \bar{\ell}_{i,d}$. In addition, Assumption 5.1 roots in the previous result about linear CPSs, it plays a important role in maintaining the controllability of the systems subject to the unknown deception attacks. To be specific, the aim of Assumption 5.1 is to exclude the situation $1 + a_i(t) = 0$, which means that the the compromised state $\hat{x}_i = (1 + a_i(t))x_i = 0$ such that the designed controller will lose efficacy.

Lemma 5.1 *[9]: If the smooth positive-definite function $W(t)$ conforms to*

$$\dot{W}(t) \leq -aW(t) - bW^q(t), \forall t \geq t_0,$$

where $0 < q < 1$, a, b are positive constants, then, $W(t)$ will converge to the equilibrium point over finite time T, in which $T \leq t_0 + (1/(a(1-q))) \ln(aV^{1-q}(t_0) + b)/b$.

Lemma 5.2 *Tong et al. [5]: Based on the approximation property of FLSs, it is widely applied in the design of adaptive controller. A FLS mainly includes fuzzy inference engine, knowledge base, defuzzifier, and fuzzifier. For any consecutive function $g(\varpi)$, which is defined on the compact set ϱ, FLSs can achieve the following performance:*

$$g(\varpi) = \lambda^T \psi(\varpi) + \varepsilon,$$

where ε denotes the bounded approximation error satisfying $|\varepsilon| \leq \epsilon$, and ϵ is an unknown constant.

5.1.2 Nussbaum Gain

This chapter employs Nussbaum gains to compensate for unknown control coefficients induced by deception attacks. A continuous function $\mathcal{N}(\Psi) : \mathbb{R} \rightarrow \mathbb{R}$ can be declared to be a function of Nussbaum type, when it satisfies

$$\lim_{s\rightarrow\infty} \sup \frac{1}{s} \int_0^s \mathcal{N}(\Psi) d\Psi = +\infty,$$

$$\lim_{s\to\infty} \inf \frac{1}{s}\int_{0}^{s} \mathcal{N}(\Psi)d\Psi = -\infty.$$

Such as, the continuous functions $\mathcal{N}(\Psi) = \Psi^2\cos(\Psi)$ and $e^{\Psi^2}\cos(\pi/2\Psi)$ are applied in [6, 17].

Lemma 5.3 *Ge et al. [6]: Suppose $\mathcal{V}(\cdot) \geq 0$ and Ψ are smooth functions defined on $[0, t_f)$, $\mathcal{N}(\cdot)$ is a smooth Nussbaum-type function. When the undermentioned condition is met*

$$\mathcal{V}(t) \leq c_0 + e^{-c_1 t}\int_{0}^{t} [h(\cdot)\mathcal{N}(\Psi) + 1]\dot{\Psi}e^{c_1\varrho}d\varrho,$$

where $h(\cdot)$ is a time-varying bounded parameter which takes values in the unknown closed intervals $L = [p-, p+]$ with $0 \notin I$, and c_0 denotes a suitable constant, c_1 is a nonnegative constant. Then $\mathcal{V}(t)$, $\int_0^t h(\cdot)\mathcal{N}(\Psi)\dot{\Psi}e^{c_1\varrho}d\varrho$ must be bounded on $[0, t_f)$.

5.2 Main Results

This section presents a novel finite-time control scheme integrating adaptive fuzzy logic and event-triggered mechanisms to counteract uncertain deception attacks. we first give a coordinate transformation below:

$$\begin{cases} z_1 = x_1 \\ z_i = x_i - \ell_i\alpha_{i-1}^f, i = 2, \cdots, n, \end{cases} \tag{5.4}$$

where α_{i-1}^f denotes the filter output with the virtual controller α_{i-1} as input.Due to the inaccessibility of actual state x_i caused by unknown deception attacks, the conventional coordinate z_i becomes unavailable for controller synthesis. Therefore, we establish a modified coordinate transformation based on measurable states and virtual control inputs as follows:

$$\begin{cases} \hat{z}_1 = \hat{x}_1 \\ \hat{z}_i = \hat{x}_i - \alpha_{i-1}^f, \end{cases} \tag{5.5}$$

where $\hat{z}_i$ plays a vital role during the design of control and adaptive laws.

To avoid differentiation explosion, a finite-time command filter is introduced:

$$\begin{cases} \dot{\varphi}_{i0} = l_{i0} \\ i0 = -\eta_{i0}|\varphi_{i0} - \alpha_i|^{\frac{1}{2}}\text{sign}(\varphi_{i0} - \alpha_i) + \varphi_{i1} \\ \dot{\varphi}_{i1} = -\eta_{i1}\text{sign}(\varphi_{i1} - l_{i0}), \end{cases} \tag{5.6}$$

where η_{i0}, η_{i1} are the positive design parameters, α_i is the virtual control signal. From [18], the terms $|\varphi_{i0} - \alpha_i|^{\frac{1}{2}} \text{sign}(\varphi_{i0} - \alpha_i)$ and $\text{sign}(\varphi_{i1} - l_{i0})$ in (5.6) can ensure that the filter can remain finite-time stable, when the parameters η_{i0} and η_{i1} are chosen properly.

Higher-order system dynamics amplify filtering errors, significantly degrading control performance. In order to eliminate the filtering error, let $e_i = \alpha_i^f - \alpha_i$, we define the following filtering error compensating signal ℓ:

$$\begin{cases} \dot{\ell}_i = r_{i1}\ell_i - r_{i2}\ell_i^{2q-1} - W_i\ell_i + e_i \\ \dot{\ell}_n = r_{n1}\ell_n - r_{n2}\ell_n^{2q-1}, i = 1, \cdots, n-1 \end{cases} \tag{5.7}$$

where $q = (2\bar{p} - 1)/(2\bar{p} + 1)$ and r_{ij} are positive design parameters with $j = 1, 2, 3$, $\bar{p} \geq 2$; $W_i = (|\hat{z}_i\ell_i| + r_{i3}e_i^2)/(||\ell_i||^2)$.

The following presents the recursive design framework for the event-triggered adaptive fuzzy controller.

Step 1: The first Lyapunov candidate function is constructed as follows:

$$\mathcal{V}_1 = \frac{1}{2}z_1^2 + \frac{1}{2}\tilde{\Lambda}_1^2 + \frac{1}{2}\ell_1^2, \tag{5.8}$$

where $\tilde{\Lambda}_1$ denotes the estimation error of Λ_1, and it will be given later.

Based on (5.4), the time derivative of z_i yields:

$$\begin{aligned} \dot{z}_i &= f_{1,s}(x_1) + x_2 \\ &= f_{1,s}(x_1) + z_2 + \ell_2\alpha_1^f. \end{aligned} \tag{5.9}$$

Thus, the time derivative of $\mathcal{V}_1$ is obtained as:

$$\begin{aligned} \dot{\mathcal{V}}_1 &= z_1\left(f_{1,s}(x_1) + z_2 + \ell_2\alpha_1^f\right) + \tilde{\Lambda}_1\dot{\tilde{\Lambda}}_1 + \ell_1\dot{\ell}_1 \\ &= z_1\left(z_2 + \mathcal{F}_{1,s}(\mathcal{Z}_1) + \ell_2\alpha_1\right) + \tilde{\Lambda}_1\dot{\tilde{\Lambda}}_1 + \ell_1\dot{\ell}_1 - r_{11}z_1^{2q} \\ &\quad - \left(c_1 + \frac{3}{2}\right)z_1^2 + c_1\hat{z}_1^2, \end{aligned} \tag{5.10}$$

where $\mathcal{F}_{1,s} = (c_1 + 3/2)z_1 + \ell_2 e_1 + r_{12}z_1^{2q-1} - c_1 z_1/\ell_1^2 + f_{1,s}(x_1)$, $\mathcal{Z}_1 = [z_1, x_1, \ell_1]^T$, $\hat{\mathcal{Z}}_1 = [\hat{z}_1, \hat{x}_1]^T$. Leveraging the approximation property of FLSs, we derive:

$$\begin{aligned} \mathcal{F}_{1,s} &= \lambda_{1,s}^T\psi_1(\mathcal{Z}_1) + \varepsilon_{1,s}(\mathcal{Z}_1) \\ &= \lambda_{1,s}^T\psi_1(\hat{\mathcal{Z}}_1) + \varepsilon_{1,s}(\mathcal{Z}_1) + \chi_{1,s}, \end{aligned} \tag{5.11}$$

where $\chi_{1,s} = \lambda_{1,s}^T(\psi_1(\mathcal{Z}_1) - \psi_1(\hat{\mathcal{Z}}_1))$ is bounded according to Lemma 1.2, which satisfies $|\chi_1| \leq \bar{\chi}_1$.

Applying Young's inequality yields:

$$
\begin{aligned}
z_1\lambda_{1,s}^T\psi_1(\hat{\mathcal{Z}}_1) \leq& \frac{||\lambda_{1,s}||^2 z_1^2\psi_1^T(\hat{\mathcal{Z}}_1)\psi_1(\hat{\mathcal{Z}}_1)}{2k_1^2} + \frac{1}{2}k_1^2 \\
\leq& \frac{\Lambda_1\hat{z}_1^2\psi_1^T(\hat{\mathcal{Z}}_1)\psi_1(\hat{\mathcal{Z}}_1)}{2k_1^2} + \frac{1}{2}k_1^2, &(5.12)\\
z_1(\varepsilon_{1,s} + \chi_{1,s}) \leq& z_1^2 + \frac{1}{2}\bar{\varepsilon}_1^2 + \frac{1}{2}\bar{\chi}_1^2, &(5.13)\\
-W_1\ell_1 \leq& -|\hat{z}_1\ell_1| - r_{13}e_1^2, &(5.14)\\
\ell_1 e_1 \leq& \frac{1}{2}\ell_1^2 + \frac{1}{2}e_1^2, &(5.15)\\
z_1 z_2 \leq& \frac{1}{2}z_1^2 + \frac{1}{2}z_2^2, &(5.16)
\end{aligned}
$$

where k_1 is a constant, $\Lambda_1 = \max_{s\in\varsigma(t)} \bar{\ell}_1^2||\lambda_{1,s}||^2$, $\hat{\Lambda}_1 = \Lambda_1 - \tilde{\Lambda}_1$ is its estimation value, $\bar{\varepsilon}_1^2 = \max_{s\in\varsigma(t)} ||\bar{\varepsilon}_{1,s}||^2$, and $\bar{\chi}_1^2 = \max_{s\in\varsigma(t)} ||\chi_{1,s}||^2$.

By combining (5.12)–(5.16) with (5.10), we obtain:

$$
\begin{aligned}
\dot{\mathcal{V}}_1 \leq& -r_{12}z_1^{2q} - c_1 z_1^2 + c_1\hat{z}_1^2 + \ell_2 z_1\alpha_1 + \frac{1}{2}z_2^2 - |\hat{z}_1\ell_1| + \Delta_1^{'} \\
&+ \frac{\hat{\Lambda}_1\hat{z}_1^2\psi_1^T(\hat{\mathcal{Z}}_1)\psi_1(\hat{\mathcal{Z}}_1)}{2k_1^2} - r_{11}\ell_1^2 - r_{12}\ell_1^{2q} - \bar{r}_{13}e_1^2 \\
&+ \tilde{\Lambda}_1(\frac{\hat{z}_1^2\psi_1^T(\hat{\mathcal{Z}}_1)\psi_1(\hat{\mathcal{Z}}_1)}{2k_1^2} - \dot{\hat{\Lambda}}_1), &(5.17)
\end{aligned}
$$

where $\bar{r}_{13} = r_{13} - 1/2$, $\Delta_1^{'} = (1/2)k_1^2 + (1/2)\varepsilon_1^2 + (1/2)\bar{\chi}_1^2$.

Design the first controller α_1, adaptive law $\hat{\Lambda}_1$, and auxiliary variable Φ_1 as

$$
\begin{aligned}
\alpha_1 =& \mathcal{N}(\Phi_1)(c_1\hat{z}_1 + \frac{\hat{\Lambda}_1\hat{z}_1\psi_1^T(\hat{\mathcal{Z}}_1)\psi_1(\hat{\mathcal{Z}}_1)}{2k_1^2} - \ell_1), &(5.18)\\
\dot{\Phi}_1 =& c_1\hat{z}_1^2 + \frac{\hat{\Lambda}_1\hat{z}_1^2\psi_1^T(\hat{\mathcal{Z}}_1)\psi_1(\hat{\mathcal{Z}}_1)}{2k_1^2} - \hat{z}_1\ell_1, &(5.19)\\
\dot{\hat{\Lambda}}_1 =& \frac{\hat{z}_1^2\psi_1^T(\hat{\mathcal{Z}}_1)\psi_1(\hat{\mathcal{Z}}_1)}{2k_1^2} - \varsigma_1\hat{\Lambda}_1, &(5.20)
\end{aligned}
$$

where c_1, ς_1 and k_1 represent tunable positive parameters.

Based on the preceding derivation, (5.17) can be reformulated as:

$$
\begin{aligned}
\dot{\mathcal{V}}_1 \leq& -r_{12}z_1^{2q} - c_1 z_1^2 - r_{11}\ell_1^2 - r_{12}\ell_1^{2q} - \bar{r}_{13}e_1^2 + \frac{1}{2}z_2^2 \\
&+ (g_1\mathcal{N}(\Phi_1) + 1)\dot{\Phi}_1 + \varsigma_1\hat{\Lambda}_1\tilde{\Lambda}_1 + \Delta_1^{'}, &(5.21)
\end{aligned}
$$

where $g_1 = \ell_1(t)\ell_2(t)$.

Applying Lemma 1.1 and the Young's inequality, we have:

$$\begin{aligned} \varsigma_1 \hat{\Lambda}_1 \tilde{\Lambda}_1 \leq& \bar{\varsigma}_1 \Lambda_1^2 - \bar{\varsigma}_1 \tilde{\Lambda}_1^2, \quad (5.22)\\ \bar{\varsigma}_1 (\frac{1}{2}\tilde{\Lambda}_1^2)^q \leq& \frac{\bar{\varsigma}_1}{2}\tilde{\Lambda}_1^2 + \bar{\varsigma}_1 (1 - q) q^{1-q}, \quad (5.23) \end{aligned}$$

where $\bar{\varsigma}_1 = (1/2)\varsigma_1$.

By inserting (5.22) and (5.23) into (5.21), we derive:

$$\begin{aligned} \dot{\mathcal{V}}_1 \leq& - r_{12} z_1^{2q} - c_1 z_1^2 - r_{11}\ell_1^2 - r_{12}\ell_1^{2q} - \bar{r}_{13} e_1^2 + \frac{1}{2} z_2^2 \\ &- \frac{\bar{\varsigma}_1}{2}\tilde{\Lambda}_1^2 - \bar{\varsigma}_1 (\frac{1}{2}\tilde{\Lambda}_1^2)^q + (g_1 \mathcal{N}(\Phi_1) + 1)\dot{\Phi}_1 + \Delta_1, \quad (5.24) \end{aligned}$$

where $\Delta_1 = \Delta_1^{'} + \bar{\varsigma}_1 \Lambda_1^2 + \bar{\varsigma}_1 (1 - q) q^{\frac{q}{1-q}}$.

Step 2: Selecting the 2th Lyapunov candidate function

$$\mathcal{V}_2 = \mathcal{V}_1 + \frac{1}{2} z_2^2 + \frac{1}{2}\tilde{\Lambda}_2^2 + \frac{1}{2}\ell_2^2, \quad (5.25)$$

The estimation error $\tilde{\Lambda}_2$ of parameter Λ_2 will be formally defined in subsequent sections. Based on (5.4), the time derivative $\dot{z}_2$ is expressed as:

$$\begin{aligned} \dot{z}_2 =& \dot{x}_2 - \dot{\ell}_2 \alpha_1^f - \ell_2 \dot{\alpha}_1^f \\ =& z_3 + \ell_3 \alpha_2^f + f_{2,s}(\bar{x}_2) - \dot{\ell}_2 \alpha_1^f - \ell \dot{\alpha}_1^f. \quad (5.26) \end{aligned}$$

Consequently, $\dot{\mathcal{V}}_2$ is given by:

$$\begin{aligned} \dot{\mathcal{V}}_2 =& \dot{\mathcal{V}}_1 + z_2(z_3 + \ell_3 \alpha_2^f + f_{2,s}(\bar{x}_2) - \dot{\ell}_2 \alpha_1^f - \ell \dot{\alpha}_1^f) \\ &+ \tilde{\Lambda}_2 \dot{\hat{\Lambda}}_2 + \ell_2 \dot{\ell}_2 \\ =& \dot{\mathcal{V}}_1 - r_{22} z_2^{2q-1} - c_2 z_2^2 + z_2(z_3 + \ell_3 \alpha_2 + \mathcal{F}_{2,s}(\mathcal{Z}_2)) \\ &+ \tilde{\Lambda}_2 \dot{\hat{\Lambda}}_2 + \ell_2 \dot{\ell}_2 + c_1 \hat{z}_2^2 - 2 z_2^2, \quad (5.27) \end{aligned}$$

where $\mathcal{F}_{2,s}(\mathcal{Z}_2) = r_{22} z_2^{2q-1} + c_2 z_2 + \ell_3 e_2 + f_{2,s}(\bar{x}_2) - \dot{\ell}_2 \alpha_1^f - \ell \dot{\alpha}_1^f + 2 z_2 - c_2 z_2/\ell_2^2$, c_2 is a positive constant.

Following the methodology of **Step 1**, we derive:

$$\mathcal{F}_{2,s}(\mathcal{Z}_2) = \lambda_{2,s}^T \psi_2(\hat{\mathcal{Z}}_2) + \varepsilon_{2,s} + \chi_{2,s}, \quad (5.28)$$

where $\chi_{2,s} = \lambda_{2,s}(\psi_2(\mathcal{Z}_2) - \psi_2(\hat{\mathcal{Z}}_2))$, $\mathcal{Z}_2 = [z_2, x_2, x_1, \ell_2, \hat{\Lambda}_1]^T$ is not available, $\hat{\mathcal{Z}}_2 = [\hat{z}_2, \hat{x}_2, \hat{x}_1, \hat{\Lambda}_1]^T$ can be obtained.

Applying Young's inequality leads to:

$$
\begin{aligned}
z_2(\mathcal{F}_{2,s}(\hat{\mathcal{Z}}_2)+\varepsilon_{2,s}+\chi_{2,s}) \leq & \frac{||\lambda_{2,s}||^2 z_2^2 \psi_2^T(\hat{\mathcal{Z}}_2)\psi_2(\hat{\mathcal{Z}}_2)}{2k_2^2} \\
& + \frac{1}{2}k_2^2 + \frac{1}{2}\bar{\varepsilon}_{2,s}^2 + \frac{1}{2}\bar{\chi}_{2,s}^2 + z_2^2 \\
\leq & \frac{\Lambda_2 \hat{z}_2^2 \psi_2^T(\hat{\mathcal{Z}}_2)\psi_2(\hat{\mathcal{Z}}_2)}{2k_2^2} + \frac{1}{2}k_2^2 \\
& + \frac{1}{2}\bar{\varepsilon}_2^2 + \frac{1}{2}\bar{\chi}_2^2 + z_2^2, \qquad (5.29) \\
-W_2\ell_2 \leq & -|\hat{z}_2\ell_2| - r_{23}e_2^2, \qquad (5.30) \\
\ell_2 e_2 \leq & \frac{1}{2}\ell_2^2 + \frac{1}{2}e_2^2, \qquad (5.31) \\
z_2 z_3 \leq & \frac{1}{2}z_2^2 + \frac{1}{2}z_3^2, \qquad (5.32)
\end{aligned}
$$

where $\Lambda_2 = \max_{s\in\varsigma(t)} \bar{\ell}_2^2||\lambda_{2,s}||^2$, $\hat{\Lambda}_2 = \Lambda_2 - \tilde{\Lambda}_2$ denotes the estimation value of Λ_2, $\bar{\varepsilon}_2^2 = \max_{s\in\varsigma(t)} ||\bar{\varepsilon}_{2,s}||^2$, $\bar{\chi}_2^2 = \max_{s\in\varsigma(t)} ||\chi_{2,s}||^2$, and $k_2 > 0$ is a constant.

The substitution of (5.29)–(5.32) into (5.27) produces:

$$
\begin{aligned}
\dot{\mathcal{V}}_2 \leq & \dot{\mathcal{V}}_1 - r_{22}z_2^{2q-1} - c_2 z_2^2 + c_2\hat{z}_2^2 + \ell_3 z_2\alpha_2 + \frac{1}{2}z_3^2 - |\hat{z}_2\ell_2| \\
& + \frac{\hat{\Lambda}_2\hat{z}_2^2\psi_2^T(\hat{\mathcal{Z}}_2)\psi_2(\hat{\mathcal{Z}}_2)}{2k_2^2} - r_{21}\ell_2^2 - r_{22}\ell_2^{2q} - \bar{r}_{23}e_2^2 \\
& + \tilde{\Lambda}_2(\frac{\hat{z}_2^2\psi_2^T(\hat{\mathcal{Z}}_2)\psi_1(\hat{\mathcal{Z}}_2)}{2k_2^2} - \dot{\hat{\Lambda}}_2) + \frac{1}{2}k_2^2 + \frac{1}{2}\bar{\varepsilon}_{2,s}^2 \\
& + \frac{1}{2}\bar{\chi}_{2,s}^2 - \frac{1}{2}z_2^2, \qquad (5.33)
\end{aligned}
$$

where $\bar{r}_{23} = r_{23} - (1/2)$.

The design of virtual controller α_2, auxiliary variable Φ_2, and adaptive law $\hat{\Lambda}_2$ proceeds as follows:

$$
\alpha_2 = \mathcal{N}(\Phi_2)(c_2\hat{z}_2 + \frac{\hat{\Lambda}_2\hat{z}_2\psi_2^T(\hat{\mathcal{Z}}_2)\psi_2(\hat{\mathcal{Z}}_2)}{2k_2^2} - \ell_2), \qquad (5.34)
$$

$$
\dot{\Phi}_2 = c_2\hat{z}_2^2 + \frac{\hat{\Lambda}_2\hat{z}_2^2\psi_2^T(\hat{\mathcal{Z}}_2)\psi_2(\hat{\mathcal{Z}}_2)}{2k_1^2} - \hat{z}_2\ell_2, \qquad (5.35)
$$

$$
\dot{\hat{\Lambda}}_2 = \frac{\hat{z}_2^2\psi_2^T(\hat{\mathcal{Z}}_2)\psi_2(\hat{\mathcal{Z}}_2)}{2k_2^2} - \varsigma_2\hat{\Lambda}_2, \qquad (5.36)
$$

where c_2, ς_2, and k_2 represent tunable positive constants.

By combining (5.34)–(5.36) with (5.33), we obtain:

$$
\begin{aligned}
\dot{\mathcal{V}}_2 \le & \dot{\mathcal{V}}_1 - r_{22} z_2^{2q} - c_2 z_2^2 - r_{21}\ell_2^2 - r_{22}\ell_2^{2q} - \bar{r}_{23} e_2^2 + \frac{1}{2} z_3^2 \\
& + (g_2 \mathcal{N}(\Phi_2) + 1)\dot{\Phi}_2 + \varsigma_2 \hat{\Lambda}_2 \tilde{\Lambda}_2 + \frac{1}{2} k_2^2 + \frac{1}{2} \bar{\varepsilon}_{2,s}^2 \\
& + \frac{1}{2} \bar{\chi}_{2,s}^2 - \frac{1}{2} z_2^2,
\end{aligned} \tag{5.37}
$$

where $g_2 = \ell_2(t)\ell_3(t)$.

By incorporating (5.21) and applying Young's inequality (cf. (5.22) and (5.23)), we derive:

$$
\begin{aligned}
\dot{\mathcal{V}}_2 \le & \sum_{l=1}^{2} (-r_{21} z_l^{2q} - c_l z_l^2 - r_{l1}\ell_l^2 - r_{l2}\ell_l^{2q} - \bar{r}_{l3} e_l^2 - \frac{\bar{\varsigma}_l}{2} \tilde{\Lambda}_l^2 \\
& - \bar{\varsigma}_l (\frac{1}{2} \tilde{\Lambda}_l^2)^q + (g_l \mathcal{N}(\Phi_l) + 1)\dot{\Phi}_l) + \frac{1}{2} z_2^3 + \Delta_2,
\end{aligned} \tag{5.38}
$$

where $\Delta_2 = \Delta_1 + (1/2)k_2^2 + (1/2)\varepsilon_2^2 + (1/2)\bar{\chi}_2^2 + \bar{\varsigma}_1 \Lambda_2^2 + \bar{\varsigma}_2 (1-q) q^{\frac{q}{1-q}}$.

Step i: The ith Lyapunov candidate function is constructed as:

$$
\mathcal{V}_i = \mathcal{V}_{i-1} + \frac{1}{2} z_i^2 + \frac{1}{2} \tilde{\Lambda}_i^2 + \frac{1}{2} \ell_i^2, \tag{5.39}
$$

where $\tilde{\Lambda}_i$ is the estimation error of Λ_i.

Building upon the previous design framework, we now construct the ith iteration components: virtual control law α_i, auxiliary variable Φ_i, and adaptive law $\hat{\Lambda}_i$, as detailed below:

$$
\alpha_i = \mathcal{N}(\Phi_i)(c_i \hat{z}_i + \frac{\hat{\Lambda}_i \hat{z}_i \psi_i^T(\hat{\mathcal{Z}}_i)\psi_i(\hat{\mathcal{Z}}_i)}{2k_i^2} - \ell_i), \tag{5.40}
$$

$$
\dot{\Phi}_i = c_i \hat{z}_i^2 + \frac{\hat{\Lambda}_i \hat{z}_i^2 \psi_i^T(\hat{\mathcal{Z}}_i)\psi_i(\hat{\mathcal{Z}}_i)}{2k_i^2} - \hat{z}_i \ell_i, \tag{5.41}
$$

$$
\dot{\hat{\Lambda}}_i = \frac{\hat{z}_i^2 \psi_i^T(\hat{\mathcal{Z}}_i)\psi_i(\hat{\mathcal{Z}}_i)}{2k_i^2} - \varsigma_i \hat{\Lambda}_i, \tag{5.42}
$$

The design parameters c_i, k_i and ς_i are positive constants requiring specification.

From the preceding equation, it follows that:

$$
\begin{aligned}
\dot{\mathcal{V}}_i \le & \sum_{l=1}^{i} (-r_{l2} z_l^{2q} - c_l z_l^2 - r_{l1}\ell_l^2 - r_{l2}\ell_l^{2q} - \bar{r}_{l3} e_l^2 - \frac{\bar{\varsigma}_l}{2} \tilde{\Lambda}_l^2 \\
& - \bar{\varsigma}_l (\frac{1}{2} \tilde{\Lambda}_l^2)^q + (g_l \mathcal{N}(\Phi_l) + 1)\dot{\Phi}_l) + \frac{1}{2} z_{i+1}^2 + \Delta_i,
\end{aligned} \tag{5.43}
$$

where $\Delta_i = \Delta_{i-1} + (1/2)k_i^2 + (1/2)\varepsilon_i^2 + (1/2)\bar{\chi}_i^2 + \bar{\varsigma}_i \Lambda_i^2 + \bar{\varsigma}_i(1-q)q^{\frac{q}{1-q}}$, $\bar{\varsigma}_i = \varsigma_i/2$, and $g_i = \ell_i(t)\ell_{i+1}(t)$.

Step n: Based on (5.4), the time derivative z_n is derived as:

$$\begin{aligned}\dot{z}_n =& \dot{x}_n - \dot{\ell}_n \alpha_n^f - \ell_n \dot{\alpha}_n^f \\ =& f_{n,\varsigma(t)}(\bar{x}_n) + \phi_0 v(t) + \Upsilon_1 - \dot{\ell}_n \alpha_n^f - \ell_n \dot{\alpha}_n^f,\end{aligned} \tag{5.44}$$

where $\Upsilon_1 = g(v) + \delta(\bar{x}_n, t)$ represents an unknown bounded disturbance with $|\Upsilon_1| \le \bar{\Upsilon}_1$ and $|\dot{\Upsilon}_1| \le \bar{\Upsilon}_{1,d}$, , where $\bar{\Upsilon}_1$, $\bar{\Upsilon}_{1,d}$ are positive constants. Building on [3], we design a disturbance observer, transforming (5.44) into:

$$\begin{aligned}\dot{z}_n =& m_1^{-1}\mathcal{F}_{n1,s}(\mathcal{Z}_n) + \Upsilon_1 \\ =& m_1^{-1}(\lambda_{n1,s}^T \psi_{n1}(\hat{\mathcal{Z}}_n) + \varepsilon_{n1,s} + \chi_{n1,s}) + \Upsilon_1,\end{aligned} \tag{5.45}$$

where $\mathcal{F}_{n1,s} = f_{n,\varsigma(t)}(\bar{x}_n) + \phi_0 v(t) - \dot{\ell}_n \alpha_n^f - \ell_n \dot{\alpha}_n^f$, $\chi_{n1,s} = \lambda_{n1,s}(\psi_{n1}(\mathcal{Z}_2) - \psi_{n1}(\hat{\mathcal{Z}}_n))$, $\mathcal{Z}_n = [z_n, x_1, \cdots, x_n, \hat{\Lambda}_{n-1}, \ell_n]^T$, $\hat{\mathcal{Z}}_n = [\hat{z}_n, \hat{x}_1, \cdots, \hat{x}_n, \hat{\Lambda}_{n-1}]^T$, and m_1^{-1} is a designed constant.

We now define an auxiliary variable $\hbar_1$:

$$\hbar_1 = \hat{z}_n - \varrho_1 = \frac{z_n}{\ell_n} - \varrho_1, \tag{5.46}$$

The intermediate variable ϱ is constructed as:

$$\dot{\varrho}_1 = b_1 \hbar_1 + \frac{\hat{\varphi}_1 \hbar_1 \psi_{n1}^T(\hat{\mathcal{Z}}_{n1})\psi_{n1}(\hat{\mathcal{Z}}_{n1})}{2m_1 k_{n1}^2}, \tag{5.47}$$

where b_1, k_{n1} are positive constant, $\hat{\varphi}_1$ will be introduced later, and $\hat{\varphi}_1 = \varphi_1 - \tilde{\varphi}_1$.

Consequently, the time derivative $\hbar_1$ is expressed as:

$$\begin{aligned}\dot{\hbar}_1 =& \frac{1}{\ell_n}(m_1^{-1}(\lambda_{n1,s}^T \psi_{n1}(\hat{\mathcal{Z}}_n) + \varepsilon_{n1,s} + \chi_{n1,s}) + \Upsilon_1) \\ & - \frac{\dot{\ell}_n}{\ell_n^2} z_n - b_1 \hbar_1 - \frac{\hat{\varphi}_1 \hbar_1 \psi_{n1}^T(\hat{\mathcal{Z}}_{n1})\psi_{n1}(\hat{\mathcal{Z}}_{n1})}{2m_1 k_{n1}^2}.\end{aligned} \tag{5.48}$$

Applying Young's inequality yields the following relations:

$$\begin{aligned}\frac{\hbar_1}{m_1 \ell_n}(\lambda_{n1,s}^T \psi_{n1}(\hat{\mathcal{Z}}_n) + \varepsilon_{n1,s} + \chi_{n1,s}) \le& \hbar_1^2 + \frac{\bar{\chi}_{n1}^2}{2m_1^2 \underline{\ell}_n^2} \\ & + \frac{\bar{\varepsilon}_{n1}^2}{2m_1^2 \underline{\ell}_n^2} + \frac{k_{n1}^2}{2m_1} + \frac{\varphi_1 \hbar_1^2 \psi_{n1}^T(\hat{\mathcal{Z}}_n)\psi_{n1}(\hat{\mathcal{Z}}_n)}{2m_1 k_{n1}^2},\end{aligned} \tag{5.49}$$

$$\frac{\hbar_1 \Upsilon_1}{\ell_n} \leq \frac{1}{2}\hbar_1^2 + \frac{\bar{\Upsilon}_1^2}{2\underline{\ell}_n^2}, \tag{5.50}$$

$$-\frac{\dot{\ell}_n}{\ell_n^2} z_n \hbar_1 \leq \frac{1}{2}\hbar_1^2 + \frac{\bar{\ell}_{n,d}^2}{\underline{\ell}_n^4} z_n^2, \tag{5.51}$$

where $\varphi_1 = \max_{s \in \varsigma(t)} \frac{\|\lambda_{n1,s}\|^2}{\underline{\ell}_n^2}$, and $\bar{\varepsilon}_{n1}^2 = \bar{\chi}_{n1}^2 = \max_{s \in \varsigma(t)} \|\bar{\chi}_{n1,s}\|^2$.

Based on the preceding analysis, it follows that:

$$\begin{aligned}\hbar_1 \dot{\hbar}_1 \leq & -(b_1 - 2)\hbar_1^2 + \frac{\tilde{\varphi}_1 \hbar_1^2 \psi_{n1}^T(\hat{\mathcal{Z}}_n)\psi_{n1}(\hat{\mathcal{Z}}_n)}{2m_1 k_{n1}^2} \\ & + \frac{\bar{\ell}_{n,d}^2}{\underline{\ell}_n^4} z_n^2 + \Delta_{\hbar_1},\end{aligned} \tag{5.52}$$

where $\Delta_{\hbar_1} = \bar{\chi}_{n1}^2/(2m_1^2\underline{\ell}_n^2) + \bar{\varepsilon}_{n1}^2/(2m_1^2\underline{\ell}_n^2) + k_{n1}^2/(2m_1^2) + \bar{\Upsilon}_1^2/(2\underline{\ell}_n^2)$.

The adaptive law for parameter $\hat{\varphi}_1$ is designed as:

$$\dot{\hat{\varphi}}_1 = \begin{cases} \dfrac{\hbar_1^2 \psi_{n1}^T(\hat{\mathcal{Z}}_n)\psi_{n1}(\hat{\mathcal{Z}}_n)}{2m_1 k_{n1}^2} - \varsigma_{n1}\hat{\varphi}_1, & \iota_1 \leq 0 \\ 0, & \iota_1 \geq 0 \end{cases} \tag{5.53}$$

where $\iota_1 = (\hbar_1^2 \psi_{n1}^T(\hat{\mathcal{Z}}_n)\psi_{n1}(\hat{\mathcal{Z}}_n)/2m_1 k_{n1}^2) - \varsigma_{n1}\hat{\varphi}_1$, such that $\hat{\varphi}_1$ is nonincreasing function, and $0 < \hat{\varphi}_1 \leq \bar{\varphi}_1 = \hat{\varphi}_1(0)$, the initial condition of $\hat{\varphi}_1(0)$ can be set manually.

The disturbance observer $\hat{\Upsilon}_1$ is designed as:

$$\hat{\Upsilon}_1 = m_1(\hbar_1 - \eta_1), \tag{5.54}$$

where $\dot{\eta}_1 = -b_1\hbar_1 + \hat{\Upsilon}_1$.

Consequently, the time derivative $\hat{\Upsilon}_1$ is expressed as:

$$\begin{aligned}\dot{\hat{\Upsilon}}_1 &= m_1\dot{\hbar}_1 - m_1\dot{\eta}_1 \\ &= \frac{1}{\ell_n}((\lambda_{n1,s}^T \psi_{n1}(\hat{\mathcal{Z}}_n) + \varepsilon_{n1,s} + \chi_{n1,s}) + m_1\Upsilon_1)) \\ &\quad - \frac{m_1\dot{\ell}_n}{\ell_n^2} z_n - m_1\hat{\Upsilon}_1 - \frac{\hat{\varphi}_1 \hbar_1 \psi_{n1}^T(\hat{\mathcal{Z}}_n)\psi_{n1}(\hat{\mathcal{Z}}_n)}{2m_1^2 k_{n1}^2},\end{aligned} \tag{5.55}$$

where $\dot{\tilde{\Upsilon}}_1 = \dot{\Upsilon}_1 - \dot{\hat{\Upsilon}}_1$.

From the preceding equation, we derive:

$$\tilde{\Upsilon}_1 \dot{\tilde{\Upsilon}}_1 = \tilde{\Upsilon}_1 \dot{\Upsilon}_1 - \tilde{\Upsilon}_1 \dot{\hat{\Upsilon}}_1. \tag{5.56}$$

Young's inequality yields the following relations:

$$-\frac{\tilde{\Upsilon}_1}{\ell_n}(\lambda_{n1,s}^T\psi_{n1}(\hat{\mathcal{Z}}_n)) \leq \frac{l_1}{2}\tilde{\Upsilon}_1^2 + \frac{\varphi_1}{2l_1\underline{\ell}_n^2}, \tag{5.57}$$

$$-\frac{\tilde{\Upsilon}_1}{\ell_n}(\varepsilon_{n1,s} + \chi_{n1,s}) \leq l_1\tilde{\Upsilon}_1^2 + \frac{\bar{\varepsilon}_{n1}^2}{2\underline{\ell}_n^2} + \frac{\bar{\chi}_{n1}^2}{2\underline{\ell}_n^2}, \tag{5.58}$$

$$\begin{aligned} -\frac{m_1}{\ell_n}\Upsilon_1\tilde{\Upsilon}_1 + m_1\hat{\Upsilon}_1\tilde{\Upsilon}_1 \leq& \frac{m_1^2}{l_1\underline{\ell}_n^2}\bar{\Upsilon}_1 + \frac{l_1}{2}\tilde{\Upsilon}^2 \\ &+ \frac{m_1}{2}\bar{\Upsilon}_1^2 - \frac{m_1}{2}\tilde{\Upsilon}_1^2, \end{aligned} \tag{5.59}$$

$$\frac{m_1\dot{\ell}_n}{\ell_n^2}z_n\tilde{\Upsilon}_1 \leq \frac{l_1}{2}\tilde{\Upsilon}_1^2 + \frac{m_1^2\bar{\ell}_{n,d}^2}{2l_1\underline{\ell}_n^4}z_n^2, \tag{5.60}$$

$$\frac{\hat{\varphi}_1\hbar_1\psi_{n1}^T(\hat{\mathcal{Z}}_n)\psi_{n1}(\hat{\mathcal{Z}}_n)}{2k_{n1}^2}\tilde{\Upsilon}_1 \leq \frac{l_1}{2}\tilde{\Upsilon}_1^2 + \frac{\bar{\varphi}_1^2}{2l_1k_{n1}^4}\hbar_1^2, \tag{5.61}$$

$$\tilde{\Upsilon}_1\dot{\Upsilon}_1 \leq \frac{l_1}{2}\tilde{\Upsilon}_1 + \frac{1}{2l_1}\bar{\Upsilon}_{1,d}^2. \tag{5.62}$$

By combining (5.57)–(5.62) with (5.56), we obtain:

$$\tilde{\Upsilon}_1\dot{\tilde{\Upsilon}}_1 \leq -(\frac{m_1}{2} - \frac{7}{2}l_1)\tilde{\Upsilon}_1^2 + \frac{m_1^2\bar{\ell}_{n,d}^2}{2l_1\underline{\ell}_n^4}z_n^2 + \frac{\bar{\varphi}_1^2}{2l_1k_{n1}^4}\hbar_1^2 + \Delta_{\Upsilon_1}, \tag{5.63}$$

where $\Delta_{\Upsilon_1} = \bar{\Upsilon}_{1,d}^2/(2l_1) + \varphi_1/(2l_1) + m_1^2\bar{\Upsilon}_1^2/(l_1\underline{\ell}_n^2) + m_1\bar{\Upsilon}_1^2/2 + \bar{\varepsilon}_{n1}^2/(2\underline{\ell}_n^2) + \bar{\chi}_{n1}^2/(2\underline{\ell}_n^2)$.

The nth Lyapunov function candidate is selected as:

$$\mathcal{V}_n = \mathcal{V}_{n-1} + \frac{1}{2}z_n^2 + \frac{1}{2}\tilde{\Lambda}_n^2 + \frac{1}{2}\hbar_1^2 + \frac{1}{2}\tilde{\varphi}_1^2 + \frac{1}{2}\tilde{\Upsilon}_1^2 + \frac{1}{2}\ell_n^2. \tag{5.64}$$

$\tilde{\Lambda}_n$ denotes the estimation error of Λ_n, where Λ_n will be formally defined in subsequent analysis.

From (5.44), the time derivative $\dot{\mathcal{V}}_n$ is expressed as:

$$\begin{aligned} \dot{\mathcal{V}}_n =& \dot{\mathcal{V}}_{n-1} + z_n(f_{n,s}(\bar{x}_n) + \phi_0 v(t) + \Upsilon_1 - \dot{\ell}_n\alpha_n^f - \ell_n\dot{\alpha}_n^f) \\ &+ \tilde{\Lambda}_n\dot{\tilde{\Lambda}}_n + \hbar_1\dot{\hbar}_1 + \tilde{\varphi}_1\dot{\tilde{\varphi}}_1 + \tilde{\Upsilon}_1\dot{\tilde{\Upsilon}}_1 + \ell_n\dot{\ell}_n. \end{aligned} \tag{5.65}$$

Inserting (5.52) and (5.63) into (5.65) yields:

$$\begin{aligned} \dot{\mathcal{V}}_n \leq& \dot{\mathcal{V}}_{n-1} - r_{n2}z_n^{2q} - c_nz_n^2 + z_n(\mathcal{F}_{n,s}(\mathcal{Z}_n) + \phi_0 v(t) + \Upsilon_1) \\ &- (b_1 - 2 - \frac{\bar{\varphi}_1^2}{2l_1k_{n1}^4})\hbar_1^2 - (\frac{m_1}{2} - \frac{7}{2}l_1)\tilde{\Upsilon}_1^2 + \Delta_{\Upsilon_1} + \Delta_{\hbar_1} \end{aligned}$$

$$
\begin{aligned}
&+\tilde{\varphi}_1\Big(\frac{\hbar_1^2\psi_{n1}^T(\hat{\mathcal{Z}}_n)\psi_{n1}(\hat{\mathcal{Z}}_n)}{2k_{n1}^2}-\dot{\hat{\varphi}}_1\Big)+\ell_n\dot{\ell}_n+c_n\hat{z}_n^2\\
&-2z_n^2+\tilde{\Lambda}_n\dot{\hat{\Lambda}}_n\\
&\le \dot{\mathcal{V}}_{n-1}-r_{n2}z^{2q}-c_nz_n^2+z_n(\mathcal{F}_{n,s}(\mathcal{Z}_n)+\phi_0 v(t)+\Upsilon_1)\\
&-\bar{b}_1\hbar_1^2-\bar{m}_1\tilde{\Upsilon}_1^2+\Delta_{\Upsilon_1}+\Delta_{\hbar_1}-r_{n1}\ell_n^2-r_{n2}\ell_1^{2q}\\
&+\tilde{\varphi}_1\Big(\frac{\hbar_1^2\psi_{n1}^T(\hat{\mathcal{Z}}_n)\psi_{n1}(\hat{\mathcal{Z}}_n)}{2k_{n1}^2}-\dot{\hat{\varphi}}_1\Big)+c_n\hat{z}_n^2\\
&-2z_n^2+\tilde{\Lambda}_n\dot{\hat{\Lambda}}_n,
\end{aligned}
\tag{5.66}
$$

where $\mathcal{F}_{n,s}(\mathcal{Z}_n)=r_{n2}z_n^{2q-1}+c_nz_n-\dot{\ell}_n\alpha_{n-1}^f-\ell_n\dot{\alpha}_{n-1}^f+2z_n-c_nz_n/\ell_n^2+\bar{\ell}_{n,d}^2z_n/\underline{\ell}_n^4+m_1^2\bar{\ell}_{n,d}^2z_n/(2l_1\underline{\ell}_n^4)$, $\bar{b}_1=b_1-2-\bar{\varphi}_1^2/(2l_1k_{n1}^4)$, $\bar{m}_1=m_1/2-7l_1/2$.

Leveraging the approximation characteristic of the FLSs, we derive:

$$
\mathcal{F}_{n,s}(\mathcal{Z}_n)=\lambda_{n,s}^T\psi_n(\hat{z}_n)+\varepsilon_{n,s}+\chi_{n,s},
\tag{5.67}
$$

where $\chi_{n,s}=\lambda_{n,s}^T(\psi_n(z_n)-\psi_n(\hat{z}_n))$.

Clearly, the following inequality is satisfied:

$$
\begin{aligned}
z_n(\lambda_{n,s}^T\psi_n(\hat{\mathcal{Z}}_n)+\varepsilon_{n,s}+\chi_{n,s})\le{}&\frac{\Lambda_n\hat{z}_n^2\psi_n^T(\hat{\mathcal{Z}}_n)\psi_n(\hat{\mathcal{Z}}_n)}{2k_n^2}\\
&+\frac{1}{2}k_n^2+z_n^2+\frac{1}{2}\bar{\varepsilon}_n^2+\frac{1}{2}\bar{\chi}_n^2,
\end{aligned}
\tag{5.68}
$$

where $\Lambda_n=\max_{s\in\varsigma(t)}\bar{\ell}_n^2||\lambda_{n,s}||^2$, $\hat{\Lambda}_n=\Lambda_n-\tilde{\Lambda}_n$ is its estimation value, $\bar{\varepsilon}_n=\max_{s\in\varsigma(t)}||\varepsilon_{n,s}||^2$, $\bar{\chi}_n=\max_{s\in\varsigma(t)}||\chi_{n,s}||^2$, and k_n represents positive parameter.

Substituting (5.68) into (5.66), we obtain the reformulated expression:

$$
\begin{aligned}
\dot{\mathcal{V}}_n\le{}&\dot{\mathcal{V}}_{n-1}-r_{n1}z_n^{2q}-c_nz_n^2+z_n\phi_0v(t)+\frac{1}{2}\bar{\varepsilon}_n^2+\frac{1}{2}\bar{\chi}_n^2\\
&-\bar{b}_1\hbar_1^2-\bar{m}_1\tilde{\Upsilon}_1^2+\Delta_{\Upsilon_1}+\Delta_{\hbar_1}-r_{n1}\ell_n^2-r_{n2}\ell_1^{2q}\\
&+\tilde{\varphi}_1\Big(\frac{\hbar_1^2\psi_{n1}^T(\hat{\mathcal{Z}}_n)\psi_{n1}(\hat{\mathcal{Z}}_n)}{2m_1k_{n1}^2}-\dot{\hat{\varphi}}_1\Big)+c_n\hat{z}_n^2-z_n^2+\frac{1}{2}k_n^2\\
&+\frac{\hat{\Lambda}_n\hat{z}_n^2\psi_n^T(\hat{\mathcal{Z}}_n)\psi_n(\hat{\mathcal{Z}}_n)}{2k_n^2}+\tilde{\Lambda}_n\Big(\frac{\hat{z}_n^2\psi_n^T(\hat{\mathcal{Z}}_n)\psi_n(\hat{\mathcal{Z}}_n)}{2k_n^2}\\
&+z_n\Upsilon_1-\dot{\hat{\Lambda}}_n\Big).
\end{aligned}
\tag{5.69}
$$

The switching threshold-based event-triggered mechanism (ETM) is designed as:

$$
v(t)=\nu(t_k),\quad \forall t\in[t_k,t_{k+1}),
\tag{5.70}
$$

$$t_{k+1} = \begin{cases} \inf\{t > t_k \,|\, |\epsilon(t)| \geq \vartheta_1 |v(t)| + \mu_1\}, \\ \text{if} \quad |u_s(t)| \leq M_1; \\ \inf\{t > t_k \,|\, |\epsilon(t)| \geq \mu_2\}, \\ \text{if} \quad |u_s(t)| > M_1; \end{cases} \tag{5.71}$$

where $\epsilon(t) = \nu(t) - v(t)$ represents measurement error, t_k, t_{k+1} are the kth update and the $(k+1)$th update of the control signal, respectively. Once $|\epsilon(t)|$ satisfies the preset condition after $t > t_k$, the actual control signal changes, i.e., $v(t) = \nu(t_{k+1})$. Different from the fixed and relative threshold ETM, the switching threshold ETM is capable of balancing system performance and communication burden. When the input signal is large, i.e., the case $u_s > M_1$, the fixed threshold ETM is activated to achieve a fast response. When the input signal satisfies that the case $u_s \leq M_1$, the relative threshold ETM can achieve precision control. Thus, in this chapter, the proposed control strategy can provide more flexibility in system response and control performance.

The subsequent parameters are defined as:

$$\vartheta = \begin{cases} \vartheta_1, & |u_s(t)| \leq M_1; \\ 0, & |u_s(t)| > M_1; \end{cases} \qquad \mu = \begin{cases} \mu_1, & |u_s(t)| \leq M_1; \\ \mu_2, & |u_s(t)| > M_1; \end{cases} \tag{5.72}$$

where $0 < \vartheta_1 < 1, M_1 > 0, \mu_1 > 0, \mu_2 > 0$ denote design constants. $\forall t \in [t_k, t_{k+1})$, assume the time-varying function $o_j(t)(j = 1, 2)$ satisfies $o_j(t) \in [-1, 1], o_j(t_k) = 0$ and $o_j(t_{k+1}) = \pm 1$, it yields

$$\nu(t) = (1 + \vartheta o_1(t))v(t) + \mu o_2(t), \tag{5.73}$$

$$v(t) = \frac{\nu(t)}{1 + o_1(t)\vartheta} - \frac{\mu o_2(t)}{1 + o_1(t)\vartheta}. \tag{5.74}$$

The control components are designed as: actual controller $\nu(t)$, nth adaptive law $\hat{\Lambda}_n$ and auxiliary variable Φ_n specified by:

$$\nu(t) = -(1 + \vartheta)(\alpha_n \tanh(\frac{\alpha_n \hat{z}_n}{\gamma_1}) + \mu \tanh(\frac{\mu \hat{z}_n}{\gamma_1})), \tag{5.75}$$

$$\alpha_n = \mathcal{N}(\Phi_n)(c_n \hat{z}_n + \frac{\hat{\Lambda}_n \hat{z}_n \psi_n^T(\hat{\mathcal{Z}}_n)\psi_n(\hat{\mathcal{Z}}_n)}{2k_n^2}) - \frac{1}{\phi_0}\hat{\Upsilon}_1, \tag{5.76}$$

$$\dot{\Phi}_n = c_n \hat{z}_n^2 + \frac{\hat{\Lambda}_n \hat{z}_n^2 \psi_n^T(\hat{\mathcal{Z}}_n)\psi_n(\hat{\mathcal{Z}}_n)}{2k_n^2}, \tag{5.77}$$

$$\dot{\hat{\Lambda}}_n = \frac{\hat{z}_n^2 \psi_n^T(\hat{\mathcal{Z}}_n)\psi_n(\hat{\mathcal{Z}}_n)}{2k_n^2} - \varsigma_n \hat{\Lambda}_n, \tag{5.78}$$

where c_n, ς_n and γ_1 represent tunable positive constants, yielding: Thus, we have

$$\begin{aligned}\phi_0 z_n v(t) =&\phi_0 z_n(\frac{\nu(t)}{1+o_1(t)\vartheta} - \frac{\mu o_2(t)}{1+o_1(t)\vartheta})\\ &\leq \phi_0 \ell_n(-\alpha_n \hat{z}_n \tanh(\frac{\alpha_n \hat{z}_n}{\gamma_1}) + \alpha_n \hat{z}_n\\ &+ |\alpha_n \hat{z}_n| + 0.2785\gamma_1)\\ &\leq \phi_0 \ell_n(\alpha_n \hat{z}_n + 0.557\gamma_1)\\ &\leq \phi_0 \ell_n \mathcal{N}(\Phi_n)\dot{\Phi}_n - z_n \hat{\Upsilon}_1 + 0.557\gamma_1 \bar{\ell}_n. \end{aligned} \tag{5.79}$$

By incorporating (5.75)–(5.79) into (5.69), we derive:

$$\begin{aligned}\dot{\mathcal{V}}_n \leq& \dot{\mathcal{V}}_{n-1} - r_{n1} z_n^{2q} - c_n z_n^2 + \frac{1}{2}\bar{\varepsilon}_n^2 + \frac{1}{2}\bar{\chi}_n^2 + z_n \tilde{\Upsilon}_1 - \bar{b}_1 \hbar_1^2\\ &- \bar{m}_1 \tilde{\Upsilon}_1^2 + \Delta_{\Upsilon_1} + \Delta_{\hbar_1} - r_{n1}\ell_n^2 - r_{n2}\ell_1^{2q} + \varsigma_n \tilde{\Lambda}_n \hat{\Lambda}_n\\ &+ \tilde{\varphi}_1(\frac{\hbar_1^2 \psi_{n1}^T(\hat{\mathcal{Z}}_n)\psi_{n1}(\hat{\mathcal{Z}}_n)}{2k_{n1}^2} - \dot{\hat{\varphi}}_1) + c_n \hat{z}_n^2 - z_n^2 + \frac{1}{2}k_n^2\\ &+ (g_n \mathcal{N}(\Phi_n) + 1)\dot{\Phi}_n + 0.557\gamma_1 \bar{\ell}_n, \end{aligned} \tag{5.80}$$

where $g_n = \phi_0 \ell_n$.

Based on the preceding analysis, Theorem 5.1 encapsulates the key results:

Theorem 5.1 *Consider the switched cyberphysical system* (5.1) *with uncertain deception attacks and backlash-like hysteresis under Assumption 5.1. By designing the virtual controllers* ((5.18), (5.34), (5.40))*, adaptive laws* ((5.20), (5.36), (5.42), (5.78))*, and actual controller* (5.75)*, when these design parameters satisfy the following conditions:* $\bar{r}_{i3} > 0\,(i = 1, \cdots, n)$*,* $\bar{b}_1 > 0$ *and* $\bar{m}_1 > 1/2$*, the whole variables in* (5.1) *are bounded.*

Remark 5.2 : Here are some guidelines for parameter selection. For example, based on the aforementioned stability analysis, we can design large $c_1, \cdots, c_n, r_{11}, r_{12}, \cdots, r_{n1}, r_{n2}, \varsigma_1, \cdots, \varsigma_n, \varsigma_{n1}, m_1$, and b_1 to obtain a good control performance. Moreover, increasing $c_1, \cdots, c_n, r_{11}, r_{12}, \cdots, r_{n1}, r_{n2}, \varsigma_1, \cdots, \varsigma_n, \varsigma_{n1}, m_1$ and b_1 can accelerate the convergence rate of system, and the corresponding control cost will augment. The designer should choose the appropriate parameters according to the specific situation in the engineering systems.

5.3 Simulation Results

This section presents a simulation to validate the proposed control scheme's effectiveness.

We consider a second-order nonlinear cyber-physical system characterized by:

$$\begin{aligned}\dot{x}_1 &= x_2 + f_{\varsigma(t),1}(x_1)\\ \dot{x}_3 &= u_s + f_{\varsigma(t),2}(\bar{x}_2) + \rho_u(t,\bar{x}_2) + d_1(t,\bar{x}_2)\end{aligned}$$

with $\varsigma(t) \in \mathcal{M} = \{1, 2\}$, $f_{1,1} = x_1\sin(x_1)$, $f_{1,2} = x_2\cos(x_1)$, $f_{2,1} = x_1\cos(x_1)$, $f_{2,2} = x_1x_2 + x_2\sin(x_1)$, the actuator deception attack is set as $\rho_u = \cos(t)x_1\sin(x_2)$, the external disturbance is defined as $d_1 = 0.5\sin(x_1x_2)\sin(3t)$, and the sensor deception attacks denote $\rho_x(x_1,t) = 0.3x_1\cos(t)$, $\rho_x(x_2,t) = x_2(0.2 + 0.5\sin(t))$.

Assume that the backlash-like hysteresis of actuator is modeled as $\frac{du_s}{dt} = \varpi|\frac{dv}{dt}|(\phi v(t) - u_s) + \varrho\frac{dv}{dt}$ with $\varpi = 2$, $\phi = 0.5$, $\varrho = 1$. We define the following design constants: $c_1 = 22, c_2 = 25; \varsigma_1 = 3, \varsigma_2 = 2, \varsigma_{21} = 3; k_1 = 6, k_2 = 6, k_{31} = 6;$ $r_{11} = 2$, $r_{12} = 2$, $r_{13} = 1$; $r_{21} = 5$, $r_{22} = 3$; $m_1 = 15$, $b_1 = 20$, $\gamma_1 = 0.25$, $\phi_0 = 5$. Besides, the filtering parameters are defined as: $\eta_{10} = 5, \eta_{11} = 1; \eta_{10} = 10, \eta_{11} = 3$.

The design parameters about switching event-triggered mechanism are designed as: $\vartheta = \begin{cases} 0.15, & |u_s(t)| \le 2; \\ 0, & |u_s(t)| > 2; \end{cases}$, $\mu = \begin{cases} 0.5, & |u_s(t)| \le 2; \\ 1.5, & |u_s(t)| > 2; \end{cases}$. In addition, the initial state variable of the closed-loop system are defined as: $x(0) = [0.5, -0.5]^T$; $\hat{\Lambda}(0) = [0.8, 0.8]^T$; $\hat{\varphi}_1(0) = 0.8$. Figures 5.1 and 5.2 demonstrate the system's stability with state variables converging near the origin, while comparative analysis with [19] confirms superior transient performance of the proposed strategy. Figure 5.3 displays the adaptive parameter evolution trajectories. Figure 5.4 shows the control input signal and event-triggered intervals, with data transmissions reduced from 40,000 to 264 (a ratio of approximately 1:152). Figure 5.5 presents the temporal evolution of both disturbance estimates and auxiliary variables. Figure 5.6 illustrates the switching law $\varsigma(t)$. Figure 5.7 displays the state trajectories x_i without the proposed controller.

5.4 Conclusion

5.4.1 *Contribution*

This work investigates finite-time adaptive control for switched cyber-physical systems (CPSs) subject to uncertain deception attacks via a novel event-triggered scheme. To address unknown control gains induced by deception attacks, Nussbaum gain techniques are employed alongside finite-time command filters with error

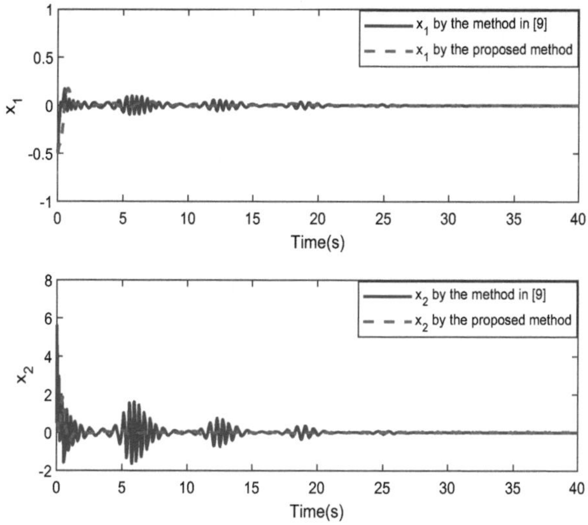

Fig. 5.1 Trajectories of state $x_i(i = 1, 2)$

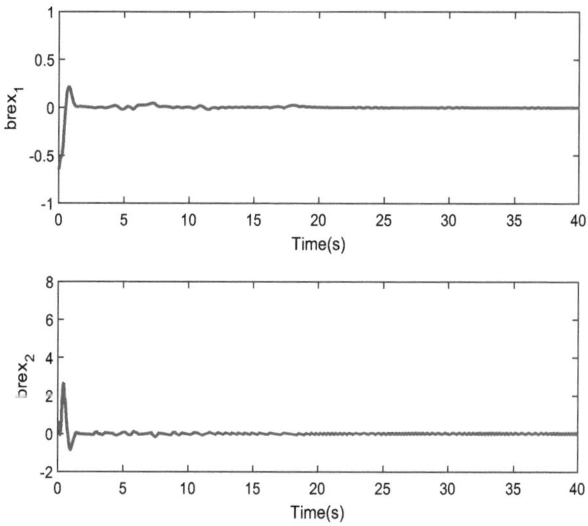

Fig. 5.2 Trajectories of state $\hat{x}_i$ under unknown deception attacks (i = 1,2)

compensation mechanisms, extending prior results in [20]. A novel compound disturbance observer is further developed to enhance system robustness by counteracting input hysteresis and mitigating deception attacks across subsystems [21]. Additionally, a threshold-dependent event-triggered strategy is introduced to optimize the trade-off between transmission efficiency and system performance, where control signals transmit only when prescribed triggering conditions are met. To circumvent computational explosion from repeated virtual control differentiation, finite-time command filters are utilized. Rigorous Lyapunov-based stability proofs

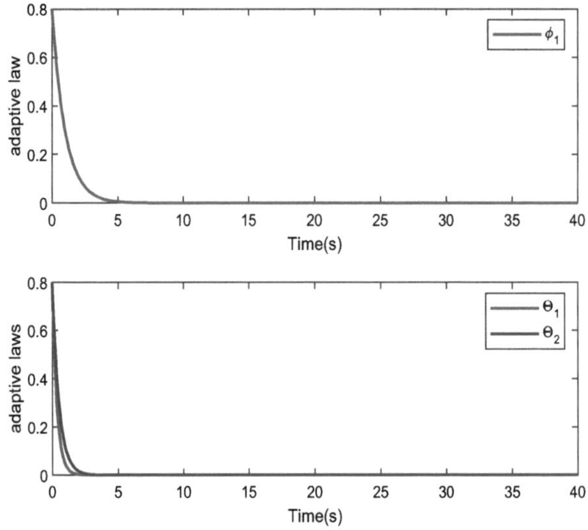

Fig. 5.3 Adaptive laws φ_1, $\Lambda_i (i = 1, 2)$

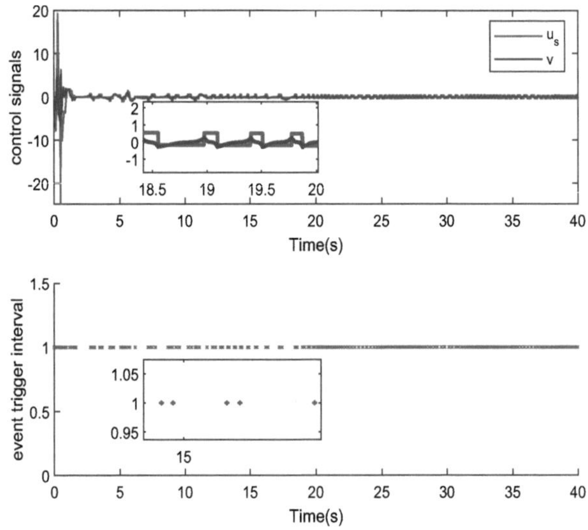

Fig. 5.4 Actual controller and triggered intervals

establish uniform finite-time boundedness for all CPS states. Simulation results validate the scheme's effectiveness, with future research directions targeting improved convergence rates and extensions to time-delay scenarios.

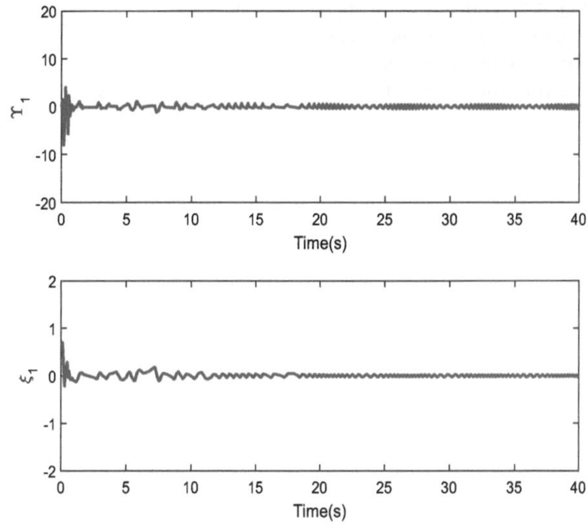

Fig. 5.5 Trajectories of disturbance observer Υ_1 and intermitted variable $\hbar_1$

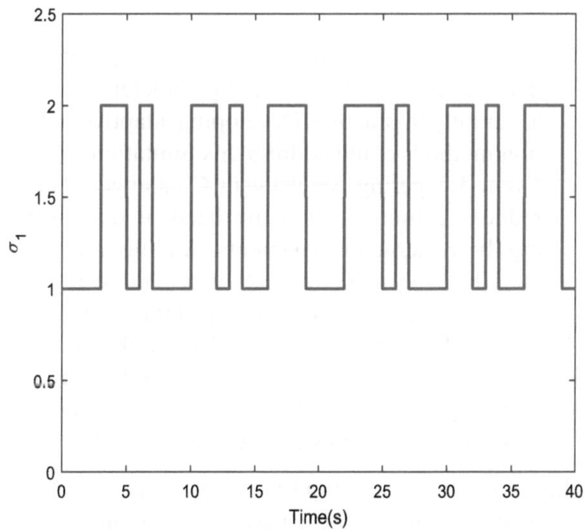

Fig. 5.6 Switching signal $\varsigma(t)$

5.4.2 *Limitations of the Current Work*

Although the proposed control strategy demonstrates strong theoretical and simulation performance, several limitations warrant attention:

(1) Simplistic Attack Modeling. The current approach assumes linear deception attacks, which may not capture complex, nonlinear attack patterns in real-world cyber-physical systems. Future work should explore more sophisticated attack models to enhance applicability.

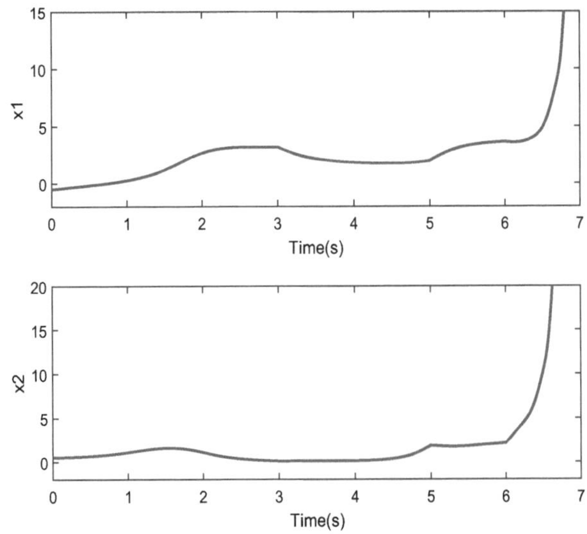

Fig. 5.7 Trajectories of state x_i under unknown deception attacks without resilient control strategy ($i = 1, 2$)

(2) Parameter Dependency. The control algorithm's effectiveness relies on prior knowledge of certain system parameters, which may be challenging to obtain accurately in practice. Developing adaptive or learning-based parameter estimation methods could address this limitation.
(3) Event-Triggering Mechanism Challenges. While the event-triggered strategy reduces communication overhead, it may suffer from frequent triggering in highly dynamic environments. Additionally, the selection of threshold parameters lacks systematic guidelines. Future research should investigate adaptive triggering schemes and optimized threshold selection.
(4) Limited System Complexity. The simulation validation focuses on second-order systems, leaving the method's scalability to higher-dimensional or strongly nonlinear systems unverified. Extending the approach to more complex system architectures remains an open challenge.

These limitations highlight key directions for future research, including enhanced attack modeling, adaptive parameter tuning, dynamic event-triggering optimization, and validation in more complex scenarios. Addressing these challenges will strengthen the method's practical applicability and performance.

5.4.3 Future Directions

Future research should focus on developing more comprehensive and practical control solutions for cyber-physical systems. A key direction involves extending the current framework to handle more sophisticated nonlinear attack patterns beyond the assumed linear models, potentially through advanced machine learning techniques for attack detection and compensation. The parameter dependency issue could be addressed by incorporating online estimation algorithms and adaptive control schemes that require less prior knowledge of system parameters.

For the event-triggered mechanism, research should explore dynamic threshold adjustment strategies and hybrid triggering approaches to better balance communication efficiency and control performance in highly dynamic environments. The scalability of the method needs to be verified through extension to higher-order systems and more complex nonlinear scenarios, possibly through distributed control architectures and time-delay compensation techniques.

Practical implementation studies should be prioritized, particularly focusing on real-world communication constraints like network delays and packet loss that were not considered in the current work. This includes hardware-in-the-loop testing and field validation in industrial settings to evaluate the method's robustness under realistic conditions. The integration of control algorithms with communication protocol design represents another important direction to enhance overall system performance.

In summary, while the current study provides a solid theoretical foundation, future work should aim to bridge the gap between theoretical advances and practical deployment by addressing the identified limitations through more adaptive, scalable, and experimentally validated solutions. This will be crucial for developing truly resilient cyber-physical systems capable of operating effectively in complex real-world environments.

References

1. Liu W, Huang J (2017) Cooperative global robust output regulation for nonlinear output feedback multiagent systems under directed switching networks. IEEE Trans Autom Control 62(12):6339–6352
2. Zhang M, Shi P, Shen C, Wu ZG (2019) Static output feedback control of switched nonlinear systems with actuator faults. IEEE Trans Fuzzy Syst 28(8):1600–1609
3. Chen M, Ge SS (2015) Adaptive neural output feedback control of uncertain nonlinear systems with unknown hysteresis using disturbance observer. IEEE Trans Industr Electron 62(12):7706–7716
4. Wang F, Chen B, Liu X, Lin C (2017) Finite-time adaptive fuzzy tracking control design for nonlinear systems. IEEE Trans Fuzzy Syst 26(3):1207–1216
5. Tong S, Li Y, Sui S (2016) Adaptive fuzzy tracking control design for SISO uncertain nonstrict feedback nonlinear systems. IEEE Trans Fuzzy Syst 24(6):1441–1454
6. Ge SS, Hong F, Lee TH (2004) Adaptive neural control of nonlinear time-delay systems with unknown virtual control coefficients. IEEE Trans Syst Man Cybern B Cybern 34(1):499–516
7. Ma L, Xu N, Zhao X, Zong G, Huo X (2020) Small-gain technique-based adaptive neural output-feedback fault-tolerant control of switched nonlinear systems with unmodeled dynamics. IEEE Trans Syst Man Cybern Syst 51(11):7051–7062
8. Chen L, Fu S, Zhao Y, Liu M, Qiu J (2019) State and fault observer design for switched systems via an adaptive fuzzy approach. IEEE Trans Fuzzy Syst 28(9):2107–2118
9. Yu S, Yu X, Shirinzadeh B, Man Z (2005) Continuous finite-time control for robotic manipulators with terminal sliding mode. Automatica 41(11):1957–1964

10. Li Y, Qu F, Tong S (2020) Observer-based fuzzy adaptive finite-time containment control of nonlinear multiagent systems with input delay. IEEE Trans Cybern 51(1):126–137
11. Zhou P, Zhang L, Zhang S, Alkhateeb AF (2020) Observer-based adaptive fuzzy finite-time control design with prescribed performance for switched pure-feedback nonlinear systems. IEEE Access 9:69481–69491
12. Zhao L, Yu J, Wang QG (2020) Finite-time tracking control for nonlinear systems via adaptive neural output feedback and command filtered backstepping. IEEE Trans Neural Netw Learn Syst 32(4):1474–1485
13. Li F, Liu Y (2019) Event-triggered stabilization for continuous-time stochastic systems. IEEE Trans Autom Control 65(10):4031–4046
14. Cao L, Li H, Dong G, Lu R (2019) Event-triggered control for multiagent systems with sensor faults and input saturation. IEEE Trans Syst Man Cybern Syst 51(6):3855–3866
15. Liang H, Liu G, Zhang H, Huang T (2020) Neural-network-based event-triggered adaptive control of nonaffine nonlinear multiagent systems with dynamic uncertainties. IEEE Trans Neural Netw Learn Syst 32(5):2239–2250
16. Zhu Z, Pan Y, Zhou Q, Lu C (2020) Event-triggered adaptive fuzzy control for stochastic nonlinear systems with unmeasured states and unknown backlash-like hysteresis. IEEE Trans Fuzzy Syst 29(5):1273–1283
17. Zhang T, Ge SS (2009) Adaptive neural network tracking control of MIMO nonlinear systems with unknown dead zones and control directions. IEEE Trans Neural Netw 20(3):483–497
18. Levant A (2003) Higher-order sliding modes, differentiation and output-feedback control. Int J Control 76(9–10):924–941
19. Yoo SJ (2019) Neural-network-based adaptive resilient dynamic surface control against unknown deception attacks of uncertain nonlinear time-delay cyberphysical systems. IEEE Trans Neural Netw Learn Syst 31(10):4341–4353
20. Li Z, Zhao J (2021) Resilient adaptive control of switched nonlinear cyber-physical systems under uncertain deception attacks. Inf Sci 543:398–409
21. Song S, Park JH, Zhang B, Song X (2021) Adaptive NN finite-time resilient control for nonlinear time-delay systems with unknown false data injection and actuator faults. IEEE Trans Neural Netw Learn Syst 33(10):5416–5428

Chapter 6
Adaptive Control for Uncertain Nonlinear Systems With Dynamic Full State Constraints: The SMDO Approach

Recent years have seen numerous adaptive controllers emerge through the incorporation of sliding-mode observers (SMOs). The core concept of SMOs involves applying nonlinear high-gain feedback to drive estimated states toward a predefined hyperplane. Consequently, SMOs exhibit strong robustness against external disturbances while providing reliable estimation of unmeasured parameters. Conversely, the growing emphasis on resource efficiency and environmental sustainability highlights limitations in time-sampled control strategies. These strategies periodically update control signals regardless of actual performance, which proves inefficient. In contrast, event-sampled strategies, where updates correlate with performance-based triggering conditions, have gained prominence by significantly reducing control signal update frequency. Zou et al. [1] applied an event-triggered strategy to consensus control in stochastic switched multiagent systems. Li et al. [2] developed an asynchronous event-based communication scheme to reduce network burdens in both reference-to-controller and plant-to-controller channels.

This chapter therefore addresses adaptive control for uncertain nonlinear systems subject to time-varying full-state constraints. First, a novel neural backstepping control approach based on Integral Barrier Lyapunov Functions (IBLFs) is designed, circumventing the state transformation requirements of traditional Barrier Lyapunov Functions (BLFs). Subsequently, sliding-mode disturbance observers (SMDOs) are established to estimate immeasurable disturbances affecting each state within the constrained system. Additionally, a dynamic threshold-based event-triggering mechanism is constructed to manage resource scarcity and reduce computational burden. Finally, an event-triggered adaptive controller is developed. This controller ensures finite-time convergence of disturbance observation errors to zero and guarantees all closed-loop signals are semiglobally uniformly ultimately bounded (SGUUB). Simulation studies verify the approach's validity.

X. Wang et al., *Optimal Adaptive Cooperative Control for Nonlinear Multiagent Systems*,
https://doi.org/10.1007/978-981-95-2964-3_6

6.1 System Descriptions

Consider an n-order uncertain nonlinear system under time-varying full state constraints:

$$\begin{aligned}\dot{x}_1 &= q_1(x_1)x_2 + g_1(x_1) + \Upsilon_1(t, x_1)\\ &\vdots\\ \dot{x}_{n-1} &= q_{n-1}(\bar{x}_{n-1})x_n + g_{n-1}(\bar{x}_{n-1}) + \Upsilon_{n-1}(t, \bar{x}_{n-1})\\ \dot{x}_n &= q_n(\bar{x}_n)u + g_n(\bar{x}_n) + \Upsilon_n(t, \bar{x}_n)\\ y &= x_1\end{aligned} \tag{6.1}$$

where $\bar{x}_r = [x_1, \ldots, x_r]^T \in \mathbb{R}^r$ denotes the partial state vector constrained by $|x_r| < \varsigma_{\varphi_r}(t)$. The known positive smooth function $q_r(\bar{x}_r) : \mathbb{R}^r \to \mathbb{R}$ characterizes system dynamics, while uncertain terms are modeled by smooth nonlinear functions $g_r(\bar{x}_r)$: $\mathbb{R}^r \to \mathbb{R}$ and unknown disturbances $\Upsilon_r(t, \bar{x}_n) \in \mathbb{R}$, $r = 1, \ldots, n$. Further, $x_d(t)$ represents the reference signal, with y and u denoting system output and control input, respectively. The control objective is defined as follows:

Control Objective: For the uncertain nonlinear systems (6.1) with time-varying full state constraints and unknown external disturbances, we develop an adaptive event-triggered controller using IBLFs and backstepping techniques to ensure:

(1) The system output can track the desired reference signal, all states remain within the constraints, and are uniformly ultimately bounded.
(2) The SMDOs can detect the disturbances in each order, with observation errors achieving theoretical finite-time convergence.
(3) The event-triggered mechanism reduces controller update frequency and avoids Zeno behavior.

To facilitate the derivation, the following assumptions and lemmas are necessary.

Assumption 6.1 : The first and second-time derivatives of the disturbances in (1) are bounded. That is, there exist unknown positive constants Γ_{r1} and known positive constants Γ_{r2} such that $\left|\dot{\Upsilon}_r\right| < \Gamma_{r1}$ and $\left|\ddot{\Upsilon}_r\right| < \Gamma_{r2}$ hold for $r = 1, \ldots, n$.

Assumption 6.2 : [3, 4]: There exist positive constants B_r and function $B_0(t)$ such that for all $\forall t \geq 0$, the desired reference signal and its time derivatives satisfy $\left|x_d^{(r)}(t)\right| < B_r$ and $|x_d(t)| \leq B_0(t) < \varsigma_{\varphi_1}(t)$.

Assumption 6.3 : Tee et al. [3]: It is assumed that $\varsigma_{\varphi_r}(t)$ and its m-th time derivatives are bounded. That is, for all $t \geq 0$, $\varsigma_{\varphi_r}(t) \leq \bar{\varsigma}^0_{\varphi_r}$ and $\varsigma^{(m)}_{\varphi_r}(t) \leq \bar{\varsigma}^m_{\varphi_r}$ hold, where $\bar{\varsigma}^0_{\varphi_r}$ and $\bar{\varsigma}^m_{\varphi_r}$ are upper bounds of constraints, for $r = 1, \ldots, n$ and $m = 1, \ldots, n$.

Lemma 6.1 *Rabiee et al. [5]: Consider the system modeled as*

$$\dot{x} = -\kappa_1 \text{sign}(x)|x|^{\iota_1} - \kappa_2 \text{sign}(x)|x|^{\iota_2}, \ x(0) = x_0, \tag{6.2}$$

then the equilibrium point of the system is finite-time stable, and the equilibrium settling time t_{eq} satisfies

$$t_{eq} < \frac{1}{\kappa_1(1-\iota_1)} + \frac{1}{\kappa_2(\iota_2 - 1)} \tag{6.3}$$

where $0 < \iota_1 < 1$, $\iota_2 > 1$, κ_1 *and* κ_2 *are positive constants, and* x_0 *is the initial state.*

Remark 6.1 : The assumptions in this chapter are standard and commonly used in previous literature. To ensure the stability of the proposed SMDOs, Assumption 6.1 presents restrictions on the properties of disturbances, and Assumption 6.2 ensures the reference signal is within the restricted boundary and its derivatives are bounded in [3, 4]. Assumption 6.3 guarantees the rationality and effectiveness of the state constraints. Assumptions 6.2 and 6.3 are crucial for proving the boundedness of the designed control laws.

6.2 SMDOs Design

Given that the systems under consideration (6.1) are influenced by unmeasurable external disturbances, we need to construct observers to estimate them. To proceed, we first define the auxiliary variables as $\Lambda_r = f_r - x_r$. The dynamic equation for f_r is given by:

$$\dot{f}_r = g_r(\bar{x}_r) - \kappa_{r1}\text{sign}(\Lambda_r)|\Lambda_r|^{\iota_{r1}} - \kappa_{r2}\text{sign}(\Lambda_r)|\Lambda_r|^{\iota_{r2}} + q_r(\bar{x}_r)x_{r+1} + \Lambda_r, \tag{6.4}$$

where $\iota_{r1} < 1, \iota_{r2} > 1$, and κ_{r1}, κ_{r2} are positive design constants. Note that $x_{n+1} = u$.

Subsequently, the sliding surfaces are designed as follows:

$$\Theta_r = \dot{\Lambda}_r + \kappa_{r1}\text{sign}(\Lambda_r)|\Lambda_r|^{\iota_{r1}} + \kappa_{r2}\text{sign}(\Lambda_r)|\Lambda_r|^{\iota_{r2}}. \tag{6.5}$$

The dynamic behavior of Δ_r is described by:

$$\dot{\Delta}_r = -\delta_r \text{sign}(\Theta_r) - \gamma_r \text{sign}(\Theta_r), \ \Delta_r(0) = 0, \tag{6.6}$$

where δ_r is a positive design constant, and γ_r is updated according to the following laws:

$$\dot{\gamma}_r(t) = -\alpha_r(t)\text{sign}(\mathfrak{E}_r(t)). \tag{6.7}$$

Here, $\alpha_r(t) = \bar{\mathcal{J}}_r + \mathcal{J}_r(t)$, $\mathfrak{E}_r(t) = \gamma_r(t) - \varpi_{dr} - \frac{1}{\beta_r}|\mathcal{M}_r|$, $\dot{\mathcal{M}}_r = -\frac{1}{\sigma_r}\Big(\mathcal{M}_r + \delta_r \text{sign}(\Theta_r) + \gamma_r \text{sign}(\Theta_r)\Big)$, and

$$\dot{\mathcal{J}}_r(t) = \begin{cases} \theta_r |\mathfrak{E}_r(t)|, & \text{if} |\mathfrak{E}_r(t)| > \bar{\mathfrak{E}}_r \\ 0, & \text{if} |\mathfrak{E}_r(t)| \le \bar{\mathfrak{E}}_r. \end{cases}$$

The positive design constants $\bar{\mathcal{J}}_r, 0 < \beta_r < 1, k_r > \max\left\{\left|\frac{d}{dt}\frac{\mathcal{M}_r(t)}{\Gamma_{r2}}\right|, 1\right\}, \bar{\mathfrak{E}}_r, \theta_r$, and ϖ_{dr} are appropriately selected to satisfy the following inequality:

$$\frac{1}{4}\varpi_{dr}^2 > \frac{1}{\theta_r}\left(\frac{k_r \Gamma_{r2}}{\beta_r}\right)^2 + \bar{\mathfrak{E}}_r^2. \tag{6.8}$$

Based on the preceding analysis, we present a proposition to demonstrate the effectiveness of the proposed SMDOs.

Proposition 6.1 : *For the SMDOs (6.6), the auxiliary variables Λ_r will converge to zero, and the finite-time disturbance estimation is given by*

$$\hat{\Upsilon}_r = \Delta_r, \quad r = 1, \ldots, n. \tag{6.9}$$

Moreover, the disturbance estimation errors $\tilde{\Upsilon}_r = \Upsilon_r - \hat{\Upsilon}_r$ will converge to zero in finite time.

Proof. : Initially, suppose γ_r is fixed and the following holds:

$$\gamma_r > \Gamma_{r1}. \tag{6.10}$$

Combining (6.1) and (6.4), the time derivative of Λ_r is

$$\begin{aligned} \dot{\Lambda}_r &= \dot{f}_r - \dot{x}_r \\ &= \Delta_r - \kappa_{r1}\text{sign}(\Lambda_r)|\Lambda_r|^{\iota_{r1}} - \kappa_{r2}\text{sign}(\Lambda_r)|\Lambda_r|^{\iota_{r2}} - \Upsilon_r, \end{aligned} \tag{6.11}$$

leading to

$$\Theta_r = \Delta_r - \Upsilon_r. \tag{6.12}$$

The derivative of the sliding surfaces is

$$\dot{\Theta}_r = \dot{\Delta}_r - \dot{\Upsilon}_r = -\delta_r \text{sign}(\Theta_r) - \gamma_r \text{sign}(\Theta_r) - \dot{\Upsilon}_r. \tag{6.13}$$

Select the Lyapunov function candidate as $V_r^D = \Theta_r^2/2$. Its time derivative is

$$\dot{V}_r^D = (-\delta_r - \gamma_r)|\Theta_r| - \dot{\Upsilon}_r \Theta_r. \tag{6.14}$$

Given $\delta_r > 0$ and $\gamma_r > \Gamma_{r1} > \left|\dot{\Upsilon}_r\right|$, from (6.10) and (6.14), we have

$$\dot{V}_r^D \leq (-\delta_r - \gamma_r)\left|\Theta_r\right| + \left|\dot{\Upsilon}_r\right|\left|\Theta_r\right| \leq -\delta_r\left|\Theta_r\right| < 0. \tag{6.15}$$

The inequality (6.15) shows that $\Theta_r = 0$ is achieved in finite time. Subsequently,

$$\dot{\Lambda}_r = -\kappa_{r1}\text{sign}(\Lambda_r)|\Lambda_r|^{\iota_{r1}} - \kappa_{r2}\text{sign}(\Lambda_r)|\Lambda_r|^{\iota_{r2}}. \tag{6.16}$$

By Lemma 6.1, the equilibrium of (6.16) at $\Lambda_r = 0$ is finite-time stable.

From (6.9) and (6.12), the condition $\tilde{\Upsilon}_r = \Upsilon_r - \hat{\Upsilon}_r = 0$ is achieved in finite time, indicating that the disturbance estimation error in each order converges to zero.

It should be noted that the above analysis shows that the SMDOs can estimate system disturbances under the condition $\gamma_r > |\Gamma_{r1}|$. However, this condition is hard to meet directly in industrial applications due to the immeasurability and uncertainty of Γ_{r1}. Therefore, the parameter updating laws (6.7) are established to overcome this limitation. According to the theorem in [5], condition (6.10) is satisfied in finite time with time-varying γ_r, meaning the proposed SMDOs do not require known upper bounds of the first derivative of disturbances. Compared to some disturbance compensation control approaches [6–8], the proposed method relaxes the strict assumptions on disturbances and their first derivative upper bounds.

It is demonstrated that the proposed SMDOs detect unknown system disturbances within finite time, thus completing the proof.

Remark 6.2 : To date, the SMDOs for monitoring system disturbances have been established. It is important to note that some related studies, such as [5], only focus on the immeasurable disturbance in the n-th order. In contrast, our work generalizes this result by fully considering the dynamic disturbances that the system may encounter in each order, thereby enhancing the universality of the proposed control approach.

6.3 Adaptive Controller Design and Analyze

In this section, we introduce the design procedure of the adaptive event-triggered controller by utilizing IBLFs and backstepping techniques.

To begin with, the tracking errors are defined as follows:

$$\Xi_1 = x_1 - x_d \tag{6.17}$$

$$\Xi_r = x_r - v_{r-1},\ r = 2, \ldots n \tag{6.18}$$

where x_d represents the desired reference signal, and v_r (for $r = 1, \ldots n$) are the virtual controllers.

Lemma 6.2 *Cheng et al. [9]: For* $\forall t \geq 0$, *under the condition* $|x_r| < \varsigma_{\varphi_r}(t)$, *the IBLFs satisfy*

$$V_r^I \leq \frac{\Xi_r^2 \varsigma_{\varphi_r}^2}{\varsigma_{\varphi_r}^2 - x_r^2}, \quad r = 1, \ldots, n. \tag{6.19}$$

Step 1: From (6.1) and (6.17), we have

$$\dot{\Xi}_1 = q_1(x_1)x_2 + g_1(x_1) + \Upsilon_1 - \dot{x}_d. \tag{6.20}$$

The IBLF candidate is chosen as

$$V_1^I = \int_0^{\Xi_1} \frac{\xi \varsigma_{\varphi_1}^2(t)}{\varsigma_{\varphi_1}^2(t) - (\xi + x_d)^2} d\xi = \int_0^{\Xi_1} \wp_1(t) d\xi \tag{6.21}$$

where $\xi = \psi \Xi_1$, $\wp_1(t) = \xi \varsigma_{\varphi_1}^2(t)/(\varsigma_{\varphi_1}^2(t) - (\xi + x_d)^2)$, with ψ lying in the integration interval [0,1]. It can be easily observed that V_1^I and $\wp_1(t)$ are continuously differentiable within the domain $|x_1| < \varsigma_{\varphi_1}(t)$. Sequently, the following inequality can be derived:

$$\frac{1}{2}\Xi_1^2 \leq V_1^I \leq \Xi_1^2 \int_0^1 \frac{\psi \varsigma_{\varphi_1}^2(t)}{\varsigma_{\varphi_1}^2(t) - (\psi \Xi_1 + B_0(t) sign(\Xi_1))^2} d\psi. \tag{6.22}$$

Based on the integral mean value theorem, the time derivative of the chosen V_1^I can be expressed as

$$\begin{aligned} \dot{V}_1^I =& \Xi_1 \mathcal{N}_1 \left(x_d, \Xi_1, \varsigma_{\varphi_1}\right) \dot{\varsigma}_{\varphi_1}(t) - \Xi_1 \mathcal{Q}_1 \left(x_d, \Xi_1, \varsigma_{\varphi_1}\right) \dot{x}_d \\ &+ \frac{\Xi_1 \varsigma_{\varphi_1}^2(t)}{\varsigma_{\varphi_1}^2(t) - x_1^2} \left((v_1 + \Xi_2) q_1(x_1) + g_1(x_1) + \hat{\Upsilon}_1 \right) \\ &+ \frac{\Xi_1 \varsigma_{\varphi_1}^2(t) \tilde{\Upsilon}_1}{\varsigma_{\varphi_1}^2(t) - x_1^2} - \frac{\Xi_1^2 \varsigma_{\varphi_1}(t) \dot{\varsigma}_{\varphi_1}(t)}{\varsigma_{\varphi_1}^2(t) - x_1^2}. \end{aligned} \tag{6.23}$$

We employ NNs to approximate the unknown nonlinear function $\mathcal{Z}_1(\chi_1)$. Correspondingly, there exists a positive scalar $\iota_1(\chi_1)$ satisfying

$$\mathcal{Z}_1(\chi_1) = W_1^{*T} E_1(\chi_1) + \iota_1(\chi_1) \tag{6.24}$$

where $E_1(\chi_1) = [e_1(\chi_1), \ldots, e_d(\chi_1)]^T$ is the basis function vector, d is the number of network nodes, and $e_i(\chi_i)$ for $i = 1, \ldots, d$ are Gaussian functions. The neural approximation error $\iota_1(\chi_1)$ is bounded, so there exists a positive constant $\bar{\iota}_1$ such that $|\iota_1(\chi_1)| < \bar{\iota}_1$. W_1^* is the ideal weight vector, $\hat{W}_1$ is the estimated weight vector, and the weight estimation error is defined as $\tilde{W}_1 = \hat{W}_1 - W_1^*$. The network input vector χ_1 is chosen as $[\hat{\Upsilon}_1, x_d, \dot{x}_d, x_1, \varsigma_{\varphi_1}, \dot{\varsigma}_{\varphi_1}]^T$.

The Lyapunov function candidate is constructed as

$$V_1^W = \frac{1}{2}\tilde{W}_1^T C_1^{-1}\tilde{W}_1 \tag{6.25}$$

where C_1 denotes a designed positive definite matrix.

Subsequently, by merging (6.21) and (6.25), the Lyapunov function for step 1 is formulated as

$$V_1 = V_1^W + V_1^I. \tag{6.26}$$

The neural weight updating law and virtual controller are established as

$$\dot{\hat{W}}_1 = \frac{\Xi_1 \varsigma_{\varphi_1}^2(t) E_1(\chi_1)}{\varsigma_{\varphi_1}^2(t) - x_1^2} C_1 - \varpi_1 \hat{W}_1 C_1 \tag{6.27}$$

$$\begin{aligned} v_1 =& \frac{1}{q_1(x_1)}\Bigg(- \hat{W}_1^T E_1(\chi_1) - \phi_1 \Xi_1 - \Xi_1 \sqrt{\left(\frac{\dot{\varsigma}_{\varphi_1}(t)}{\varsigma_{\varphi_1}(t)}\right)^2 + \zeta_1} \\ & - \frac{\Xi_1 \varsigma_{\varphi_1}^2(t)}{2\left(\varsigma_{\varphi_1}^2(t) - x_1^2\right)} - \hat{\Upsilon}_1 \Bigg). \end{aligned} \tag{6.28}$$

where ϖ_1, ϕ_1, and ζ_1 are positive design parameters.

By applying Young's inequality and considering (6.28), the following inequalities can be derived:

$$\frac{\Xi_1 \varsigma_{\varphi_1}^2(t)}{\varsigma_{\varphi_1}^2(t) - x_1^2} \iota_1(\chi_1) \le \frac{\Xi_1^2 \varsigma_{\varphi_1}^4(t)}{2(\varsigma_{\varphi_1}^2(t) - x_1^2)^2} + \frac{1}{2}\bar{\iota}_1^2 \tag{6.29}$$

$$\varsigma_{\varphi_1}(t)\sqrt{\left(\frac{\dot{\varsigma}_{\varphi_1}(t)}{\varsigma_{\varphi_1}(t)}\right)^2 + \zeta_1} + \dot{\varsigma}_{\varphi_1}(t) \ge 0. \tag{6.30}$$

Furthermore, based on (6.23)–(6.30), it follows that

$$\begin{aligned} \dot{V}_1 =& \frac{\Xi_1 \varsigma_{\varphi_1}^2(t)}{\varsigma_{\varphi_1}^2(t) - x_1^2}\left(v_1 q_1(x_1) + W_1^{*T} E_1(\chi_1) + \iota_1(\chi_1)\right) \\ & + \frac{\Xi_1 \varsigma_{\varphi_1}^2(t)\tilde{\Upsilon}_1}{\varsigma_{\varphi_1}^2(t) - x_1^2} - \frac{\Xi_1^2 \varsigma_{\varphi_1}(t)\dot{\varsigma}_{\varphi_1}(t)}{\varsigma_{\varphi_1}^2(t) - x_1^2} + \tilde{W}_1^T C_1^{-1}\dot{\hat{W}}_1 \\ \le & \frac{1}{2}\bar{\iota}_1^2 - \varpi_1 \tilde{W}_1^T \hat{W}_1 - \frac{\phi_1 \Xi_1^2 \varsigma_{\varphi_1}^2(t)}{\varsigma_{\varphi_1}^2(t) - x_1^2}. \end{aligned} \tag{6.31}$$

For Step r ($r = 2, \ldots, n-1$), from (6.1) and (6.18), it can be deduced that

$$\dot{\Xi}_r = q_r(\bar{x}_r)x_{r+1} + g_r(\bar{x}_r) + \Upsilon_r - \dot{v}_{r-1}. \tag{6.32}$$

Assume there exist positive constants K_{r-1} such that $|v_{r-1}| \le K_{r-1} \le \varsigma_{\varphi_r}(t)$ with $r = 2, \ldots, n$, and

$$\begin{aligned}\dot{v}_{r-1} =& \sum_{i=1}^{r-1} \frac{\partial v_{r-1}}{\partial x_i}(q_i(\bar{x}_i)x_{i+1} + g_i(\bar{x}_i) + \Upsilon_i) + \sum_{i=1}^{r-1} \frac{\partial v_{r-1}}{\partial \hat{W}_i}\dot{\hat{W}}_i \\ &+ \sum_{i=0}^{r-1} \frac{\partial v_{r-1}}{\partial x_d^{(i)}} x_d^{(i+1)} + \sum_{i=1}^{r-1}\sum_{m=0}^{r-i} \frac{\partial v_{r-1}}{\partial \varsigma_{\varphi_i}^{(m)}} \varsigma_{\varphi_i}^{(m+1)} + \sum_{i=1}^{r-1} \frac{\partial v_{r-1}}{\partial \hat{\Upsilon}_i}\dot{\hat{\Upsilon}}_i. \end{aligned} \tag{6.33}$$

The IBLF candidate is given by

$$V_r^I = \int_0^{\Xi_r} \frac{\xi \varsigma_{\varphi_r}^2(t)}{\varsigma_{\varphi_r}^2(t) - (\xi + v_{r-1})^2} d\xi \tag{6.34}$$

where $\xi = \psi \Xi_r$.

It can be shown that

$$\frac{1}{2}\Xi_r^2 \le V_r^I \le \Xi_r^2 \int_0^1 \frac{\psi \varsigma_{\varphi_r}^2(t)}{\varsigma_{\varphi_r}^2(t) - (\psi \Xi_r + K_{r-1} sign(\Xi_r))^2} d\psi. \tag{6.35}$$

Similar to step 1, the time derivative of (6.34) is calculated as

$$\begin{aligned}\dot{V}_r^I =& \dot{v}_{r-1} \int_0^{\Xi_r} \frac{\partial}{\partial v_{r-1}} \frac{\varsigma_{\varphi_r}^2(t)\xi}{\varsigma_{\varphi_r}^2(t) - (\xi + v_{r-1})^2} d\xi \\ &+ \dot{\varsigma}_{\varphi_r}(t) \int_0^{\Xi_r} \frac{\partial}{\partial \varsigma_{\varphi_r}(t)} \frac{\varsigma_{\varphi_r}^2(t)\xi}{\varsigma_{\varphi_r}^2(t) - (\xi + v_{r-1})^2} d\xi + \frac{\Xi_r \dot{\Xi}_r \varsigma_{\varphi_r}^2(t)}{\varsigma_{\varphi_r}^2(t) - x_r^2} \end{aligned} \tag{6.36}$$

Following a similar approach, NNs are used to approximate the unknown nonlinear function $\mathcal{Z}_r(\chi_r)$. There exists a positive scalar $\iota_r(\chi_r)$ such that

$$\begin{aligned}\mathcal{Z}_r(\chi_r) =& g_r(\bar{x}_r) + \tilde{\Upsilon}_r + \Xi_{r+1} q_r(\bar{x}_r) + \frac{\varsigma_{\varphi_r}^2(t) - x_r^2}{\varsigma_{\varphi_r}^2(t)} \mathcal{N}_r\left(v_{r-1}, \Xi_r, \varsigma_{\varphi_r}\right) \dot{\varsigma}_{\varphi_r}(t) \\ &- \frac{\varsigma_{\varphi_r}^2(t) - x_r^2}{\varsigma_{\varphi_r}^2(t)} \mathcal{Q}_r\left(v_{r-1}, \Xi_r, \varsigma_{\varphi_r}\right) \dot{v}_{r-1} \\ =& W_r^{*T} E_r(\chi_r) + \iota_r(\chi_r) \end{aligned} \tag{6.37}$$

where χ_r is defined as $\left[\hat{\Upsilon}_1, \ldots, \hat{\Upsilon}_r, x_d, \ldots, x_d^{(r)}, x_1, \ldots, x_r, \varsigma_{\varphi_1}, \ldots, \varsigma_{\varphi_1}^{(r)}, \varsigma_{\varphi_2}, \ldots, \varsigma_{\varphi_2}^{(r-1)}, \ldots, \varsigma_{\varphi_r}, \dot{\varsigma}_{\varphi_r}, \hat{W}_1, \ldots, \hat{W}_{r-1}\right]^T$. There exist constants $\bar{\iota}_r > 0$ such that $|\iota_r(\chi_r)| < \bar{\iota}_r$. W_r^* represents the ideal weight vector, and $\hat{W}_r$ is the estimated weight vector. The weight estimation error is defined as $\tilde{W}_r = \hat{W}_r - W_r^*$.

The Lyapunov function for step r is then constructed as

$$V_r = V_r^W + V_r^I \tag{6.38}$$

where $V_r^W = \frac{1}{2}\tilde{W}_r^T C_r^{-1} \tilde{W}_r$, and C_r is a designed positive definite matrix.

The neural weight updating law and virtual controller are established as

$$\dot{\hat{W}}_r = \frac{\Xi_r \varsigma_{\varphi_r}^2(t) E_r(\chi_r)}{\varsigma_{\varphi_r}^2(t) - x_r^2} C_r - \varpi_r \hat{W}_r C_r \tag{6.39}$$

$$\begin{aligned} \upsilon_r = & \frac{1}{q_r(\bar{x}_r)}\Bigg(- \hat{W}_r^T E_r(\chi_r) - \phi_r \Xi_r - \Xi_r \sqrt{\left(\frac{\dot{\varsigma}_{\varphi_r}(t)}{\varsigma_{\varphi_r}(t)}\right)^2 + \zeta_r} \\ & - \frac{\Xi_r \varsigma_{\varphi_r}^2(t)}{2\left(\varsigma_{\varphi_r}^2(t) - x_r^2\right)} - \hat{\Upsilon}_r \Bigg) \end{aligned} \tag{6.40}$$

where ϖ_r, ϕ_r, and ζ_r are positive design parameters.

Similarly, from (6.36)–(6.40), it can be deduced that

$$\begin{aligned} \dot{V}_r = & \frac{\Xi_r \varsigma_{\varphi_r}^2(t)}{\varsigma_{\varphi_r}^2(t) - x_r^2} \left(\upsilon_r q_r(\bar{x}_r) + W_r^{*T} E_r(\chi_r) + \iota_r(\chi_r)\right) \\ & + \frac{\Xi_r \varsigma_{\varphi_r}^2(t)\tilde{\Upsilon}_r}{\varsigma_{\varphi_r}^2(t) - x_r^2} - \frac{\Xi_r^2 \varsigma_{\varphi_r}(t)\dot{\varsigma}_{\varphi_r}(t)}{\varsigma_{\varphi_r}^2(t) - x_r^2} + \tilde{W}_r^T C_r^{-1} \dot{\hat{W}}_r \\ \le & \frac{1}{2}\bar{\iota}_r^2 - \varpi_r \tilde{W}_r^T \tilde{W}_r - \frac{\phi_r \Xi_r^2 \varsigma_{\varphi_r}^2(t)}{\varsigma_{\varphi_r}^2(t) - x_r^2}. \end{aligned} \tag{6.41}$$

Step n: In view of (6.1) and (6.18), we have

$$\dot{\Xi}_n = q_n(\bar{x}_n)u + g_n(\bar{x}_n) + \Upsilon_n - \dot{\upsilon}_{n-1}. \tag{6.42}$$

Similar to step r, the time derivative of υ_{n-1} is expressed as

$$\begin{aligned} \dot{\upsilon}_{n-1} = & \sum_{i=1}^{n-1} \frac{\partial \upsilon_{n-1}}{\partial x_i} \left(q_i(\bar{x}_i)x_{i+1} + g_i(\bar{x}_i) + \Upsilon_i\right) + \sum_{i=1}^{n-1} \frac{\partial \upsilon_{n-1}}{\partial \hat{W}_i} \dot{\hat{W}}_i \\ & + \sum_{i=0}^{n-1} \frac{\partial \upsilon_{n-1}}{\partial x_d^{(i)}} x_d^{(i+1)} + \sum_{i=1}^{n-1}\sum_{m=0}^{n-i} \frac{\partial \upsilon_{n-1}}{\partial \varsigma_{\varphi_i}^{(m)}} \varsigma_{\varphi_i}^{(m+1)} + \sum_{i=1}^{n-1} \frac{\partial \upsilon_{n-1}}{\partial \hat{\Upsilon}_i} \dot{\hat{\Upsilon}}_i. \end{aligned} \tag{6.43}$$

The IBLF candidate is given by

$$V_n^I = \int_0^{\Xi_n} \frac{\xi \varsigma_{\varphi_n}^2(t)}{\varsigma_{\varphi_n}^2(t) - (\xi + \upsilon_{n-1})^2} d\xi \tag{6.44}$$

where $\xi = \psi \Xi_n$.

Subsequently, we derive

$$\frac{1}{2}\Xi_n^2 \le V_n^I \le \Xi_n^2 \int_0^1 \frac{\psi \varsigma_{\varphi_n}^2(t)}{\varsigma_{\varphi_n}^2(t) - (\psi \Xi_n + K_{n-1}\mathrm{sign}(\Xi_n))^2} d\psi. \tag{6.45}$$

Taking the time derivative of (6.44) yields

$$\begin{aligned}\dot{V}_n^I =& \dot{\upsilon}_{n-1} \int_0^{\Xi_n} \frac{\partial}{\partial \upsilon_{n-1}} \frac{\varsigma_{\varphi_n}^2(t)\xi}{\varsigma_{\varphi_n}^2(t) - (\xi + \upsilon_{n-1})^2} d\xi \\ &+ \dot{\varsigma}_{\varphi_n}(t) \int_0^{\Xi_n} \frac{\partial}{\partial \varsigma_{\varphi_n}(t)} \frac{\varsigma_{\varphi_n}^2(t)\xi}{\varsigma_{\varphi_n}^2(t) - (\xi + \upsilon_{n-1})^2} d\xi + \frac{\Xi_n \dot{\Xi}_n \varsigma_{\varphi_n}^2(t)}{\varsigma_{\varphi_n}^2(t) - x_n^2}\end{aligned} \tag{6.46}$$

where

$$\int_0^{\Xi_n} \frac{\partial}{\partial \upsilon_{n-1}} \frac{\varsigma_{\varphi_n}^2(t)\xi}{\varsigma_{\varphi_n}^2(t) - (\xi + \upsilon_{n-1})^2} d\xi = \frac{\Xi_n \varsigma_{\varphi_n}^2(t)}{\varsigma_{\varphi_n}^2(t) - x_n^2} - \Xi_n \mathcal{Q}_n\left(\upsilon_{n-1}, \Xi_n, \varsigma_{\varphi_n}\right)$$

and

$$\begin{aligned}&\int_0^{\Xi_n} \frac{\partial}{\partial \varsigma_{\varphi_n}(t)} \frac{\varsigma_{\varphi_n}^2(t)\xi}{\varsigma_{\varphi_n}^2(t) - (\xi + \upsilon_{n-1})^2} d\xi \\ &= \Xi_n \mathcal{N}_n\left(\upsilon_{n-1}, \Xi_n, \varsigma_{\varphi_n}\right) - \frac{\Xi_n^2 \varsigma_{\varphi_n}(t)}{\varsigma_{\varphi_n}^2(t) - (\Xi_n + \upsilon_{n-1})^2}\end{aligned}$$

with

$$\begin{aligned}\mathcal{Q}_n\left(\upsilon_{n-1}, \Xi_n, \varsigma_{\varphi_n}\right) =& \int_0^1 \frac{\varsigma_{\varphi_n}^2(t)}{\varsigma_{\varphi_n}^2(t) - (\psi \Xi_n + \upsilon_{n-1})^2} d\psi \\ =& \frac{\varsigma_{\varphi_n}(t)}{2\Xi_n} \ln \frac{\left(\varsigma_{\varphi_n}(t) + x_n\right)\left(\varsigma_{\varphi_n}(t) - \upsilon_{n-1}\right)}{\left(\varsigma_{\varphi_n}(t) - x_n\right)\left(\varsigma_{\varphi_n}(t) + \upsilon_{n-1}\right)}\end{aligned}$$

$$\mathcal{N}_n\left(v_{n-1}, \Xi_n, \varsigma_{\varphi_n}\right) = \int_0^1 \frac{2\psi \Xi_n \varsigma_{\varphi_n}(t) + v_{n-1}\varsigma_{\varphi_n}(t)}{\varsigma_{\varphi_n}^2(t) - (\psi \Xi_n + v_{n-1})^2} d\psi$$
$$- \frac{\varsigma_{\varphi_n}(t) v_{n-1}}{\varsigma_{\varphi_n}^2(t) - (\Xi_n + v_{n-1})^2}$$
$$= \frac{\varsigma_{\varphi_n}(t)}{\Xi_n} \ln \frac{\varsigma_{\varphi_n}^2(t) - x_n^2}{\varsigma_{\varphi_n}^2(t) - v_{n-1}^2} - \frac{\varsigma_{\varphi_n}(t) v_{n-1}}{\varsigma_{\varphi_n}^2(t) - x_n^2}$$
$$+ \frac{v_{n-1}}{2\Xi_n} \ln \frac{\varsigma_{\varphi_n}^2(t) - v_{n-1}^2}{\varsigma_{\varphi_n}^2(t) - x_n^2}.$$

By applying L'Hospital's rule, it is evident that $\mathcal{N}_n(v_{n-1}, \Xi_n, \varsigma_{\varphi_n})$ is bounded. In a similar vein, $\mathcal{Q}_n(v_{n-1}, \Xi_n, \varsigma_{\varphi_n})$ is also bounded.

To approximate the unknown nonlinear function $\mathcal{Z}_n(\chi_n)$, we employ NNs. There exists a positive scalar $\iota_n(\chi_n)$ such that

$$\mathcal{Z}_n(\chi_n) = W_n^{*T} E_n(\chi_n) + \iota_n(\chi_n) \tag{6.47}$$

where $\chi_n = \left[\hat{\Upsilon}_1, \ldots, \hat{\Upsilon}_n, x_d, \ldots, x_d^{(n)}, x_1, \ldots, x_n, \varsigma_{\varphi_1}, \ldots, \varsigma_{\varphi_1}^{(n)}, \varsigma_{\varphi_2}, \ldots, \varsigma_{\varphi_2}^{(n-1)}, \ldots, \varsigma_{\varphi_n}, \dot{\varsigma}_{\varphi_n}, \hat{W}_1, \ldots, \hat{W}_{n-1}\right]^T$, and there exist constants $\bar{\iota}_n > 0$ such that the inequality $|\iota_n(\chi_n)| < \bar{\iota}_n$ holds. W_n^* represents the ideal weight vector and $\hat{W}_n$ represents the estimated weight vector. Define $\tilde{W}_n = \hat{W}_n - W_n^*$. Subsequently, the Lyapunov function for step n is constructed as

$$V_n = V_n^W + V_n^I \tag{6.48}$$

where $V_n^W = \frac{1}{2}\tilde{W}_n^T C_n^{-1} \tilde{W}_n$, with C_n being a designed positive definite matrix.

To alleviate the system burden and conserve resources, the dynamic threshold based event-sampled triggering mechanism is employed, which is presented as

$$u = w(t_{\mathcal{A}}), \ \forall \mathrm{t} \in [\mathrm{t}_{\mathcal{A}}, \mathrm{t}_{\mathcal{A}+1}) \tag{6.49}$$
$$\mathrm{t}_{\mathcal{A}+1} = \inf\left\{t \in \mathbb{R} \,|\varrho\, |u(t)| + c \le |D(t)| \right\} \tag{6.50}$$
$$D(t) = w(t) - u(t) \tag{6.51}$$

where

$$w(t) = -(1+\varrho)\left[v_n \tanh\left(\frac{\Xi_n v_n}{\mu}\right) + \bar{c}\tanh\left(\frac{\Xi_n \bar{c}}{\mu}\right)\right] \tag{6.52}$$
$$v_n = \frac{1}{q_n(\bar{x}_n)}\Bigg(- \phi_n \Xi_n - \hat{W}_n^T E_n(\chi_n) - \hat{\Upsilon}_n$$

$$-\frac{\Xi_n \varsigma_{\varphi_n}^2(t)}{2\left(\varsigma_{\varphi_n}^2(t)-x_n^2\right)}-\Xi_n\sqrt{\left(\left(\frac{\dot{\varsigma}_{\varphi_n}(t)}{\varsigma_{\varphi_n}(t)}\right)^2+\zeta_n\right)} \tag{6.53}$$

where $D(t)$ is termed as the trigger error, $0<\varrho<1$, μ, c and $\bar{c}$ are positive design parameters with $\bar{c}>c/(1-\varrho)$.

Subsequently, we have

$$u(t)=\frac{w(t)}{1+\varrho\iota_1(t)}-\frac{c\iota_2(t)}{1+\varrho\iota_1(t)} \tag{6.54}$$

with time-varying parameters $|\iota_i(t)|\le 1$ $(i=1,2)$.

The neural weight updating law is established as

$$\dot{\hat{W}}_n=\frac{\Xi_n\varsigma_{\varphi_n}^2(t)E_n(\chi_n)}{\varsigma_{\varphi_n}^2(t)-x_n^2}C_n-\varpi_n\hat{W}_nC_n \tag{6.55}$$

where ϖ_n, ϕ_n, and ζ_n are positive design parameters.

According to Lemma 1.2, since $x\in\mathbb{R}$ and $\mu>0$, it follows that $-x\tanh(x/\mu)\le 0$. Hence, $\Xi_n w\le 0$ is obtained. Moreover, the following relations can be derived: $|L_2(t)c/(1+\theta L_1(t))|\le c/(1-\varrho)$ and $\Xi_n w/(1+\varrho\iota_1(t))\le\Xi_n w/(1+\varrho)$. Combining (6.48)–(6.55), we obtain

$$\begin{aligned}\dot{V}_n\le&\frac{\Xi_n\varsigma_{\varphi_n}^2(t)}{\varsigma_{\varphi_n}^2(t)-x_n^2}\left(q_n(\bar{x}_n)u+W_n^{*T}E_n(\chi_n)+\iota_n(\chi_n)\right)\\&+\tilde{W}_n^TC_n^{-1}\dot{\hat{W}}_n+\frac{\Xi_n\varsigma_{\varphi_n}^2(t)\Upsilon_n}{\varsigma_{\varphi_n}^2(t)-x_n^2}-\frac{\Xi_n^2\varsigma_{\varphi_n}(t)\dot{\varsigma}_{\varphi_n}(t)}{\varsigma_{\varphi_n}^2(t)-x_n^2}\\&+\frac{q_n\varsigma_{\varphi_n}^2(t)}{\varsigma_{\varphi_n}^2(t)-x_n^2}\left|\frac{\Xi_nc}{1-\varrho}\right|-\frac{q_n\varsigma_{\varphi_n}^2(t)}{\varsigma_{\varphi_n}^2(t)-x_n^2}|\Xi_n\bar{c}|+0.577\mu\\\le&\frac{1}{2}\bar{\iota}_n^2-\frac{\phi_n\Xi_n^2\varsigma_{\varphi_n}^2(t)}{\varsigma_{\varphi_n}^2(t)-x_n^2}-\varpi_n\tilde{W}_n^T\hat{W}_n+0.577\mu.\end{aligned} \tag{6.56}$$

Remark 6.3 : Regarding the virtual control laws (6.28), (6.40) and (6.53), unlike [4], it should be highlighted that the terms $\Xi_{r+1}q_r(\bar{x}_r)$ $(r=1,\dots,n-1)$ are treated as unknown functions and estimated using the NN technique as in (6.24), (6.37) and (6.47). This approach effectively simplifies the control protocol, meaning that in the control protocol of the $(r+1)$-th order, it is unnecessary to specifically design a nonlinear term to counteract $\Xi_{r+1}q_r(\bar{x}_r)$ from the r-th order. Moreover, by integrating disturbance estimation information from the developed SMDOs into the control laws, the proposed method's anti-interference capability is significantly enhanced.

Remark 6.4 : Comparisons with other approaches are outlined as follows:

(1) Recent adaptive nonlinear disturbance observer designs [6, 10, 11] have been limited to asymptotic stability. In contrast, the SMDOs presented in this chapter

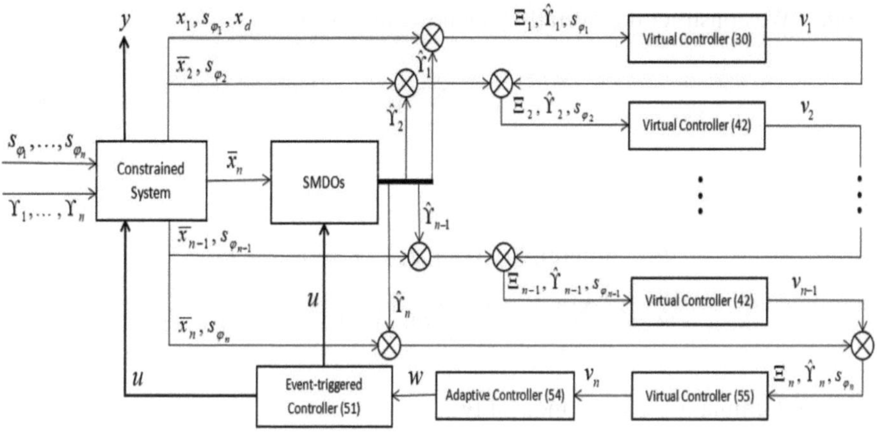

Fig. 6.1 Block diagram of the proposed adaptive control structure

achieve finite-time stability. SMDOs are more complex than adaptive nonlinear disturbance observers. With proper parameter selection, SMDOs can offer a smoother disturbance estimation process and greater optimization potential.

(2) Our design employs the time-varying IBLFs method. Compared to other approaches for addressing state constraints, such as constant IBLFs and conventional BLFs [12, 13], our method can handle dynamic full state constraints. This avoids conservative limitations and conversion operations, thereby expanding the scope of applications.

(3) Unlike event-triggered mechanisms with fixed thresholds, the dynamic threshold used here can flexibly balance system performance. When system states deviate from the desired condition, the larger control signal results in a higher threshold, effectively reducing the frequency of signal updates. Conversely, when the control signal amplitude is small, the threshold is adaptively lowered to provide finer control performance.

Figure 6.1 illustrates the block diagram of the proposed adaptive control structure based on SMDOs. Taking all the above deductions into account, the main result of the proposed control approach is presented in Theorem 6.1.

Theorem 6.1 : *For the system defined in (6.1) and the event-triggered controller specified in (6.49), under the initial conditions that* $|x_r(0)| < \varsigma_{\varphi_r}(0)$ *and with positive design parameters* ϖ_r, ϕ_r, ζ_r, C_r, ϱ, μ, $\bar{c}$ *and* c *for each* $r = 1, \ldots, n$, *it is established that the system's tracking errors are ultimately confined within a compact set around the origin. All signals within the closed-loop system remain bounded. Furthermore, the system state signals at each order never exceed the time-varying constraint boundaries. Importantly, the control mechanism effectively prevents the occurrence of the Zeno phenomenon.*

Proof. : We construct the overall Lyapunov function as

$$V = \sum_{r=1}^{n} V_r. \tag{6.57}$$

Define

$$\Phi = \min\left\{\varpi_r \lambda_{\min}(C_r), \phi_r\right\}, \ r = 1, \ldots, n \tag{6.58}$$

$$\Psi = \frac{1}{2}\sum_{r=1}^{n} \bar{\iota}_r^2 + \frac{1}{2}\sum_{r=1}^{n} \varpi_r \left\| W_r^* \right\|^2 + 0.577\mu. \tag{6.59}$$

Using Lemma 6.1 and the inequality

$$\tilde{W}_r^T \hat{W}_r \leq \frac{1}{2}\left\| W_r^* \right\|^2 - \frac{1}{2}\left\| \tilde{W}_r \right\|^2, \tag{6.60}$$

we derive

$$\dot{V} \leq -\Phi V + \Psi. \tag{6.61}$$

From (6.61), we obtain

$$V(t) \leq \frac{\Psi}{\Phi} + V(0)e^{-\Phi t}. \tag{6.62}$$

It follows that the tracking errors are bounded, specifically,

$$\left\| \Xi_r \right\| \leq \sqrt{2V(0)e^{-\Phi t} + 2\Psi/\Phi}. \tag{6.63}$$

Thus, the tracking error $\| \Xi_1 \|$ can be made arbitrarily small.
The weight estimate errors satisfy

$$\left\| \tilde{W}_r \right\| \leq \sqrt{2\left(\Psi/\Phi + V(0)e^{-\Phi t}\right)\lambda_{\max}(C_r)}, \tag{6.64}$$

indicating that $\tilde{W}_r$ are bounded for $r = 1, \ldots, n$.

As proven in Proposition 6.1, the disturbances estimation $\tilde{\Upsilon}_r$ are bounded. By Assumptions 6.2 and 6.3, the virtual control input and actual control input are bounded. Hence, all closed-loop signals are uniformly bounded.

Moreover, (6.62) holds provided $|x_r| < \varsigma_{\varphi_r}$ with initial information $|x_r(0)| < \varsigma_{\varphi_r}(0)$, ensuring the system state never violates the desired restricted boundary.

To examine the avoidance of the Zeno behavior, inspired by the event-based strategy [14], we note that $w(t)$ is continuous and differentiable. We also have

$$\frac{d}{dt}|D(t)| = \text{sign}(D(t))\,\dot{D}(t) \leq |\dot{w}(t)|. \tag{6.65}$$

From (6.52), we get

$$\dot{w}(t) = -(1+\varrho)\left[\frac{\bar{c}^2\dot{\Xi}_n}{\mu\cosh^2\left(\frac{\bar{c}\Xi_n}{\mu}\right)} + v_n\frac{\dot{v}_n\Xi_n + v_n\dot{\Xi}_n}{\mu\cosh^2\left(\frac{v_n\Xi_n}{\mu}\right)} + \dot{v}_n\left(\tanh\left(\frac{\Xi_n v_n}{\mu}\right)\right)\right]. \tag{6.66}$$

As all closed-loop signals are uniformly bounded, $\dot{w}(t)$ is bounded, implying a positive constant $\mathcal{H}$ exists such that $|\dot{w}(t)| \leq \mathcal{H}$. Based on the event-triggered approach, $D(t_{\mathcal{A}}) = 0$ and $\lim_{t\to t_{\mathcal{A}+1}} |D(t)| = \varrho\,|u(t)| + c$. Thus, the lower bound of inter-execution intervals is $\underset{\rightarrow}{t} \geq \left[(\varrho\,|u(t)| + c)\,/\mathcal{H}\right]$, ensuring the exclusion of Zeno behavior. This completes the proof of Theorem 1.

It is worth noting that the developed control approach can be extended to various practical systems, such as unmanned ground vehicles [15], two-wheeled mobile robots [16], and flapping-wing aerial vehicles [17]. Another promising avenue for future research is the application of our theoretical findings to robotic fish target tracking problems [18].

Remark 6.5 : The design parameters provide sufficient conditions for system stability. According to (6.63), reducing ϖ_r and μ or increasing $\lambda_{\min}(C_r)$ and ϕ_r helps narrow the tracking error but may increase control input amplitude, potentially causing strong signal impulses. Therefore, a balance between system safety and tracking accuracy should be carefully considered.

6.4 Simulation

This section presents a numerical simulation to validate the proposed adaptive tracking method's effectiveness and reliability. The simulation involves a nonlinear system described by

$$\begin{aligned}\dot{x}_1 &= q_1(x_1)x_2 + g_1(x_1) + \Upsilon_1(t, x_1)\\ \dot{x}_2 &= q_2(\bar{x}_2)u + g_2(\bar{x}_2) + \Upsilon_2(t, \bar{x}_2)\\ y &= x_1\end{aligned} \tag{6.67}$$

The system parameters are defined as $q_1(x_1) = 1 + \cos(x_1 + x_2)$, $q_2(\bar{x}_2) = 1.8 + \sin(3.6x_1x_2)$, $g_1(x_1) = 0.2\sin(x_1^2 + 0.5x_2)$, $g_2(\bar{x}_2) = 0.25\cos(x_1^6)$, $\Upsilon_1(t, x_1) = 0.2\sin(2.5t) - 0.15\cos(0.25t)$, and $\Upsilon_2(t, \bar{x}_2) = 0.5 + 0.2\sin(x_1 + x_2)$. The desired trajectory is set to $x_d = 0.6\cos(2t)$. The state constraints are defined as $|x_1| \leq \varsigma_{\varphi_1} = 1 + 0.1\cos(7t)$ and $|x_2| \leq \varsigma_{\varphi_2} = 1.86 + 0.4\sin(3t)$. The simulation parameters are initialized to $\kappa_{11} = 0.2$, $\kappa_{12} = 0.2$, $\kappa_{21} = 0.25$, $\kappa_{22} = 0.4$, $\varpi_{d1} = \varpi_{d2} =$

0.6, $\bar{\mathcal{J}}_1 = \bar{\mathcal{J}}_2 = 0.1$, $\delta_1 = \delta_2 = 0.02$, $\sigma_1 = \sigma_2 = 0.1$, $\phi_1 = 100$, $\phi_2 = 276$, $\mu = 0.3$, $\varrho = 0.2$, $c = 0.15$, $\bar{c} = 3$, $\zeta_1 = \zeta_2 = 0.05$, $\varpi_1 = 0.8$, $\varpi_2 = 0.6$, $C_1 = diag\{0.8\}$, $C_2 = diag\{0.5\}$, $\imath_{11} = 0.5$, $\imath_{12} = 1.5$, $\imath_{21} = 0.36$, $\imath_{22} = 1.6$, $\theta_1 = 999$, and $\theta_2 = 666$.

The simulation results are presented in Figs. 6.2, 6.3, 6.4, 6.5, 6.6, 6.7, 6.8, 6.9 and 6.10. Figure 6.2 illustrates the trajectories of the system output and the reference signal. Figure 6.3 depicts the second-order state and the virtual control input v_1. The tracking performance is satisfactory, as the state signals are maintained within

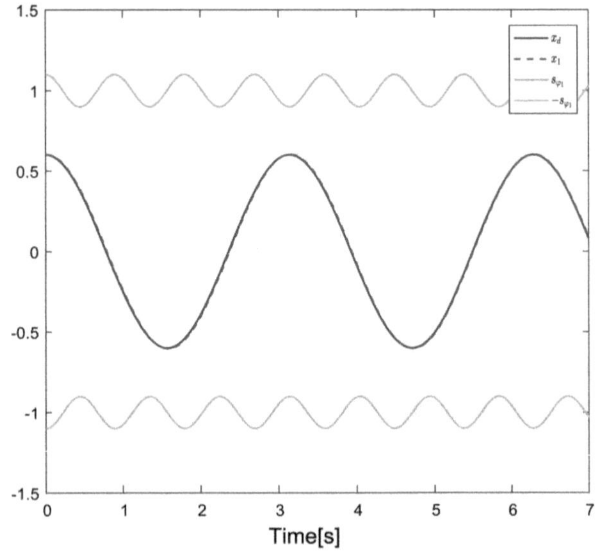

Fig. 6.2 Trajectories of first-order signals and constraints

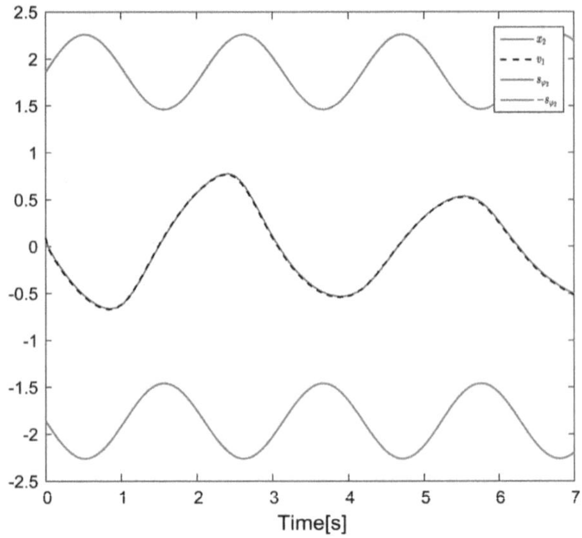

Fig. 6.3 Trajectories of second-order signals and constraints

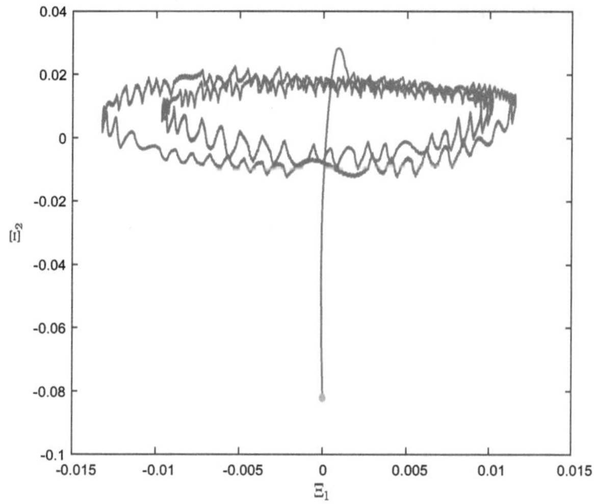

Fig. 6.4 Phase portrait of tracking error

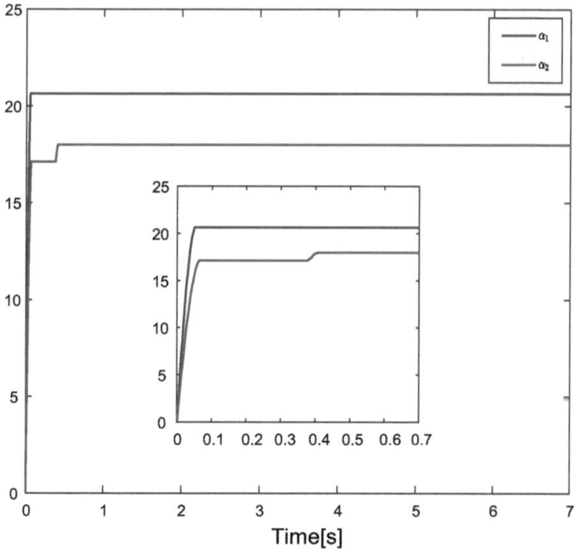

Fig. 6.5 Trajectories of α_i $(i = 1, 2)$

specified ranges, ensuring that dynamic boundary constraints are not violated. The phase portrait of the tracking errors in Fig. 6.4 indicates that the errors converge to a bounded set around the origin. The evolution of α_1 and α_2 is shown in Fig. 6.5, which stabilizes in finite time. The difference between γ_i and the absolute value of $\dot{\Upsilon}_i$ $(i = 1, 2)$ is presented in Fig. 6.6, confirming that $\gamma_i > \left|\dot{\Upsilon}_i\right|$ is achieved in finite time, ensuring persistent sliding motion. A comparison of actual disturbances with SMDOs estimations is shown in Fig. 6.7, indicating that SMDOs can adaptively detect external

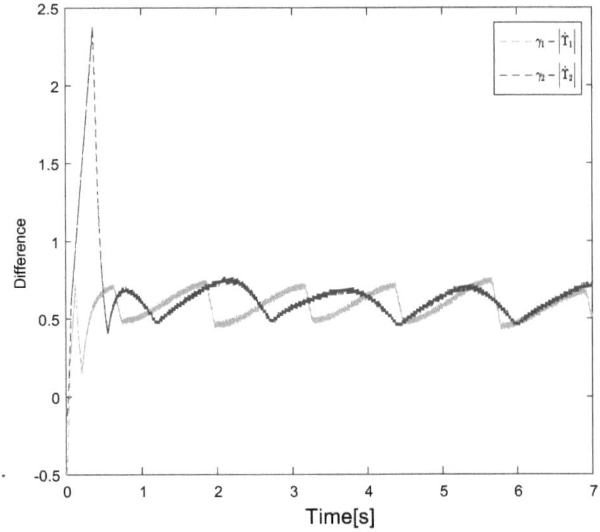

Fig. 6.6 Difference

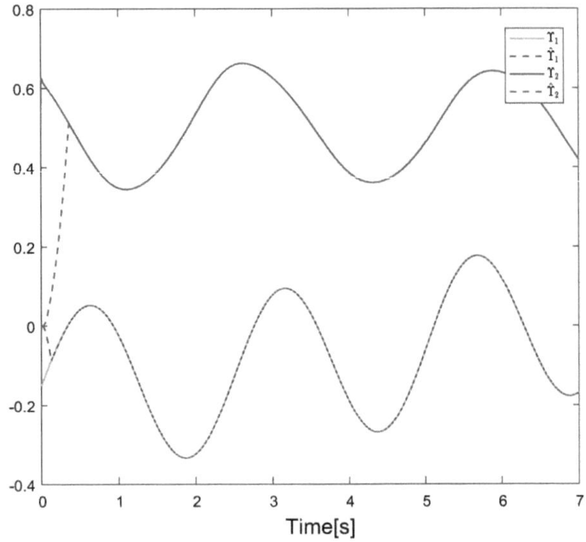

Fig. 6.7 Disturbances and their estimation

disturbances after a short interval. The phase portrait of the disturbances estimation errors is shown in Fig. 6.8, with highlighted initial positions. The time-triggered and event-triggered control inputs are presented in Fig. 6.9, and the trigger times and intervals are shown in Fig. 6.10, with 381 control signal updates recorded.

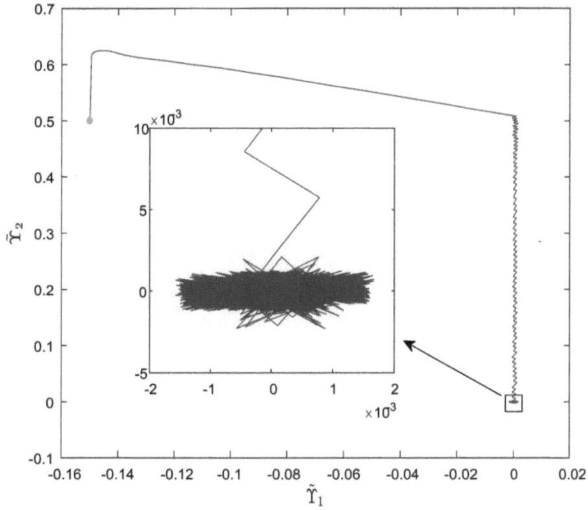

Fig. 6.8 Phase portrait of disturbances estimation error

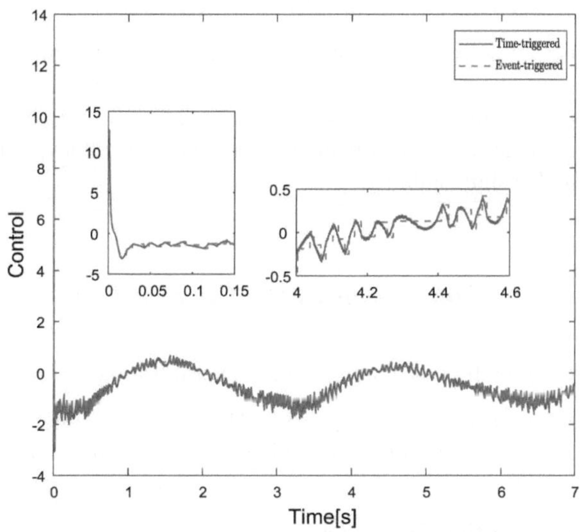

Fig. 6.9 System control input

6.5 Conclusion

6.5.1 Contribution

This work proposes an event-based adaptive control strategy for uncertain nonlinear systems with dynamic full-state constraints and unknown external disturbances. Sliding mode disturbance observers (SMDOs) are utilized to track and compensate for disturbances. An event-triggered controller is designed based on disturbance and

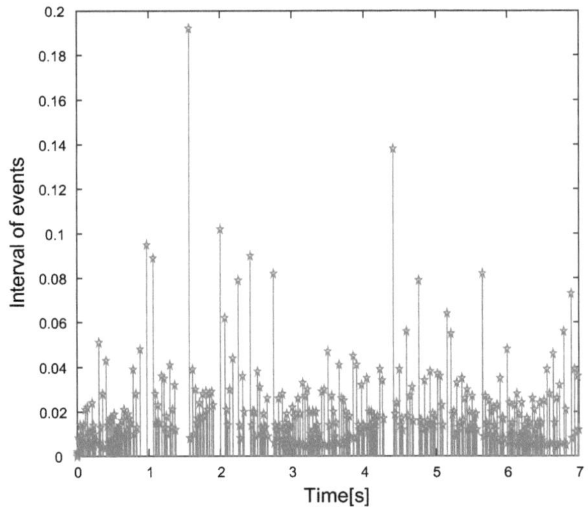

Fig. 6.10 Interval of events

neural network estimations. Integral barrier Lyapunov functions (IBLFs) guarantee system states stay within constraints. Theoretical analysis shows all closed - loop signals are uniformly bounded. A simulation case verifies the approach's effectiveness, demonstrating its advantages in control accuracy, control signal update frequency, and system stability.

6.5.2 *Limitations of the Current Work*

Despite the promising results, several limitations remain:

Computational Complexity: The triggering condition necessitates continuous control signal monitoring, which may consume significant resources. Additionally, incorporating the RNFNN also increases computational demands.

Assumption Dependency: Assumption 6.1 assumes the first and second derivatives of disturbances are bounded, which may not apply to scenarios with abrupt disturbances. Assumption 6.2 requires the reference signal and its derivatives to be bounded by certain constants and functions, which may be challenging in dynamic environments. Assumption 6.3 assumes the boundedness of constraints and their derivatives, but in practice, these may exceed predefined bounds due to unforeseen factors.

Parameter Sensitivity: The control performance is sensitive to design parameters like $\iota_{r1} < 1$, $\iota_{r2} > 1$, κ_{r1}, κ_{r2}, and neural network parameters. Using inappropriate parameters may lower the tracking precision or lead to system instability.

6.5.3 Future Directions

To address these limitations, future research could focus on:

Advanced Event-Triggering Schemes: Future work could focus on developing a hybrid multi-layer self-triggered mechanism to balance energy consumption, system safety, and control performance.

Reduced-Order Observers: Developing lightweight observer designs to lower computational overhead while maintaining estimation accuracy, and enhancing the observation of external disturbances.

Machine Learning Integration: Utilizing reinforcement learning or metaheuristic algorithms to automatically adjust parameters and enhance adaptability.

Experimental Validation: Testing the proposed event-based adaptive control strategy on physical platforms (e.g., robotic swarms or smart grids) to evaluate scalability and robustness under real-world constraints, especially in the presence of dynamic full-state constraints and unknown external disturbances.

By addressing these challenges, the framework could be extended to broader applications like autonomous vehicles, distributed energy systems, or bio-inspired collective behaviors. This work establishes a foundation for adaptive control in complex multiagent systems. However, bridging the gap between theoretical guarantees and practical deployment remains an open frontier.

Future research should focus on enhancing scalability, robustness, and real-world feasibility to unlock the full potential of neural network-based control in distributed systems.

References

1. Zou WC, Shi P, Xiang ZR, Shi Y (2020) Consensus tracking control of switched stochastic nonlinear multiagent systems via event-triggered strategy. IEEE Transactions on Neural Networks and Learning Systems 31(3):1036–1045
2. Li ZM, Chang XH, Park JH (2021) Quantized static output feedback fuzzy tracking control for discrete-time nonlinear networked systems with asynchronous event-triggered constraints. IEEE Transactions on Systems Man Cybernetics-Systems 51(6):3820–3831
3. Tee KP, Ren B, Ge SS (2011) Control of nonlinear systems with time-varying output constraints. Automatica 47(11):2511–2516
4. Liu L, Gao T T, Liu Y J, Tong S C, Chen C L P, Ma L (2021) Time-varying IBLFs-based adaptive control of uncertain nonlinear systems with full state constraints. Automatica 129
5. Rabiee H, Ataei M, Ataei M (2019) Continuous nonsingular terminal sliding mode control based on adaptive sliding mode disturbance observer for uncertain nonlinear systems. Automatica 109
6. Zhang YH, Sun J, Liang HJ, Li HY (2020) Event-triggered adaptive tracking control for multiagent systems with unknown disturbances. IEEE Transactions on Cybernetics 50(3):890–901
7. Ginoya D, Shendge PD, Phadke SB (2014) Sliding mode control for mismatched uncertain systems using an extended disturbance observer. IEEE Trans Industr Electron 61(4):1983–1992

8. Wang X, Wang H, Huang TW, Kurths J (2023) Neural-network-based adaptive tracking control for nonlinear multiagent systems: The observer case. IEEE Transactions on Cybernetics 53(1):138–150
9. Cheng FB, Wang HQ, Zong GD, Niu B, Zhao XD (2023) Adaptive finite-time command-filtered control for switched nonlinear systems with input quantization and output constraints. Circuits Systems Signal Process 42(1):147–172
10. Chen M, Ge SS (2015) Adaptive neural output feedback control of uncertain nonlinear systems with unknown hysteresis using disturbance observer. IEEE Trans Industr Electron 62(12):7706–7716
11. Pang N, Wang X, Wang ZM (2022) Event-triggered adaptive control of nonlinear systems with dynamic uncertainties: The switching threshold case. IEEE Trans Circuits Syst II Express Briefs 69(8):3540–3544
12. Gao TT, Liu YJ, Li DP, Tong SC, Li TS (2021) Adaptive neural control using tangent time-varying BLFs for a class of uncertain stochastic nonlinear systems with full state constraints. IEEE Transactions on Cybernetics 51(4):1943–1953
13. Wang LJ, Chen CLP, Li HY (2020) Event-triggered adaptive control of saturated nonlinear systems with time-varying partial state constraints. IEEE Transactions on Cybernetics 50(4):1485–1497
14. Xing LT, Wen CY, Liu ZT, Su HY, Cai JP (2017) Event-triggered adaptive control for a class of uncertain nonlinear systems. IEEE Trans Autom Control 62(4):2071–2076
15. Ren HR, Karimi HR, Lu RQ, Wu YQ (2020) Synchronization of network systems via aperiodic sampled-data control with constant delay and application to unmanned ground vehicles. IEEE Trans Industr Electron 67(6):4980–4990
16. Mu B, Chen J, Shi Y, Chang Y (2017) Design and implementation of nonuniform sampling cooperative control on a group of two-wheeled mobile robots. IEEE Trans Industr Electron 64(6):5035–5044
17. He W, Mu X, Zhang L, Zou Y (2021) Modeling and trajectory tracking control for flapping-wing micro aerial vehicles. IEEE/CAA Journal of Automatica Sinica 8(1):148–156
18. Yu JZ, Wu ZX, Yang X, Yang YQ, Zhang PF (2021) Underwater target tracking control of an untethered robotic fish with a camera stabilizer. IEEE Transactions on Systems Man Cybernetics-Systems 51(10):6523–6534

Chapter 7
Observer-Based-Triggered Adaptive Control for Nonlinear Multiagent Systems With Unknown States and Disturbances

Achieving consensus in nonlinear multiagent systems (MAS) is a crucial task for a wide range of applications, including unmanned air vehicles, underwater vehicles, and various networked systems. The fundamental objective is to ensure that the outputs of all follower agents converge to the output of a designated leader agent. However, a significant obstacle in many practical industrial settings is the inability to measure system states directly and the presence of unknown external disturbances. These uncertainties pose a considerable challenge to conventional control methods for achieving consensus, highlighting the importance of developing robust control strategies to address them.

To manage unmeasurable states and disturbances, techniques involving observers have been widely investigated in existing literature. For various systems, including second-order linear MAS, state observers that utilize tools like fuzzy logic and neural networks have been essential for estimating unknown system states. In parallel, disturbance observers, first introduced by Nakao et al. [1], have a critical function in neutralizing external interferences and have been applied in fields such as motion control. More recently, disturbance observers have been incorporated into event-triggered consensus frameworks to improve overall efficiency.

Traditional control systems typically rely on time-sampling methods, which require continuous updates of the controller. This approach can lead to significant consumption of computational and communication resources, particularly when the system is not actively changing. As a resource-conscious alternative, event-triggered control has attracted substantial research interest because it activates control updates only when system performance dictates a need for intervention. A variety of strategies have been developed for nonlinear systems, such as centralized and distributed models for MAS, adaptations for discrete-time systems, and adaptive neural network methods for complex nonaffine nonlinear MAS.

Despite these developments, the specific issue of event-based consensus for high-order nonlinear multiagent systems that face both unknown states and disturbances concurrently has received limited attention, although some studies have used observer or event-triggered methods for low-order MAS [2–4]. This chapter devotes to present

X. Wang et al., *Optimal Adaptive Cooperative Control for Nonlinear Multiagent Systems*,
https://doi.org/10.1007/978-981-95-2964-3_7

an event-triggered adaptive consensus tracking scheme for a class of high-order nonlinear MAS where only agent output signals are accessible and both states and disturbances are unknown. The solution develops through a synergistic combination of backstepping procedures, filter techniques, RBF NNs and observer methods.

The core of this scheme is the systematic development of a comprehensive control framework that integrates a state observer, a disturbance observer and a low-pass filter for high-order uncertain systems. This framework is conceived as an organic whole, specifically designed to address multiple challenges of uncertainty. A key distinction of this design is the parallel operation of the state and disturbance observers. This parallel architecture allows for robust online detection and dynamic compensation of both unmeasured system states and time-varying external disturbances, effectively mitigating the risk of system failure that these uncertainties could cause and thereby enhancing the controller's practical viability.

Building upon this robust estimation, the chapter further puts forward a novel, event-based adaptive consensus tracking control method for this class of semi-strict-feedback uncertain nonlinear MAS. In comparison to existing adaptive tracking control techniques [5–7], the resulting controller not only ensures computational and communication efficiency via the event-triggered mechanism but also demonstrates superior anti-interference performance.

7.1 Preliminaries

7.1.1 Problem Formulation

The behavior of each agent within the considered multiagent system is described by the following dynamic model:

$$
\begin{aligned}
\dot{x}_{i,m} &= x_{i,m+1} + f_{i,m}\left(\bar{x}_{i,m}(t)\right) + \mathcal{D}_{i,m}(t) \\
\dot{x}_{i,n} &= u_i + f_{i,n}\left(\bar{x}_{i,n}(t)\right) + \mathcal{D}_{i,n}(t) \\
y_i(t) &= x_{i,1}(t)
\end{aligned}
\tag{7.1}
$$

where $m = 1, \ldots, n-1$, $\bar{x}_{i,j} = \left[x_{i,1}^T, \ldots, x_{i,j}^T\right]^T$ represents the state vector, $f_{i,j}(\bar{x}_{i,j})$ is an unknown smooth nonlinear function and $\mathcal{D}_{i,j}$ is an unknown external disturbance for $i = 1, \ldots, N$, $j = 1, \ldots, n$.

The objective of this chapter is to present a novel consensus tracking approach that uses neural networks, observers, filters, and backstepping techniques. This approach is designed to ensure that the output signal y_i of each follower agent can effectively track the reference signal y_r.

Assumption 7.1 The leader's desired reference signal is both measurable and smooth, with $y_r(t)$ and $\dot{y}_r(t)$ being bounded functions.

Assumption 7.2 The unmeasured external disturbance for the i-th agent at state j is a bounded function, meaning the inequality $\|\mathcal{D}_{i,j}\| \leq \mathcal{D}_{i,j}^0$ is valid, where $\mathcal{D}_{i,j}^0$ is a positive constant.

Assumption 7.3 Each agent in the system can only access information from its direct neighbors; it cannot receive communication from other agents. This limitation on information exchange places greater demands on the performance of the consensus tracking controller.

Neural networks are utilized across many fields due to their superior capability in approximating unknown continuous functions. We employ RBF NNs to estimate an unknown function $\Phi(\bar{\delta}) : R^p \to R^m$ as follows:

$$\Phi(\bar{\delta}) = \epsilon^T q(\bar{\delta}) \tag{7.2}$$

where $\bar{\delta}$ is the input vector,, $\epsilon = [\epsilon_1, \ldots, \epsilon_d]^T \in R^d$ is the weights vector and $q(\bar{\delta}) = \left[q_1(\bar{\delta}), \ldots, q_d(\bar{\delta})\right]^T$ denotes the basis function vector, d stands for the number of RBF NN nodes. Let $q_i(\bar{\delta}) = \exp\left[-(x-\sigma_i)^T(x-\sigma_i)/{\varpi_i}^2\right]$, where $\sigma_i = \left[\sigma_{i,1}, \ldots, \sigma_{i,p}\right]^T$ stands for the center of the receptive field, ϖ_i is known as the width of gaussian function, $i = 1, 2, \ldots, d$.

By applying RBF NNs to approximate the unknown smooth nonlinear function in the multiagent system (7.1), an ideal weight vector $\epsilon_{i,l}^*$ exists that satisfies:

$$f_{i,l}(\bar{x}_{i,l}) = \epsilon_{i,l}^{*T} q_{i,l}\left(\hat{\bar{x}}_{i,l}(t)\right) + \rho_{i,l}(t) \tag{7.3}$$

where $\hat{\bar{x}}_{i,l} = \left[\hat{x}_{i,1}^T, \ldots, \hat{x}_{i,l}^T\right]^T$ is the estimate of the true state, $\rho_{i,l}(t)$ denotes the approximation error and it is bounded, i.e., existing a constant $\rho_{0,l} > 0$ such that $\left|\rho_{i,l}(t)\right| \leq \rho_{0,l}$.

Letting $\hat{\epsilon}_{i,l}^T$ be the estimate of $\epsilon_{i,l}^{*T}$, the estimation of the unknown smooth nonlinear function is:

$$\hat{f}_{i,l}(\hat{\bar{x}}_{i,l}) = \hat{\epsilon}_{i,l}^T q_{i,l}\left(\hat{\bar{x}}_{i,l}(t)\right). \tag{7.4}$$

The ideal weight $\epsilon_{i,l}^*$ is then selected as:

$$\epsilon_{i,l}^* = \arg\min_{\hat{\epsilon}_{i,l} \in \breve{\Omega}} \sup_{\substack{\bar{x}_{i,l} \in \Omega_{i,l} \\ \hat{\bar{x}}_{i,l} \in \hat{\Omega}_{i,l}}} \left|\hat{f}_{i,l}(\hat{\bar{x}}_{i,l}) - f(\bar{x}_{i,l})\right| \tag{7.5}$$

where $\breve{\Omega}$, $\Omega_{i,l}$ and $\hat{\Omega}_{i,l}$ are compact sets of $\hat{\epsilon}_{i,l}$, $\bar{x}_{i,l}$, $\hat{\bar{x}}_{i,l}$, respectively. Furthermore, define $\vartheta_i^* = \max\left\{\left\|\epsilon_{i,l}^*\right\|\right\}$, where $\epsilon_{i,l}^*$ denotes the ideal weight vector of RBF NNs, $\hat{\vartheta}_i$ denotes the estimation of ϑ_i^* with $\tilde{\vartheta}_i = \vartheta_i^* - \hat{\vartheta}_i$, $i = 1, \ldots, M, l = 1, \ldots, n$.

7.1.2 Observer Design

To estimate the states that cannot be directly measured, the following state observers are introduced:

$$\begin{aligned}\dot{\hat{\bar{x}}}_{i,n}(t) &= (B_i \otimes I_m)\,\hat{\bar{x}}_{i,n}(t) + (G_i \otimes y_i(t)) \\ &\quad + \sum_{l=1}^{n}\left(A_{i,l} \otimes \hat{f}_{i,l}\left(\hat{\bar{x}}_{i,l}(t)\right)\right) + (\varsigma_i \otimes u_i(t)) + \hat{\mathcal{D}}_i \\ y_i(t) &= \left(C_i^T \otimes I_m\right)\hat{\bar{x}}_n(t),\ i = 1, \ldots, M \end{aligned} \tag{7.6}$$

where $\otimes$ denotes the Kronecker product, $A_{i,l} = \Big[\underbrace{0, \ldots, 1}_{l}, \ldots, 0\Big]^T_{n\times 1}$, $G_i = \left[g_{i,1}, \ldots, g_{i,n}\right]^T$, $B_i = \begin{bmatrix} -g_{i,1} & \\ \vdots & I_{(n-1)\times(n-1)} \\ -g_{i,n} & 0 \ldots 0 \end{bmatrix}$, $C_i = [1, 0, \ldots, 0]^T_{n\times 1}$, $\varsigma_i = [0, \ldots, 0, 1]^T_{n\times 1}$, $\hat{\mathcal{D}}_i = \left[\hat{\mathcal{D}}_{i,1}^T, \ldots, \hat{\mathcal{D}}_{i,n}^T\right]^T$ denotes the estimate of external disturbance. The vector $G_i = \left[g_{i,1}, \ldots, g_{i,n}\right]^T$ satisfies that B_i is a strict Hurwitz matrix.

Define $\Pi_i(t) = \left[(x_{i,1} - \hat{x}_{i,1})^T, \ldots, (x_{i,n} - \hat{x}_{i,n})^T\right]^T$ as the state observation error, we can easily get

$$\dot{\Pi}_i(t) = (B_i \otimes I_m)\,\Pi_i(t) + \tilde{\mathcal{D}}_i + \sum_{l=1}^{n} A_{i,l} \otimes \left(f_{i,l}\left(\bar{x}_{i,l}(t)\right) - \hat{f}_{i,l}\left(\hat{\bar{x}}_{i,l}(t)\right)\right).$$

Consider $\xi_i(t) = \sum_{l=1}^{n} A_{i,l} \otimes \left(f_{i,l}\left(\bar{x}_{i,l}(t)\right) - \hat{f}_{i,l}\left(\hat{\bar{x}}_{i,l}(t)\right)\right)$ as the function approximation error and $\xi_i(t) = \left[\xi_{i,1}^T(t), \ldots, \xi_{i,n}^T(t)\right]^T$, suppose that $\xi_i(t)$ is bounded, i.e., existing an unknown constant $\xi_i^0 > 0$ satisfies that $\|\xi_i(t)\| \le \xi_i^0$.

Then, the state observers can be rewritten as

$$\begin{aligned}\dot{\hat{x}}_{i,m} &= \hat{x}_{i,m+1} + \hat{\epsilon}_{i,m}^T q_{i,m}\left(\hat{\bar{x}}_{i,m}(t)\right) + g_{i,m}\Pi_{i,1}(t) + \hat{\mathcal{D}}_{i,m}(t) \\ \dot{\hat{x}}_{i,n} &= u_i + \hat{\epsilon}_{i,n}^T q_{i,n}\left(\hat{\bar{x}}_{i,n}(t)\right) + g_{i,n}\Pi_{i,1}(t) + \hat{\mathcal{D}}_{i,n}(t)\end{aligned} \tag{7.7}$$

with $m = 1, \ldots, n-1$.

Since the external disturbance is unknown, a new auxiliary variable is defined to facilitate consensus control

$$c_{i,l} = \mathcal{D}_{i,l} - \varphi_{i,l}x_{i,l} \tag{7.8}$$

where $\varphi_{i,l}$ is a positive design parameter. Based on this, the disturbance observer model is formulated as:

$$\begin{aligned}\hat{\mathcal{D}}_{i,l} &= \hat{c}_{i,l} + \varphi_{i,l}\hat{x}_{i,l} \\ \dot{\hat{c}}_{i,l} &= -\varphi_{i,l}\left(\hat{\epsilon}_{i,l}^{T} q_{i,l}(\hat{\bar{x}}_{i,l}) + \hat{c}_{i,l} + \varphi_{i,l}\hat{x}_{i,l} + \hat{x}_{i,l+1}\right)\end{aligned} \tag{7.9}$$

where $\hat{\epsilon}_{i,l}$ is the estimation of $\epsilon_{i,l}^{*}$ and $x_{i,n+1} = u_i$.

Using (7.1), (7.3), and (7.9), we can derive:

$$\dot{\tilde{c}}_{i,l} = \dot{\mathcal{D}}_{i,l} - \varphi_{i,l}\left(\tilde{c}_{i,l} + \Pi_{i,l+1} + \varphi_{i,l}\Pi_{i,l} + \tilde{\epsilon}_{i,l}^{T} q_{i,l}(\hat{\bar{x}}_{i,l}) + \rho_{i,l}\right)$$

where $\tilde{c}_{i,l} = c_{i,l} - \hat{c}_{i,l}$.

Lemma 7.1 *[5] For any given positive definite matrix $H_i = H_i^T > 0$, there exists a symmetric positive matrix P_i such that the Lyapunov equality*

$$B_i^T P_i + P_i B_i = -2H_i \tag{7.10}$$

is satisfied.

7.2 Distributed Consensus Tracking Controller Design

This section details the design methodology for the observer-based distributed event-triggered controller. The design process involves n steps. We first define the following errors

$$e_{i,1} = \sum_{j=1}^{N} a_{i,j}\left(y_i - y_j\right) + b_i\left(y_i - y_r\right)$$

$$e_{i,k} = \hat{x}_{i,k} - \bar{w}_{i,k} \tag{7.11}$$

$$s_{i,k} = \bar{w}_{i,k} - w_{i,k} \tag{7.12}$$

where $e_{i,k}$ denotes the consensus tracking error, $w_{i,k}$ denotes the virtual control law and $\bar{w}_{i,k}$ denotes the filtered virtual control law, $s_{i,k}$ is known as the filtered error between $w_{i,k}$ and $\bar{w}_{i,k}$, where $k = 2, \ldots, n$.

Step 1: By utilizing (7.1), the time derivative of the tracking error $e_{i,1}$ is

$$\begin{aligned}\dot{e}_{i,1} = &\left(d_i + b_i\right)\left(\hat{x}_{i,2} + \Pi_{i,2}(t) + \mathcal{D}_{i,1}(t) + f_{i,1}\left(\bar{x}_{i,1}\right)\right) \\ &- \sum_{j=1}^{N} a_{i,j}\left(\hat{x}_{j,2} + \Pi_{j,2}(t) + \mathcal{D}_{j,1}(t) + f_{j,1}\left(\bar{x}_{j,1}\right)\right) - b_i\dot{y}_r.\end{aligned} \tag{7.13}$$

The Lyapunov function for this step is selected as

$$V_{i,1} = \frac{e_{i,1}^2}{2} + \frac{\tilde{c}_{i,1}^2}{2} + \frac{1}{2\eta_{i,1}}\tilde{\epsilon}_{i,1}^T\tilde{\epsilon}_{i,1} + \frac{1}{2\mathcal{D}_i}\tilde{\vartheta}_i^2 + V_0 \tag{7.14}$$

where $V_0 = \frac{1}{2}\Pi_i^T(P_i \otimes I_m)\Pi_i$ with $\Pi_i = [\Pi_{i,1}^T, \ldots, \Pi_{i,n}^T]^T$, $\eta_{i,1}$ and $\mathcal{D}_i$ are positive constants.

A new auxiliary function is defined as

$$\begin{aligned} \bar{S}_{i,1}\left(\bar{\delta}_{i,1}\right) = &- \sum_{j=1}^{N} a_{i,j}\left(\hat{x}_{j,2} + f_{j,1}\left(\bar{x}_{j,1}\right) + \mathcal{D}_{j,1} + \Pi_{j,2}\right) \\ &+ (d_i + b_i)\left(f_{i,1}\left(\bar{x}_{i,1}\right) + \mathcal{D}_{i,1} + \Pi_{i,2}\right) - b_i\dot{y}_r. \end{aligned} \tag{7.15}$$

$\bar{S}_{i,1}\left(\bar{\delta}_{i,1}\right)$ is estimated by RBF NN:

$$\begin{aligned} \bar{S}_{i,1}\left(\bar{\delta}_{i,1}\right) =& W_{i,1}^{*T}q_{i,1}\left(\bar{\delta}_{i,1}\right) + \wp_{i,1}\left(\bar{\delta}_{i,1}\right) \\ \left|\wp_{i,1}\left(\bar{\delta}_{i,1}\right)\right| \le& \varepsilon_{i,1} \end{aligned} \tag{7.16}$$

where $\varepsilon_{i,1}$ is an unknown positive constant.

Applying Young's inequality, for any constant $q_{i,1} > 0$, we derive

$$e_{i,1}\bar{S}_{i,1}\left(\bar{\delta}_{i,1}\right) \le \frac{\vartheta_i^*}{2q_{i,1}^2}e_{i,1}^2 q_{i,1}^T\left(\bar{\delta}_{i,1}\right)q_{i,1}\left(\bar{\delta}_{i,1}\right) + \frac{q_{i,1}^2}{2} + \frac{e_{i,1}^2}{2} + \frac{\varepsilon_{i,1}^2}{2}. \tag{7.17}$$

where $\bar{\delta}_{i,1} = \left[y_r, \dot{y}_r, \hat{x}_{i,1}^T, \hat{x}_{j,1}^T, \hat{\mathcal{D}}_{i,1}^T, \hat{\mathcal{D}}_{j,1}^T\right]^T$ stands for the input of RBF NNs.

From Assumption 7.2 and Young's inequality, we obtain the following relations

$$-\varphi_{i,1}\tilde{c}_{i,1}\rho_{i,1} \le \frac{1}{2}\tilde{c}_{i,1}^2 + \frac{\varphi_{i,1}^2\rho_{0,1}^2}{2} \tag{7.18}$$

$$-\varphi_{i,1}\tilde{c}_{i,1}\Pi_{i,2} \le \frac{1}{2}\tilde{c}_{i,1}^2 + \varphi_{i,1}^2\left\|\Pi_i\right\|^2 \tag{7.19}$$

$$-\varphi_{i,1}^2\tilde{c}_{i,1}\Pi_{i,1} \le \frac{1}{2}\tilde{c}_{i,1}^2 + \varphi_{i,1}^4\left\|\Pi_i\right\|^2 \tag{7.20}$$

$$\Pi_i^T(P_i \otimes I_m)\xi_i \le \frac{1}{2}\|P_i \otimes I_m\|^2\left\|\xi_i^0\right\|^2 + \frac{1}{2}\|\Pi_i\|^2 \tag{7.21}$$

$$\Pi_i^T(P_i \otimes I_m)\tilde{\mathcal{D}}_i \le \frac{1}{2}\|\Pi_i\|^2 + \frac{1}{2}\|P_i \otimes I_m\|^2\left\|\tilde{\mathcal{D}}_i^0\right\|^2. \tag{7.22}$$

The virtual control law and adaptive law are formulated as

$$w_{i,2} = \frac{1}{d_i + b_i}\left(r_{i,1}e_{i,1} - \frac{e_{i,1}}{2} - \frac{\hat{\vartheta}_i}{2q_{i,1}^2}e_{i,1}q_{i,1}^T\left(\bar{\delta}_{i,1}\right)q_{i,1}\left(\bar{\delta}_{i,1}\right)\right) \tag{7.23}$$

$$\dot{\hat{\epsilon}}_{i,1} = -h_{i,1}\hat{c}_{i,1} - \eta_{i,1}\tilde{c}_{i,1}\varphi_{i,1}q_{i,1}\left(\hat{\bar{x}}_{i,1}\right) \tag{7.24}$$

where $h_{i,1}$ and $r_{i,1}$ are design constants.

By combining (7.14), (7.19)-(7.25), and Lemma 7.1, the derivative of the Lyapunov function is bounded as

$$\begin{aligned}\dot{V}_{i,1} \leq & r_{i,1}e_{i,1}^2 + (d_i + b_i)\,e_{i,1}\left(e_{i,2} + s_{i,2}\right) + \tilde{c}_{i,1}\dot{\mathcal{D}}_{i,1} \\ & + \left(\frac{3 - 2\varphi_{i,1}}{2}\right)\tilde{c}_{i,1}^2 + \iota_{i,1} + \frac{h_{i,1}}{\eta_{i,1}}\tilde{\epsilon}_{i,1}^T\hat{\epsilon}_{i,1} \\ & + \tilde{\vartheta}_i\left(\frac{1}{2q_{i,1}^2}e_{i,1}^2q_{i,1}^T\left(\bar{\delta}_{i,1}\right)q_{i,1}\left(\bar{\delta}_{i,1}\right) - \frac{\dot{\hat{\vartheta}}_i}{\mathcal{D}_i}\right) \\ & - \Pi_i^T\left(H_i \otimes I_m\right)\Pi_i + \left(1 + \varphi_{i,1}^2 + \varphi_{i,1}^4\right)\|\Pi_i\|^2\end{aligned} \tag{7.25}$$

where $\iota_{i,1} = \frac{q_{i,1}^2}{2} + \frac{\varepsilon_{i,1}^2}{2} + \frac{\varphi_{i,1}^2\rho_{0,1}^2}{2} + \frac{1}{2}\|P_i \otimes I_m\|^2\left\|\xi_i^0\right\|^2 + \frac{1}{2}\|P_i \otimes I_m\|^2\left\|\tilde{\mathcal{D}}_i^0\right\|^2$.

Next, we pass $w_{i,2}$ through a first-order low-pass filter with a small positive time constant $\tau_{i,2}$ to obtain $\bar{w}_{i,2}$:

$$\bar{w}_{i,2}(0) = w_{i,2}(0),\ \tau_{i,2}\dot{\bar{w}}_{i,2} + \bar{w}_{i,2} = w_{i,2}. \tag{7.26}$$

Step k: The derivative of $e_{i,k}$ is found by using (7.8), (7.11), and (7.12): $\dot{e}_{i,k} = e_{i,k+1} + s_{i,k+1} + \hat{\epsilon}_{i,k}^T q_{i,k}\left(\hat{\bar{x}}_{i,k}\right) + g_{i,k}\Pi_{i,1} + \hat{\mathcal{D}}_{i,k}(t) + w_{i,k+1} - \dot{\bar{w}}_{i,k}$, where $k = 2, \ldots, n-1$.

For this step, the Lyapunov function is $V_{i,k} = e_{i,k}^2/2 + \tilde{c}_{i,k}^2/2 + \tilde{\epsilon}_{i,k}^T\tilde{\epsilon}_{i,k}/2\eta_{i,k}$, where $\eta_{i,k}$ is a positive constant.

An RBF NN is used again for approximation:

$$\begin{aligned}\bar{S}_{i,k}\left(\bar{\delta}_{i,k}\right) = & \hat{\epsilon}_{i,k}^T q_{i,k}\left(\hat{\bar{x}}_{i,k}\right) + \hat{\mathcal{D}}_{i,k} \\ = & W_{i,k}^{*T}q_{i,k}\left(\bar{\delta}_{i,k}\right) + \wp_{i,k}\left(\bar{\delta}_{i,k}\right) \\ \left|\wp_{i,k}\left(\bar{\delta}_{i,k}\right)\right| \leq & \varepsilon_{i,k}.\end{aligned} \tag{7.27}$$

The virtual control law and adaptive law are designed as

$$\begin{aligned}w_{i,k+1} = & r_{i,k}e_{i,k} - \frac{e_{i,k}}{2} - \frac{\hat{\vartheta}_i}{2q_{i,k}^2}e_{i,k}q_{i,k}^T\left(\bar{\delta}_{i,k}\right)q_{i,k}\left(\bar{\delta}_{i,k}\right) \\ & + \frac{w_{i,k} - \bar{w}_{i,k}}{\tau_{i,k}} - g_{i,k}\Pi_{i,1}\end{aligned} \tag{7.28}$$

$$\dot{\hat{\epsilon}}_{i,k} = - h_{i,k}\hat{\epsilon}_{i,k} - \eta_{i,k}\tilde{c}_{i,k}\varphi_{i,k}q_{i,k}\left(\hat{\bar{x}}_{i,k}\right) \tag{7.29}$$

where $\bar{\delta}_{i,k} = [y_r, \hat{\vartheta}_i, \hat{\bar{x}}_{i,k}^T, \hat{\bar{x}}_{j,k}^T, \hat{\mathcal{D}}_{i,k}^T, \hat{\mathcal{D}}_{j,k}^T]^T$, $\varepsilon_{i,k}$ is an unknown positive constant, $h_{i,k}$ and $r_{i,k}$ are design constants.

Utilizing Young's inequality, the derivative of the Lyapunov function satisfies

$$\begin{aligned}\dot{V}_{i,k} \leq & r_{i,k}e_{i,k}^2 + \left(\varphi_{i,k}^2 + \varphi_{i,k}^4\right)\|\Pi_i\|^2 + e_{i,k}\left(e_{i,k+1} + s_{i,k+1}\right) \\ & + \frac{h_{i,k}}{\eta_{i,k}}\tilde{\epsilon}_{i,k}^T\hat{\epsilon}_{i,k} + \left(\frac{3 - 2\varphi_{i,k}}{2}\right)\tilde{c}_{i,k}^2 + \tilde{c}_{i,k}\dot{\mathcal{D}}_{i,k} \\ & + \frac{\tilde{\vartheta}_i}{2q_{i,k}^2}e_{i,k}^2 q_{i,k}^T\left(\bar{\delta}_{i,k}\right)q_{i,k}\left(\bar{\delta}_{i,k}\right) + \iota_{i,k}\end{aligned} \tag{7.30}$$

where $\iota_{i,k} = \frac{q_{i,k}^2}{2} + \frac{\varepsilon_{i,k}^2}{2} + \frac{\varphi_{i,k}^2\rho_{0,k}^2}{2}$, $q_{i,k} > 0$.

A low-pass filter is introduced similarly to the first step:

$$\begin{aligned}\bar{w}_{i,k+1}(0) &= w_{i,k+1}(0) \\ \tau_{i,k+1}\dot{\bar{w}}_{i,k+1} + \bar{w}_{i,k+1} &= w_{i,k+1}.\end{aligned} \tag{7.31}$$

Step n: The derivative of $e_{i,n}$ is given by $\dot{e}_{i,n} = u_i + \hat{\epsilon}_{i,n}^T q_{i,n}\left(\hat{\bar{x}}_{i,n}\right) + \hat{\mathcal{D}}_{i,n}(t) + g_{i,n}\Pi_{i,1} - \dot{\bar{w}}_{i,n}$, based on (7.8), (7.11), and (7.12).

The Lyapunov function is $V_{i,n} = e_{i,n}^2/2 + \tilde{c}_{i,n}^2/2 + \tilde{\epsilon}_{i,n}^T\tilde{\epsilon}_{i,n}/2\eta_{i,n}$ with positive constant $\eta_{i,n}$.

The RBF NN approximation is

$$\begin{aligned}\bar{S}_{i,n}\left(\bar{\delta}_{i,n}\right) =& \hat{\epsilon}_{i,n}^T q_{i,n}\left(\hat{\bar{x}}_{i,n}\right) + \hat{\mathcal{D}}_{i,n} \\ =& W_{i,n}^{*T}q_{i,n}\left(\bar{\delta}_{i,n}\right) + \wp_{i,n}\left(\bar{\delta}_{i,n}\right) \\ \left|\wp_{i,n}\left(\bar{\delta}_{i,n}\right)\right| \leq & \varepsilon_{i,n}.\end{aligned} \tag{7.32}$$

The adaptive event-triggered consensus tracking controller is constructed as

$$\alpha_i(t) = w_{i,n+1} - \bar{m}_i \tanh\left(\frac{e_{i,n}\bar{m}_i}{\aleph_i}\right) \tag{7.33}$$

where $w_{i,n+1} = \frac{w_{i,n} - \bar{w}_{i,n}}{\tau_{i,n}} + r_{i,n}e_{i,n} - g_{i,n}\Pi_{i,1} - \frac{\hat{\vartheta}_i}{2q_{i,n}^2}e_{i,n}q_{i,n}^T\left(\bar{\delta}_{i,n}\right)q_{i,n}\left(\bar{\delta}_{i,n}\right)$-$\frac{e_{i,n}}{2}$, $\bar{\delta}_{i,n} = [y_r, \hat{\vartheta}_i, \hat{\bar{x}}_{i,n}^T, \hat{\bar{x}}_{j,n}^T, \hat{\mathcal{D}}_{i,n}^T, \hat{\mathcal{D}}_{j,n}^T]^T$.

The adaptive parameters are constructed as

$$\dot{\hat{\epsilon}}_{i,n} = - h_{i,n}\hat{\epsilon}_{i,n} - \eta_{i,n}\tilde{c}_{i,n}\varphi_{i,n}q_{i,n}\left(\hat{\bar{x}}_{i,n}\right) \tag{7.34}$$

$$\dot{\hat{\vartheta}}_i = -\gamma_i \hat{\vartheta}_i + \sum_{k=1}^{n} \frac{\mathcal{D}_i}{2q_{i,k}^2} e_{i,k}^2 q_{i,k}^T \left(\bar{\delta}_{i,k}\right) q_{i,k} \left(\bar{\delta}_{i,k}\right) \tag{7.35}$$

where $\varepsilon_{i,n}$ is an unknown positive constant, $h_{i,n}$, $r_{i,n}$ and γ_i are design constants.

The event-triggering condition is defined as:

$$u_i(t) = \alpha_i\left(t_\beta\right), \ \forall \mathrm{t} \in \left[t_\beta, t_{\beta+1}\right) \tag{7.36}$$

$$t_{\beta+1} = \inf\left\{\mathrm{t} \in R \| \phi_i(t) | \geq m_i\right\}, \mathrm{t}_1 = 0 \tag{7.37}$$

where the measurement triggered error $\phi_i(t) = \alpha_i(t) - u_i(t)$, and t_β denotes the update time of the controller, $\beta \in Z$. $\bar{m}_i$, m_i and $\aleph_i$ are positive parameters, $\bar{m}_i > m_i$. The control signal $u_i\left(t_{\beta+1}\right)$ will be entered into the system when (7.38) is triggered, while during the time $\mathrm{t} \in \left[t_\beta, t_{\beta+1}\right)$, i.e. $|\alpha_i(t) - u_i(t)| \leq m_i$, the controller holds as a constant $\alpha_i\left(t_\beta\right)$. For a continuous time-varying function $\bar{o}_i(t)$ which satisfies $\bar{o}_i\left(t_\beta\right) = 0$ and $\bar{o}_i\left(t_{\beta+1}\right) = \pm 1$ with $|\bar{o}_i(t)| \leq 1$, one has $\alpha_i(t) = u_i(t) + \bar{o}_i(t) m_i$.

Based on Lemma 1.2, the inequality $-\bar{o}_i(t) m_i e_{i,n} - \bar{m}_i e_{i,n} \tanh\left(\frac{e_{i,n}\bar{m}_i}{\aleph_i}\right) \leq 0.2875\aleph_i$ holds, with this in mind, differentiate $V_{i,n}$ based on Young's inequality and (7.33)-(7.35), one gets

$$\begin{aligned}
\dot{V}_{i,n} \leq & r_{i,n} e_{i,n}^2 + \frac{\tilde{\vartheta}_i}{2q_{i,n}^2} e_{i,n}^2 q_{i,n}^T(\bar{\delta}_{i,n}) q_{i,n}(\bar{\delta}_{i,n}) - \bar{m}_i e_{i,n} \tanh\left(\frac{e_{i,n}\bar{m}_i}{\aleph_i}\right) \\
& + \tilde{c}_{i,n}\dot{\mathcal{D}}_{i,n} - \bar{o}_i(t) m_i e_{i,n} + \frac{q_{i,n}^2}{2} + \frac{\varepsilon_{i,n}^2}{2} + \frac{3}{2}\tilde{c}_{i,n}^2 - \varphi_{i,n}\tilde{c}_{i,n}^2 + \frac{\varphi_{i,n}^2 \rho_{0,n}^2}{2} \\
& + \frac{h_{i,n}}{\eta_{i,n}} \tilde{\epsilon}_{i,n}^T \hat{\epsilon}_{i,n} + \left(\varphi_{i,n}^2 + \varphi_{i,n}^4\right) \|\Pi_i\|^2 \\
\leq & r_{i,n} e_{i,n}^2 + \left(\frac{3 - 2\varphi_{i,n}}{2}\right) \tilde{c}_{i,n}^2 + \tilde{c}_{i,n}\dot{\mathcal{D}}_{i,n} + \frac{\tilde{\vartheta}_i}{2q_{i,n}^2} e_{i,n}^2 q_{i,n}^T\left(\bar{\delta}_{i,n}\right) q_{i,n}\left(\bar{\delta}_{i,n}\right) \\
& + \frac{h_{i,n}}{\eta_{i,n}} \tilde{\epsilon}_{i,n}^T \hat{\epsilon}_{i,n} + \left(\varphi_{i,n}^2 + \varphi_{i,n}^4\right) \|\Pi_i\|^2 + \iota_{i,n}
\end{aligned} \tag{7.38}$$

where $\iota_{i,n} = \frac{q_{i,n}^2}{2} + \frac{\varepsilon_{i,n}^2}{2} + \frac{\varphi_{i,n}^2 \rho_{0,n}^2}{2} + 0.2785\aleph_i$, $q_{i,n} > 0$.

Remark 7.1 In contrast to the consensus control approaches detailed in [5, 8, 9], the work in this chapter simultaneously addresses unmeasured states and disturbances, as well as the problem of limited resources. This is accomplished by utilizing an observer method and an event-triggered mechanism, which provides the proposed approach with the capability to handle more practical industrial control scenarios that require low resource consumption.

Remark 7.2 Traditionally, acquiring the actual control input (7.34 necessitates the repeated time derivative of the virtual control law (7.24, as seen in methods like [19], which leads to a significant increase in complexity and computational workload.

This problem, often termed "complexity explosion," becomes more extreme when dealing with high-order systems that have many agents, unmeasurable states, and external disturbances, rendering some traditional control methods difficult to apply in industrial settings. To overcome this limitation, the filters specified in (7.27 and (7.32 are introduced to eliminate the "repeated derivation" issue, and a novel, simpler controller design approach is subsequently proposed using the backstepping technique. Notably different from the approach in [30], the filter assumes a more critical role and its effectiveness is more pronounced in this chapter; this is attributed to the consideration of practical factors such as unmeasurable states and dynamic external disturbances.

7.3 Stability Analysis

Following the controller design, this section establishes the stability of the system. The main result is presented in the following theorem.

Theorem 7.1 *For the nonlinear multiagent system (7.1 with the event-triggered controller (7.34, if the initial condition satisfies $V(0) \leq \mu$, then the system error signals $e_{i,k}$, $\tilde{c}_{i,k}$, $\tilde{\epsilon}_{i,k}$, $\tilde{\vartheta}_i$, $s_{i,k}$, and $\Pi_{i,k}$ are uniformly bounded. Furthermore, the consensus tracking errors between the follower outputs and the leader's reference signal can be reduced to an arbitrarily small range.*

Proof. : A comprehensive Lyapunov function for the entire multiagent system (7.1 is chosen as:

$$V = \sum_{i=1}^{N}\sum_{k=1}^{n} V_{i,k} + \sum_{i=1}^{N}\sum_{k=1}^{n-1} \frac{s_{i,k+1}^2}{2}. \tag{7.39}$$

By applying Young's inequality, we have

$$e_{i,1}\left(e_{i,2}+s_{i,2}\right) \leq e_{i,1}^2 + \frac{1}{2}e_{i,2}^2 + \frac{1}{2}s_{i,2}^2 \tag{7.40}$$

$$\tilde{\epsilon}_{i,k}^T\hat{\epsilon}_{i,k} \leq -\frac{1}{2}\tilde{\epsilon}_{i,k}^T\tilde{\epsilon}_{i,k} + \frac{1}{2}\epsilon_{i,k}^{*T}\epsilon_{i,k}^* \tag{7.41}$$

$$\tilde{\vartheta}_i\hat{\vartheta}_i \leq \frac{1}{2}\left(\vartheta_i^*\right)^2 - \frac{1}{2}\tilde{\vartheta}_i^2. \tag{7.42}$$

We select the parameters as $r_{i,1} = -(d_i+b_i) + r_{i,1}^*$, $r_{i,2} = -\frac{d_i+b_i}{2} - 1 + r_{i,2}^*$, $r_{i,k} = -\frac{3}{2} + r_{i,k}^*$, $r_{i,n} = -\frac{1}{2} + r_{i,n}^*$ with $i = 1, \ldots, N$, $k = 3, \ldots, n-1$.

When $\left\|\dot{\mathcal{D}}_{i,k}\right\| \leq \varphi_{i,k}^*\left\|\tilde{c}_{i,k}\right\|$, we can show that

$$\dot{V} \leq \sum_{i=1}^{N}\frac{d_i+b_i}{2}s_{i,2}^2 + \sum_{i=1}^{N}\sum_{k=2}^{n-1}\frac{s_{i,k+1}^2}{2} + \sum_{i=1}^{N}\sum_{k=1}^{n-1}s_{i,k+1}\dot{s}_{i,k+1} + \sum_{i=1}^{N}\sum_{k=1}^{n}r_{i,k}^*e_{i,k}^2$$

$$+2\sum_{i=1}^{N}\sum_{k=1}^{n}\varphi_{i,k}^{*}\tilde{c}_{i,k}^{2}+\frac{1}{2}\sum_{i=1}^{N}\sum_{k=1}^{n}\frac{h_{i,k}}{\eta_{i,k}}\epsilon_{i,k}^{*T}\epsilon_{i,k}^{*}-\frac{1}{2}\sum_{i=1}^{N}\sum_{k=1}^{n}\frac{h_{i,k}}{\eta_{i,k}}\tilde{\epsilon}_{i,k}^{T}\tilde{\epsilon}_{i,k}$$
$$-\sum_{i=1}^{N}\Pi_i^T\left(\left(H_i-\left(1+\sum_{k=1}^{n}\left(\varphi_{i,k}^2+\varphi_{i,k}^4\right)\right)I_n\right)\otimes I_m\right)\Pi_i \tag{7.43}$$
$$+\frac{1}{2}\sum_{i=1}^{N}\frac{\gamma_i}{\mathcal{D}_i}\left(\vartheta_i^*\right)^2-\frac{1}{2}\sum_{i=1}^{N}\frac{\gamma_i}{\mathcal{D}_i}\tilde{\vartheta}_i^2+\sum_{i=1}^{N}\sum_{k=1}^{n}\iota_{i,k}$$

where $\varphi_{i,k}^* = (3-2\varphi_{i,k})/2$, $r_{i,k}^*$ and $\varphi_{i,k}^*$ are unknown constants for stability analysis, $i = 1, \ldots, M, k = 1, \ldots, n$.

Let $q(\bullet)$ denote the eigenvalue of a matrix. We choose the matrix H_i to satisfy $q_{\min}\left[H_i - \left(1+\sum_{k=1}^{n}\left(\varphi_{i,k}^2+\varphi_{i,k}^4\right)\right)I_n\right] = \vartheta_i/2q_{\max}(P_i)$, where ϑ_i is a positive constant.

Take the time derivative of $s_{i,k+1}$ and $s_{i,2}$, one gets

$$\dot{s}_{i,k+1} = -\frac{s_{i,k+1}}{\tau_{i,k+1}} + \Gamma_{i,k+1} \tag{7.44}$$
$$\dot{s}_{i,2} = -\frac{s_{i,2}}{\tau_{i,2}} + \Gamma_{i,2} \tag{7.45}$$

where $\Gamma_{i,2} = -\frac{r_{i,1}\dot{e}_{i,1}}{d_i+b_i} + \frac{\hat{\vartheta}_i}{2q_{i,1}^2(d_i+b_i)}\dot{e}_{i,1}q_{i,1}^T\left(\bar{\delta}_{i,1}\right)q_{i,1}\left(\bar{\delta}_{i,1}\right) + \frac{\dot{e}_{i,1}}{2(d_i+b_i)} + \frac{\hat{\vartheta}_i}{q_{i,1}^2(d_i+b_i)} \times e_{i,1}q_{i,1}^T\left(\bar{\delta}_{i,1}\right)\dot{q}_{i,1}\left(\bar{\delta}_{i,1}\right)$ and $\Gamma_{i,k+1} = g_{i,k}\dot{\Pi}_{i,1} + \frac{\hat{\vartheta}_i}{2q_{i,k}^2}\dot{e}_{i,k}q_{i,k}^T\left(\bar{\delta}_{i,k}\right)q_{i,k}\left(\bar{\delta}_{i,k}\right) + \frac{\dot{s}_{i,k}}{\tau_{i,k}} - r_{i,k}\dot{e}_{i,k} + \frac{\dot{e}_{i,k}}{2} + \frac{\hat{\vartheta}_i}{q_{i,k}^2}e_{i,k}q_{i,k}^T\left(\bar{\delta}_{i,k}\right)\dot{q}_{i,k}\left(\bar{\delta}_{i,k}\right)$.

Following the approach in [7], we consider the sets

$$\mathbb{C}_{i,k} = \left\{\sum_{i=1}^{N}\sum_{k=1}^{n}\left(e_{i,k}^2+\tilde{c}_{i,k}^2+\frac{1}{\eta_{i,k}}\tilde{\epsilon}_{i,k}^T\tilde{\epsilon}_{i,k}\right)+\sum_{i=1}^{N}\frac{1}{\mathcal{D}_i}\tilde{\vartheta}_i^2 + \sum_{i=1}^{N}\Pi_i^T(P_i\otimes I_m)\Pi_i+\sum_{i=1}^{N}\sum_{k=1}^{n-1}s_{i,k+1}^2\le 2\mu\right\},$$

where $\mathbb{C}_{i,k}$ is compact in $R^{\dim(\mathbb{C}_{i,k})}$, there exists a positive constant $D_{i,k+1}$ such that the inequality $\left\|\Gamma_{i,k+1}\right\| \le D_{i,k+1}$ holds.

From Young's inequality, for constant $\Xi > 0$, one gets

$$\Gamma_{i,k+1}s_{i,k+1} \le \frac{\Xi}{2} + \frac{D_{i,k+1}^2 s_{i,k+1}^2}{2\Xi} \tag{7.46}$$

where $i = 1, \ldots, N,\ k = 1, \ldots, n-1$.

Choose the parameters as $\frac{1}{\tau_{i,2}} = \frac{d_i+b_i}{2} + \frac{D_{i,2}^2}{2\Xi} + \tau_{i,2}^*$ and $\frac{1}{\tau_{i,k+1}} = \frac{1}{2} + \frac{D_{i,k+1}^2}{2\Xi} + \tau_{i,k+1}^*$, where $\tau_{i,k+1}^*$ is an unknown positive constant, $k = 2, \ldots, n-1$.

Then, set

$$\kappa = \min\left\{-2r_{i,k}^*,\ -4\varphi_{i,k}^*,\ h_{i,k},\ \vartheta_i,\ \gamma_i,\ 2\tau_{i,l}^*\right\}$$
$$l = 2, ..., n,\ k = 1, ..., n. \tag{7.47}$$

To get satisfactory tracking control performance and guarantee the stability of the overall multiagent system, the value of parameters should be appropriately chosen and the following rules should be observed: $-r_{i,k}^* > 0,\ -\varphi_{i,k}^* > 0,\ h_{i,k} > 0,\ \vartheta_i > 0,\ \gamma_i > 0$ and $\tau_{i,l}^* > 0$.

This leads to the final inequality:

$$\dot{V} \leq -\kappa V + C \tag{7.48}$$

where $C = \sum_{i=1}^N \sum_{k=1}^n \iota_{i,k} + \sum_{i=1}^N \sum_{k=1}^{n-1} \frac{\Xi}{2} + \sum_{i=1}^N \frac{\gamma_i}{2D_i} \left(\vartheta_i^*\right)^2 + \frac{1}{2} \sum_{i=1}^N \sum_{k=1}^n \frac{h_{i,k}}{\eta_{i,k}} \epsilon_{i,k}^{*T} \epsilon_{i,k}^*$.

When set $\kappa > C/\mu$, we get $\dot{V} < 0$ on $V = \mu$. Further, if $V \leq \mu$ at time $t = 0$, then $V \leq \mu$ for all time $t > 0$, which illustrates the error signals $e_{i,k}$, $\tilde{c}_{i,k}$, $\tilde{\epsilon}_{i,k}$, $\tilde{\vartheta}_i$, $s_{i,k}$, and $\Pi_{i,k}$ are uniformly bounded.

We can easily get the following one:

$$\frac{1}{2}\|e_1\|^2 \leq V(t) \leq e^{-\kappa t} V(0) + \frac{C}{\kappa}\left(1 - e^{-\kappa t}\right) \tag{7.49}$$

where $e_1 = \left[e_{1,1}^T, e_{2,1}^T, \ldots, e_{M,1}^T\right]^T$. Then, we have

$$\|e_1\|^2 \leq 2e^{-\kappa t} V(0) + \frac{2C}{\kappa}\left(1 - e^{-\kappa t}\right). \tag{7.50}$$

Therefore, all the consensus tracking errors will converge to compact set $\Im = \{e_1 \mid \|e_1\| \leq \sqrt{2C/\kappa}\}$ as time increases, i.e., all the tracking errors between agents are adjustable and can be narrow down to arbitrarily small range by increasing the value of κ.

From (7.34), we can obtain

$$\dot{\alpha}_i(t) = \dot{w}_{i,n+1} - \frac{\bar{m}_i \dot{e}_{i,n}}{\cosh^2\left(\frac{e_{i,n}\bar{m}_i}{\aleph_i}\right)}. \tag{7.51}$$

According to [23], we have

$$\begin{aligned}\frac{d}{dt}|\phi_i(t)| &= \frac{d}{dt}(\phi_i(t) \times \phi_i(t))^{\frac{1}{2}} \\ &= \mathrm{sign}(\phi_i(t))\, \dot{\phi}_i(t) \leq |\dot{\alpha}_i(t)|.\end{aligned} \tag{7.52}$$

Finally, to preclude Zeno behavior, we analyze the inter-execution time. The derivative of the controller output is bounded, i.e., $|\dot{\alpha}_i(t)| \leq \ell$ for some constant $\ell > 0$. Since the triggering error $\phi_i(t)$ goes from 0 to m_i between events, the time interval t^* between triggers must be at least $t^* \geq \frac{m_i}{\ell}$. This positive lower bound guarantees that Zeno behavior is avoided.

7.4 Simulation Studies

To validate the performance of the developed control strategy, this section presents a simulation study. The system under consideration includes one leader and four follower agents. The dynamics of each agent are given by

$$
\begin{aligned}
\dot{x}_{i,1} &= x_{i,2} + f_{i,1}\left(\bar{x}_{i,1}(t)\right) + \mathcal{D}_{i,1}(t) \\
\dot{x}_{i,2} &= u_i + f_{i,2}\left(\bar{x}_{i,2}(t)\right) + \mathcal{D}_{i,2}(t) \\
y_i(t) &= x_{i,1}(t)
\end{aligned} \tag{7.53}
$$

where the nonlinear functions are $f_{i,1}\left(\bar{x}_{i,1}(t)\right) = -0.8x_{i,1}e^{-0.5x_{i,2}^2}$, $f_{i,2}\left(\bar{x}_{i,2}(t)\right) = 1.5x_{i,1}^2\cos\left(x_{i,2}\right)$, and the disturbances are $\mathcal{D}_{i,1}(t) = 0.4x_{i,1}\cos\left(x_{i,2}\right)\cos(t)$ and $\mathcal{D}_{i,2}(t) = 0.2x_{i,2}\sin\left(x_{i,1}\right)\cos(t)$, $i = 1, \ldots, 4$.

The communication topology is shown as Fig. 7.1 and the leader's output signal is chosen as $y_r = 0.8\sin(4t)$, The connection matrix between leader and followers is represented by $D = diag\{0, 1, 0, 0\}$. The adjacent matrix and the Laplacian matrix are described as

$$
A = \begin{bmatrix} 0 & 1 & 0 & 0 \\ 0 & 0 & 0 & 0 \\ 0 & 1 & 0 & 0 \\ 1 & 0 & 0 & 0 \end{bmatrix}, L = \begin{bmatrix} 1 & -1 & 0 & 0 \\ 0 & 0 & 0 & 0 \\ 0 & -1 & 1 & 0 \\ -1 & 0 & 0 & 1 \end{bmatrix}.
$$

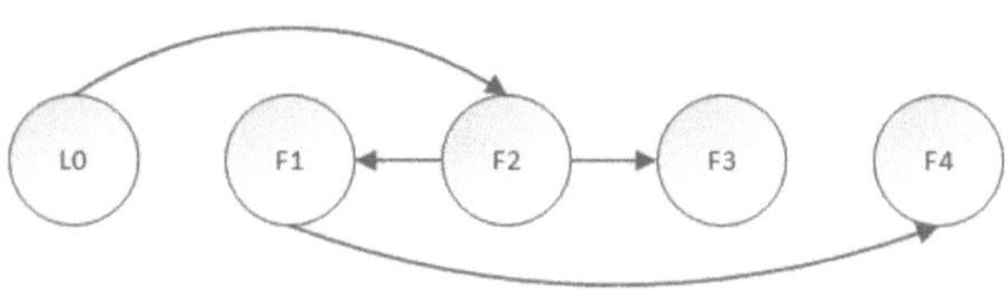

Fig. 7.1 Communication topology

The initial states for the followers and their respective observers were chosen for the simulation. The design parameters were carefully selected according to the stability analysis rules: $\bar{x}_{1,2}(0) = [0.7,\ 0]^T, \bar{x}_{2,2}(0) = [-0.8,\ 0]^T, \bar{x}_{3,2}(0) = [0.1,\ 0]^T$ and $\bar{x}_{4,2}(0) = [0.2,\ 0]^T$. $\hat{\bar{x}}_{1,2}(0) = [0.3,\ 1.7]^T$, $\hat{\bar{x}}_{2,2}(0) = [-0.5,\ 1.7]^T$, $\hat{\bar{x}}_{3,2}(0) = [0,\ \text{-}4]^T$ and $\hat{\bar{x}}_{4,2}(0) = [0, -4]^T$. Considering the parameter choosing rules stated in stability analyse, the numerical value of the design parameters are carefully chosen as $m_i = 3, \bar{m}_i = 4, \aleph_i = 5.4, \tau_i = 0.005, h_{i,1} = h_{i,2} = 50, r_{i,1} = -95, r_{i,2} = -100, q_{i,1} = q_{i,2} = 100, g_{i,1} = 350, g_{i,2} = 0.5, \eta_{i,1} = \eta_{i,2} = 0.01, \gamma_i = 120$ and $\mathcal{D}_i = 24$.

The simulation results demonstrate the controller's effectiveness. Figure 7.2 shows that all followers rapidly converge to track the leader's trajectory, achieving the adaptive consensus objective. The performance of the state observers is highlighted in Figs. 7.3 and 7.4, which show excellent estimation of the states. In Fig. 7.5, the tracking errors between the followers and the leader are shown to decrease quickly, converging to a small bounded region around zero. Finally, Fig. 7.6 displays the event-triggering instants for the controllers, with the total number of triggers being 395, 304, 403, and 486 for each agent, respectively, demonstrating efficient resource utilization.

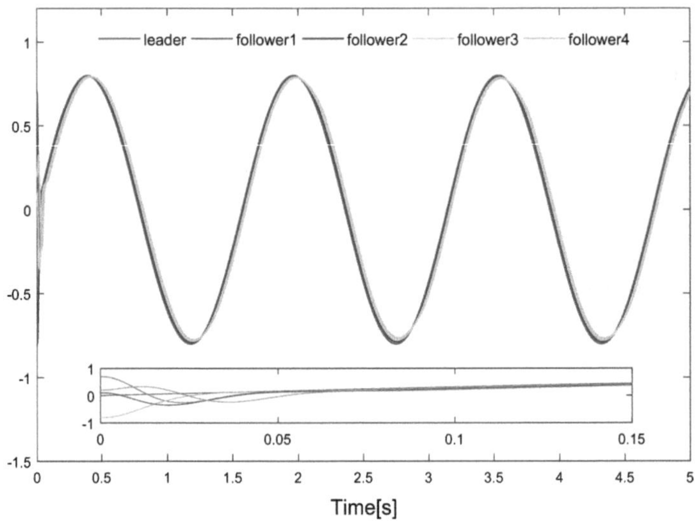

Fig. 7.2 Consensus tracking performance

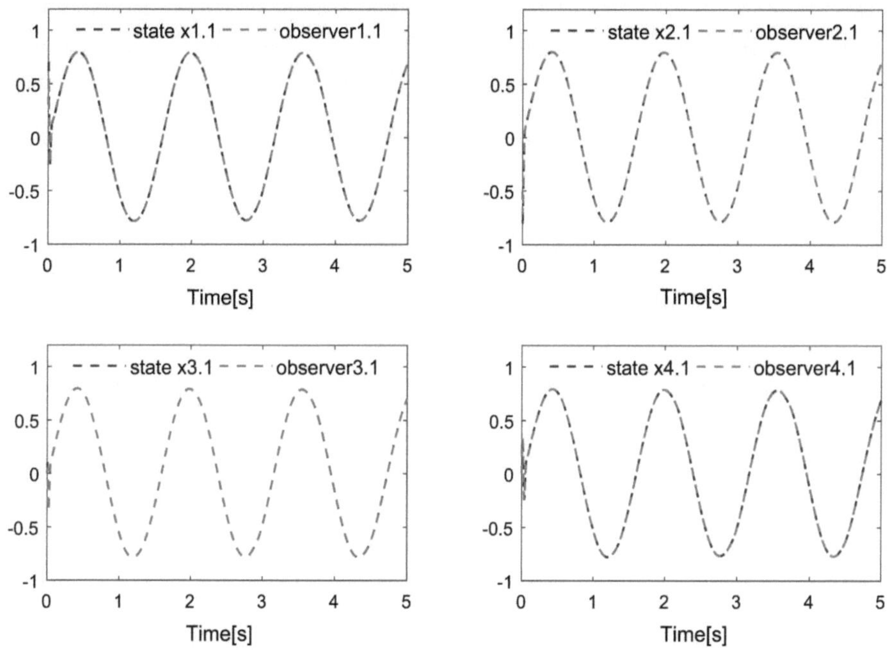

Fig. 7.3 First-order states and state observers

7.5 Conclusion

7.5.1 Contribution

This chapter investigated the problem of observer-based event-triggered consensus tracking for high-order nonlinear multiagent systems. A key contribution is the development of a synergistic control framework that cohesively integrates state observers, disturbance observers, and low-pass filters. To contend with the unmeasurable states and unknown disturbances, state observers and disturbance observers were developed to operate in parallel, allowing for robust online estimation and compensation that enhances the controller's practical viability.

To circumvent the "complexity explosion" problem inherent in traditional backstepping methods, low-pass filters were systematically employed in each design step. Furthermore, an adaptive event-triggered consensus tracking scheme was formulated to conserve limited communication and computational resources. The stability analysis guarantees that all signals in the closed-loop system are uniformly bounded, and the viability and effectiveness of the proposed tracking approach were confirmed through a simulation example.

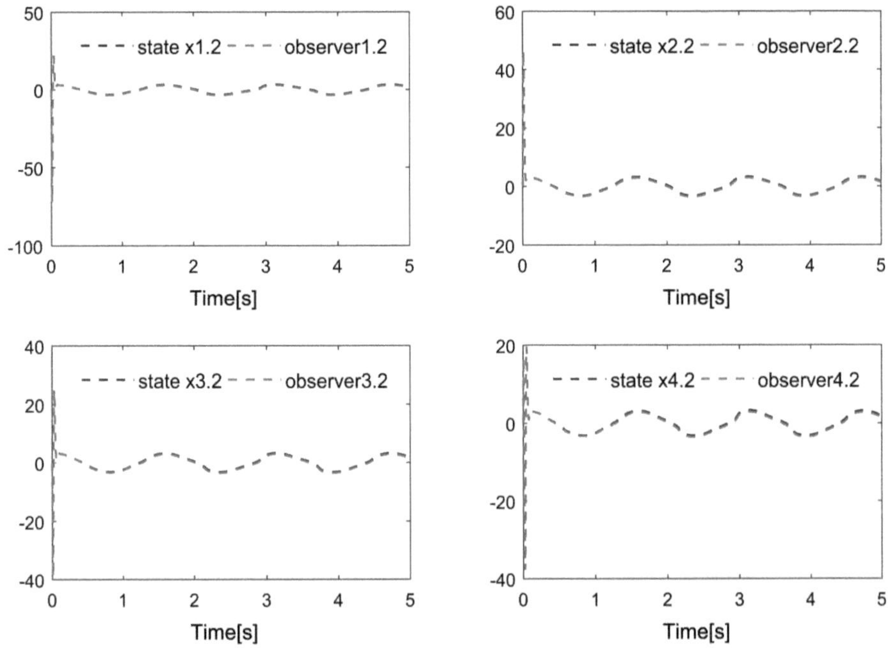

Fig. 7.4 Second-order states and state observers

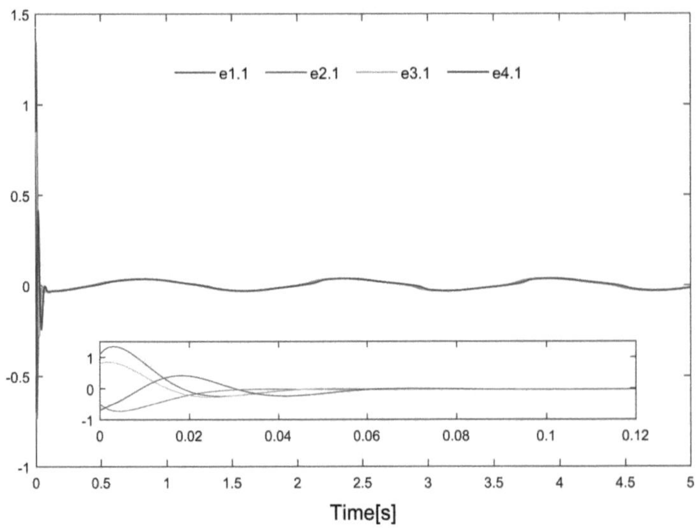

Fig. 7.5 Tracking errors

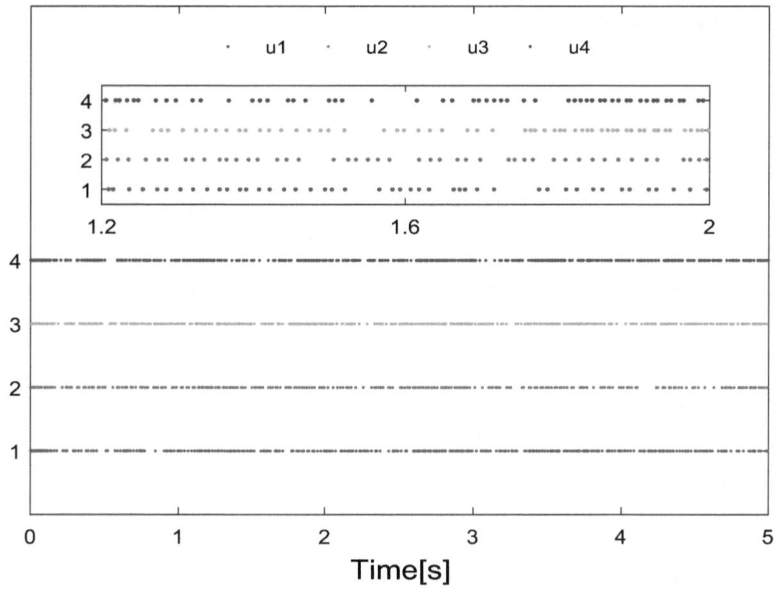

Fig. 7.6 Triggering event

7.5.2 Limitations of the Current Work

Despite the positive results, this study has several limitations:

Assumption Dependency: The stability proof relies on the assumption that external disturbances are bounded. This may not hold in scenarios with more complex, state-dependent, or unbounded disturbances. Additionally, the leader's reference signal is assumed to be smooth and bounded, which can be restrictive.

Filtering Errors: While low-pass filters effectively solve the "complexity explosion" issue, they introduce filtering errors that can affect tracking precision. The overall system performance is sensitive to the selection of the filter time constants.

Static Event-Triggering: The chapter employs an event-triggered mechanism with a static, predefined threshold. This type of trigger may not be optimal for all operating conditions, potentially leading to either unnecessary transmissions or delayed control updates when system dynamics change.

Computational Load: The use of RBF NNs for function approximation, combined with parallel state and disturbance observers, increases the overall computational burden, which could be a concern for resource-constrained agents.

7.5.3 Future Directions

To address these limitations, future research could be pursued in the following directions:

Robustness Enhancements: Future work should focus on relaxing the assumption of bounded disturbances by developing observers and control laws capable of handling unbounded or state-dependent uncertainties.

Advanced Filtering Techniques: To improve tracking accuracy, more advanced command filtering techniques with explicit error compensation mechanisms could be investigated to mitigate the impact of filtering errors.

Adaptive Event-Triggering Schemes: Exploring dynamic or adaptive event-triggering thresholds could provide a better balance between communication efficiency and control performance across a wider range of operating conditions.

Expanded System Complexities: As originally proposed, future research will focus on extending these methods to output-feedback control for nonaffine multi-agent systems that are also subject to unknown states and disturbances. Other practical considerations such as switching topologies, actuator faults, unknown hysteresis, and anti-consensus techniques will also be explored.

Experimental Validation: Testing the proposed control strategy on physical multi-agent platforms, such as mobile robots or UAVs, would be essential to evaluate its performance and robustness under real-world conditions and communication constraints.

References

1. Nakao M, Ohnishi K, Miyachi K (1987) A robust decentralized joint control based on interference estimation. In: Proceedings of the IEEE international conference on robotics and automation, pp 326–331
2. Li Z, Liu X, Lin P, Ren W (2011) Consensus of linear multi-agent systems with reduced-order observer-based protocols. Syst Control Lett 60(7):510–516
3. Henningsson T, Johannesson E, Cervin A (2008) Sporadic event-based control of first-order linear stochastic systems. Automatica 44(11):2890–2895
4. Zhu W, Pu H, Wang D, Li H (2017) Event-based consensus of second order multi-agent systems with discrete time. Automatica 79:78–83
5. Chen CLP, Wen G, Liu Y, Liu Z (2016) Observer-based adaptive backstepping consensus tracking control for high-order nonlinear semi-strict-feedback multiagent systems. IEEE Trans Cybern 46(7):1591–1601
6. Zhang Y, Sun J, Liang H, Li H (2020) Event-triggered adaptive tracking control for multiagent systems with unknown disturbances. IEEE Trans Cybern 50(3):890–901

7. Yoo SJ (2013) Distributed consensus tracking for multiple uncertain nonlinear strict-feedback systems under a directed graph. IEEE Trans Neural Netw Learn Syst 24(4):666–672
8. Zou W, Shi P, Xiang Z, Shi Y (2020) Finite-time consensus of second-order switched nonlinear multi-agent systems. IEEE Trans Neural Netw Learn Syst 31(5):1757–1762
9. Zou W, Shi P, Xiang Z, Shi Y (2020) Consensus tracking control of switched stochastic nonlinear multiagent systems via event-triggered strategy. IEEE Trans Neural Netw Learn Syst 31(3):1036–1045

Chapter 8
Optimized Adaptive Finite-Time Consensus Control for Stochastic Nonlinear Multiagent Systems With Non-affine Nonlinear Faults

In the complex and ever-changing technological and engineering fields of the present day, Multi-Agent Systems (MASs) hold tremendous potential in scenarios such as unmanned aerial vehicle (UAV) formation flight and distributed energy management in smart grids [1, 2]. These scenarios demand that MAS achieve efficient and stable cooperation in complex environments. Consensus control is the core of their cooperation. However, practical MASs encounter issues such as the system's inherent nonlinear characteristics, stochastic noise, and non-affine nonlinear faults, which are difficult to address using traditional methods. When designing consensus control strategies, it is essential to account for faults to enhance fault-tolerant capabilities. Therefore, researching consensus control for stochastic nonlinear MAS is of great significance.[3–5].

Unlike asymptotic consensus, finite-time consensus exhibits higher accuracy, stronger robustness, and faster convergence rates [6]. To address the issue of the absence of velocity measurements in nonlinear systems, a novel estimation method was introduced in [7] by employing a finite-time convergent observer, based on the application of the Lyapunov stability theorem. Considering that stochastic disturbances are inevitable in certain practical systems and may lead to instability, further a new stochastically finite-time stability criterion by utilizing the Itô differential equation and traditional finite-time theory is developed to tackle this complex problem in [8].

Event-triggered control enhances communication efficiency by intelligently scheduling data collection and controller updates based on specific triggering criteria, thereby reducing resource usage [9–11]. Furthermore, the study presented in [12] introduced an adaptive fault-tolerant control approach integrated with an ETM. This method not only ensures superior control performance but also effectively conserves communication resources. However, it is important to note that none of the previously mentioned research works adequately address the challenges posed by limitations in convergence speed and resource consumption.

X. Wang et al., *Optimal Adaptive Cooperative Control for Nonlinear Multiagent Systems*,
https://doi.org/10.1007/978-981-95-2964-3_8

Building upon the insights gained from the earlier discussion, this chapter focuses on tackling the challenge of optimized adaptive finite-time consensus control for stochastic nonlinear MASs that are plagued by non-affine nonlinear faults. In a nutshell, the main contributions of this chapter can be outlined in the following three distinct yet interrelated areas.

(1) The reinforcement learning (RL) algorithm is introduced based on the Identifier-Critic-Actor structure, which is grounded on the Optimal Backstepping (OB) technique. Specifically, the identifier is employed to estimate the unknown dynamics of the system, the critic is used to evaluate the system's performance, and the actor is responsible for implementing control actions. Moreover, the update rules for the critic and the actor are derived from the negative gradient of a simple positive function.
(2) During the actual design process of the controller, a Butterworth low-pass filter is utilized to counteract the adverse effects caused by non-affine nonlinear faults, thereby further significantly enhancing the performance of the control system.
(3) Within the framework of finite-time stability theory, by utilizing the Itô differential equation, an optimized adaptive finite-time fault-tolerant control problem is thoroughly investigated. This problem aims to ensure that all state variables of the controlled system achieve semi-globally finite-time probabilistically stable (SGFSP) status and converge to the origin within a finite time with the minimum resource consumption.

8.1 Model Formulation

The stochastic nonlinear system dynamic for followers $i (i = 1, ..., N)$ are modeled as follows:

$$\begin{aligned} dx_{i,k} =&(x_{i,k+1} + f_{i,k}(\bar{x}_{i,k}))dt + \Upsilon_{i,k}(\bar{y})d\omega_i \\ dx_{i,n} =&(u_i + f_{i,n}(\bar{x}_{i,n}) + \theta_i(t - T_0)\lambda_i(\bar{x}_{i,n}, u_i))dt + \Upsilon_{i,n}(\bar{y})d\omega_i \\ y_i =&x_{i,1}, \ 1 \le i \le N, \ 1 \le k \le n-1 \end{aligned} \tag{8.1}$$

where $\bar{x}_{i,k} = [x_{i,1}, x_{i,2}, \ldots, x_{i,k}]^{\mathrm{T}} \in \mathbb{R}^k$ represents the state vector of the i-th agent, $y_i \in \mathbb{R}$ denotes the system output of the i-th agent; and $u_i \in \mathbb{R}$ is the control input of the i-th agent. $\bar{y} = [y_1, ..., y_N]^T$ is a vector composed of the outputs of all N agents. Both $f_{i,k}(\cdot)$ and $\Upsilon_{i,k}(\cdot)$ are unknown smooth nonlinear functions that satisfy the Lipschitz continuity condition. $\lambda_i(\bar{x}_{i,n}, u_i)$ represents the nonlinear fault function of the i-th agent. In addition, $\theta_i(t - T_0)$ describes the time profile of the fault occurring at some unknown time T_0 and it is given by the following piece-wise function: $\theta_i(t - T_0) = \begin{cases} 0, & t < T_0 \\ 1 - e^{-\mu(t-T_0)}, & t \ge T_0 \end{cases}$, where $\mu > 0$ is the evolution rate of the unknown fault. ω_i is an independent r-dimensional standard Wiener process, and its incremental covariance satisfies $E\{d\omega_i \cdot d\omega_i^T\} = \sigma_i(t)\sigma_i^T(t)dt$.

The initial aim of this chapter is centered around constructing an optimized adaptive finite-time consensus control strategy for stochastic nonlinear MASs that are plagued by non-affine nonlinear faults. This strategy is designed to ensure that all signals within the closed-loop systems achieve SGFSP conditions, and that the outputs of each follower agent can precisely track the output of the leader. To achieve this control objective, we define the local synchronization tracking error e_i for each follower agent i as follows:

$$e_i(t) = \sum_{k \in \mathcal{N}_i} a_{i,k}(y_i - y_k) + p_i(y_i - y_r), \tag{8.2}$$

where y_i denotes the output of the follower, and y_r represents the output of the leader.

In the subsequent content, some reasonable assumptions, definitions, and lemmas will be presented to facilitate the following derivations.

Assumption 8.1 [13] In the context of a connected directed graphs $\mathcal{G}$, we assume that there is at least one follower capable of communicating with the leader.

Assumption 8.2 [14] Supposed the unknown nonlinear function $f_{i,n}(\bar{x}_{i,n})$ and fault function $\theta_i(t - T_0)\lambda_i(\bar{x}_{i,n}, u_i)$ as an indivisible term. Define $f_{i,n}(\bar{x}_{i,n}) + \theta_i(t - T_0)\lambda_i(\bar{x}_{i,n}, u_i)$ as the fault function, which satisfies the following inequality:

$$|f_{i,n}(\bar{x}_{i,n}) + \theta_i(t - T_0)\lambda_i(\bar{x}_{i,n}, u_i)| \leq \overline{f_i}(\bar{x}_{i,n}, u_i), \tag{8.3}$$

where $\overline{f_i}(\bar{x}_{i,n}, u_i)$ represents an unknown non-negative function.

Remark 8.1 Assumption 8.1 indicates that all follower agents can communicate with the leader agent, directly or indirectly. Assumption 8.2 gives the upper boundary of the unknown fault functions $\theta_i(t - T_0)\lambda_i(\bar{x}_{i,n}, u_i)$ and unknown functions $f_{i,n}(\bar{x}_{i,n})$. In particular, Assumption 8.2 is the mainstream method to handle the unknown boundary function on non-affine nonlinear faults. For more detail, please refer to [14, 15].

Definition 8.1 [16] Consider the system (8.1) and a differentiable function $V(x) \in \mathbb{C}^2$, the differential operator ℓ is given as

$$\ell V = \frac{\partial V(x)}{\partial x} f(x) + \frac{1}{2}\mathrm{Tr}\{\Omega^T(x)\frac{\partial^2 V(x)}{\partial x^2}\Omega(x)\}, \tag{8.4}$$

where $\mathrm{Tr}\{\cdot\}$ stands for the trace of the matrix.

Definition 8.2 [16] For the stochastic system (8.1) with initial value $x(t_0, \omega) = x_0$, the equilibrium point $x = 0$ of the nonlinear system is said to be semi-global finite-time stable in probability (SGFSP), if there is a positive constant ε and a setting time $T(\varepsilon, x_0, \omega) < \infty$ such that the equilibrium point satisfies $E[\| x(t, \omega) \|] < \varepsilon$ for all $t \geq t_0 + T$.

Lemma 8.1 *[16] Consider the system (8.1), if there is a positive-define function $V(x)$, along with $A > 0,\ 0 < L < 1,\ B > 0$ and k_∞ -functions $\overline{F}$ and $\underline{F}$, such that*

$$\overline{F}(|\ x\ |) \leq V(x) \leq \underline{F}(|\ x\ |)$$
$$\ell V(x) \leq -AV^L(x) + B, t \geq 0$$

then the system (8.1) is SGFSP.

Lemma 8.2 *[15] Let $\wp_i \in \mathbb{R},\ i = 1, ..., N$, and $0 < \iota \leq 1$. The following inequality holds:*

$$\left(\sum_{i=1}^{N} |\wp_i|\right)^\iota \leq \sum_{i=1}^{N} |\wp_i|^\iota \leq N^{1-\iota}\left(\sum_{i=1}^{N} |\wp_i|\right)^\iota.$$

Lemma 8.3 *[15] For the given variables $\Im_1$ and $\Im_2$, and any positive constants $\flat_1$, $\flat_2$, and $\flat_3$, one has*

$$|\Im_1|^{\flat_1}|\Im_2|^{\flat_2} \leq \frac{\flat_1}{\flat_1 + \flat_2}\flat_3|\Im_1|^{\flat_1+\flat_2} + \frac{\flat_2}{\flat_1 + \flat_2}\flat_3^{\frac{-\flat_1}{\flat_2}}|\Im_2|^{\flat_1+\flat_2}.$$

Lemma 8.4 *In this chapter, the terminology of graph theory is utilized to represent the communication topology among $N + 1$ agents containing N followers and one leader. The elements of the adjacency matrix $\underline{A}$ is defined as: if there is communication relationship between ith agent and kth agent, $a_{i,k} = 1$, whereas $a_{i,k} = 0$. For the agent i, $\mathcal{N}_i = \{\varphi_k|(\varphi_i, \varphi_k) \in \underline{E}\}$ represents its neighbors set. The Laplacian matrix of the directed graph satisfies $\mathcal{L} = \mathcal{D} - \underline{A}$, in which $\mathcal{D} = \text{diag}\{d_1, \ldots, d_N\}$ with $d_i = \sum_{\varphi_k \in \mathcal{N}_i} a_{i,k}$ being the degree matrix. The connection matrix among agents is denoted by $P = \text{diag}\{p_1, \ldots, p_N\}$ with $p_i \geq 0$ yielding that the ith follower can receive the information from the leader; otherwise, $p_i = 0$.*

Lemma 8.5 *As stated in [17], the neural networks (NNs) are applied to estimate the unknown continuous nonlinear function $\Psi(M)$: $\mathbb{R}^n \to \mathbb{R}^J$ defined over a compact set $\Omega_M \in \mathbb{R}^n$, which is depicted as $\Psi(M) = \Theta^T\zeta(M)$, where $\zeta(M) = [\zeta_1(M), ..., \zeta_q(M)]^T$ stands for the basis function vector, and $\Theta = [\emptyset_1, ..., \emptyset_q] \in \mathbb{R}^{q \times J}$ denotes the weight with $q > 1$ being the number of neurons. There exists the ideal parameter vectors Θ^*, which is expressed as $\Theta^* := \arg\min_{\Theta \in \mathbb{R}^{q\times J}}\{\sup_{M \in \Omega_M} ||\Psi(M) - \Theta^T\zeta(M)||\}$, in which the function $\Psi(M)$ can be reformulated as $\Psi(M) = \Theta^{*T}\zeta(M) + \Delta(M)$, where $\Delta(M)$ represents the approximation error satisfying $||\Delta(M)|| \leq \kappa$ and κ stands for a positive constant. In addition, the Gaussian function is selected as $\zeta_i(M) = \exp[-\frac{(M-\varrho_i)^T(M-\varrho_i)}{\upsilon_i^2}],\ i = 1, 2, ..., q$, where $\varrho_i = [\varrho_{i1}, \varrho_{i2}, ..., \varrho_{in}]^T$ denotes the center of the receptive field, and υ_i represents the width of Gaussian function.*

8.2 Controller Design

In this section, by resorting to the adaptive backstepping method and the RL strategy, an optimized adaptive consensus algorithm is derived. Then, the SGFSP investigated by the manners of stochastic analysis approach and finite-time stability theory. Before moving further, let us deliver the following coordinate transformation

$$z_{i,1} = e_i \tag{8.5}$$

$$z_{i,s} = x_{i,s} - \hat{\alpha}^*_{i,s-1} \tag{8.6}$$

where $\hat{\alpha}^*_{i,s-1}$ refers to the local optimal virtual control. In what follows, the adaptive backstepping approach containing n steps are used for accomplishing the optimal controller design. The elaborate configuration procedure is described as follows.

Step 1: It follows from (8.1) and (8.5) that

$$\begin{aligned} dz_{i,1} =& \{(d_i + p_i)(x_{i,2} + f_{i,1}(x_{i,1})) - \sum_{s\in\mathcal{N}_i} a_{i,s}(x_{s,2} + f_{s,1}(x_{s,1})) - p_i \dot{y}_r\}dt \\ & - \{(d_i + p_i)\Omega_{i,1}(\bar{y}) - \sum_{s\in\mathcal{N}_i} a_{i,s}\Omega_{s,1}(\bar{y})\}d\omega_i \end{aligned} \tag{8.7}$$

Note that the problems of adaptive controller design and energy consumption optimization are both considered during the finite-time consensus control process for MASs. Regarding the dynamics investigation of MASs in certain practical applications, it is always preferred that some specific performance limitations should be satisfied. In this chapter, thereby, based on the optimal principle, we develop the local performance index function for each agent as

$$\begin{aligned} J^*_{i,1}(z_{i,1}) =& \min_{\alpha_{i,1}\in\Xi(\underline{\Omega}_{i,1})} \int_t^\infty \hbar_{i,1}(z_{i,1}(\xi), \alpha_{i,1}(\xi))d\xi \\ =& \int_t^\infty \hbar_{i,1}(z_{i,1}(\xi), \alpha^*_{i,1}(\xi))d\xi \end{aligned} \tag{8.8}$$

where $\Xi(\underline{\Omega}_{i,1})$ stands for the admissible control and $\underline{\Omega}_{i,1}$ denotes a compact set containing the origin. $\hbar_{i,1}(z_{i,1}, \alpha_{i,1}) = z^2_{i,1}(t) + \alpha^2_{i,1}(z_{i,1})$ represents the cost function.

Considering $x_{i,2}$ as the optimal virtual controller $\alpha^*_{i,1}$, we can obtain the Hamilton-Jacobi-Bellman equation (HJBE) as follows:

$$\begin{aligned}H_{i,1}(z_{i,1}, \alpha^*_{i,1}, \frac{dJ^*_{i,1}(z_{i,1})}{dz_{i,1}}) =& \alpha^{*2}_{i,1} + z^2_{i,1} + \frac{dJ^*_{i,1}(z_{i,1})}{dz_{i,1}} \times \Big\{(d_i + p_i)(f_{i,1}(x_{i,1}) \\ &+ \alpha^*_{i,1}) - \sum_{s\in\mathcal{N}_i} a_{i,s}(x_{s,2} + f_{s,1}(x_{s,1})) - p_i \dot{y}_r\Big\} \\ &+ \frac{1}{2}\frac{d^2 J^*_{i,1}(z_{i,1})}{dz^2_{i,1}} \times \| (d_i + p_i)\Omega_{i,1}(\bar{y}) \\ &- \sum_{s\in\mathcal{N}_i} a_{i,s}\Omega_{s,1}(\bar{y}) \|^2 = 0 \end{aligned} \tag{8.9}$$

It is clear that the optimal virtual controller $\alpha^*_{i,1} = -\frac{d_{i,1}/2+p_{i,1}}{2}\frac{dJ^*_{i,1}(z_{i,1})}{dz_{i,1}}$ can be obtained by solving the equation $\partial H_{i,1}(z_{i,1}, \alpha^*_{i,1}, \frac{dJ^*_{i,1}(z_{i,1})}{dz_{i,1}})/\partial\alpha^*_{i,1} = 0$, and so as to fulfill the control goal, the term $dJ^*_{i,1}(z_{i,1})/dz_{i,1}$ can be decomposed into

$$\frac{dJ^*_{i,1}(z_{i,1})}{dz_{i,1}} = \frac{1}{(d_i + p_i)^2}\Big\{2\rho_{i,1}z^{4L-3}_{i,1} + \frac{z^3_{i,1}}{2\gamma_{i,1}} + 2F_{i,1}(\chi_{i,1}) + J^0_{i,1}(X_{i,1})\Big\}, \tag{8.10}$$

where $\rho_{i,1}$ and $\gamma_{i,1}$ denote positive design parameters, $F_{i,1}(\chi_{i,1}) = (d_i + p_i)f_{i,1} - \sum_{s\in\mathcal{N}_i} a_{i,s}(x_{s,2} + f_{s,1}(x_{s,1})) + (1/2)z^3_{i,1} + (3/4)(d_i + p_i)z_{i,1} - p_i\dot{y}_r + z_{i,1}\| G \|^4$ with $\chi_{i,1} = [x_{i,1}, y_r, x_{s,2}, z_{i,1}, \bar{y}, \dot{y}_r]^T$ being a unknown function and $J^0_{i,1}(X_{i,1}) = (d_i + p_i)^2(dJ^*_{i,1}(z_{i,1})/dz_{i,1}) - 2\rho_{i,1}z^{4L-3}_{i,1} - 2F_{i,1}(\chi_{i,1}) - z^3_{i,1}/2\gamma_{i,1}$ with $X_{i,1} = [x_{i,1}, z_{i,1}]^T$ and $L = [(2\mu - 1)/(2\mu + 1)](\mu \geq 2,\ \mu \in \mathbb{N})$.

Based on (8.10), it is easy to obtain

$$\alpha^*_{i,1} = \frac{1}{d_i + p_i}\Big\{ - \rho_{i,1}z^{4L-3}_{i,1} - \frac{z^3_{i,1}}{4\gamma_{i,1}} - F_{i,1}(\chi_{i,1}) - \frac{J^0_{i,1}(X_{i,1})}{2}\Big\}. \tag{8.11}$$

Considering the challenge in finite-time optimal controller design brought by the unknown functions $F_{i,1}(\chi_{i,1})$ and $J^0_{i,1}(X_{i,1})$, the universal approximation capability of NNs is adopted here to estimate them. Thus, it yields

$$F_{i,1}(\chi_{i,1}) = \Theta^{*T}_{Fi,1}S_{Fi,1}(\chi_{i,1}) + \Lambda_{Fi,1}(\chi_{i,1}) \tag{8.12}$$

$$J^0_{i,1}(X_{i,1}) = \Theta^{*T}_{Ji,1}S_{Ji,1}(X_{i,1}) + \Lambda_{Ji,1}(X_{i,1}) \tag{8.13}$$

where $\Theta^*_{Fi,1}$ and $\Theta^*_{Ji,1}$ stand for the ideal weights, $S_{Fi,1}(\chi_{i,1})$ and $S_{Ji,1}(X_{i,1})$ represent the activation function vectors, and $\Lambda_{Fi,1}(\chi_{i,1})$ and $\Lambda_{Ji,1}(X_{i,1})$ denote NNs approximation errors.

Based on the above analysis, one obtains

$$\begin{aligned}\frac{dJ_{i,1}^*}{dz_{i,1}} =&\frac{1}{(d_i+p_i)^2}\bigg\{2\rho_{i,1}z_{i,1}^{4L-3}+\Theta_{Ji,1}^{*T}S_{Ji,1}(X_{i,1})\\&+\frac{z_{i,1}^3}{2\gamma_{i,1}}+2\Theta_{Fi,1}^{*T}S_{Fi,1}(\chi_{i,1})+\Lambda_{i,1}\bigg\}\end{aligned} \tag{8.14}$$

$$\begin{aligned}\alpha_{i,1}^* =&\frac{1}{d_i+p_i}\bigg\{-\rho_{i,1}z_{i,1}^{4L-3}-\frac{1}{2}\Theta_{Ji,1}^{*T}S_{Ji,1}(X_{i,1})\\&-\frac{z_{i,1}^3}{4\gamma_{i,1}}-\Theta_{Fi,1}^{*T}S_{Fi,1}(\chi_{i,1})-\frac{1}{2}\Lambda_{i,1}\bigg\}\end{aligned} \tag{8.15}$$

with $\Lambda_{i,1}=2\Lambda_{Fi,1}+\Lambda_{Ji,1}$.

Significantly, the optimal virtual controller $\alpha_{i,1}^*$ in (8.15) contains the unknown ideal weights $\Theta_{Fi,1}^*$ and $\Theta_{Ji,1}^*$, so that the controller cannot be directly applied in subsequent analysis. So as to overcome this dilemma, the NN-based RL algorithm is introduced to approximate $\alpha_{i,1}^*$, $F_{i,1}(\chi_{i,1})$ and $J_{i,1}^0(X_{i,1})$ with identifier, critic and actor NNs, respectively. In addition, the identifier NN is employed to approximate the unknown function, the critic NN is applied to evaluate the control performance, and the actor NN is utilized to perform the control behavior, which are designed as

$$\hat{F}_{i,1}(\chi_{i,1}) =\hat{\Theta}_{Fi,1}^T S_{Fi,1}(\chi_{i,1}) \tag{8.16}$$

$$\begin{aligned}\frac{d\hat{J}_{i,1}^*(z_{i,1})}{dz_{i,1}} =&\frac{1}{(d_{i,1}+p_{i,1})^2}\{\frac{z_{i,1}^3}{2\gamma_{i,1}}+2\hat{\Theta}_{Fi,1}^T S_{Fi,1}(\chi_{i,1})\\&+2\rho_{i,1}z_{i,1}^{4L-3}+\hat{\Theta}_{ci,1}^T S_{Ji,1}(X_{i,1})\}\end{aligned} \tag{8.17}$$

$$\begin{aligned}\hat{\alpha}_{i,1}^* =&\frac{1}{d_{i,1}+p_{i,1}}\{-\frac{1}{4\gamma_{i,1}}z_{i,1}^3-\hat{\Theta}_{Fi,1}^T S_{Fi,1}(\chi_{i,1})\\&-\rho_{i,1}z_{i,1}^{4L-3}-\frac{1}{2}\hat{\Theta}_{ai,1}^T S_{Ji,1}(X_{i,1})\}\end{aligned} \tag{8.18}$$

where $\hat{F}_{i,1}$, $d\hat{J}_{i,1}^*/dz_{i,1}$, and $\hat{\alpha}_{i,1}^*$, respectively, represent the estimation of $F_{i,1}$, $dJ_{i,1}^*/dz_{i,1}$, and $\alpha_{i,1}^*$. $\hat{\Theta}_{Fi,1}$, $\hat{\Theta}_{ci,1}$, and $\hat{\Theta}_{ai,1}$, respectively, denote the identifier, critic, and actor NNs weights. The adaptive laws of the identifier, critic, and actor NNs weights are constructed as follows:

$$\dot{\hat{\Theta}}_{Fi,1} =\eta_{i,1}(S_{Fi,1}(\chi_{i,1})z_{i,1}^3-\phi_{i,1}\hat{\Theta}_{Fi,1}) \tag{8.19}$$

$$\dot{\hat{\Theta}}_{ci,1} =-\pi_{ci,1}S_{Ji,1}(X_{i,1})S_{Ji,1}^T(X_{i,1})\hat{\Theta}_{ci,1} \tag{8.20}$$

$$\dot{\hat{\Theta}}_{ai,1} =-S_{Ji,1}(X_{i,1})S_{Ji,1}^T(X_{i,1})(\pi_{ai,1}(\hat{\Theta}_{ai,1}-\hat{\Theta}_{ci,1})+\pi_{ci,1}\hat{\Theta}_{ci,1}) \tag{8.21}$$

where $\eta_{i,1}$ represents a positive definite matrix. $\phi_{i,1}$, $\pi_{ci,1}$, and $\pi_{ai,1}$ denote positive design parameters. For subsequent analysis, $\phi_{i,1}$, $\pi_{ci,1}$, and $\pi_{ai,1}$ should satisfy the following conditions:

$$\pi_{ai,1} > \frac{1}{2}, \ \pi_{ai,1} > \frac{\phi_{i,1}}{2}, \ \pi_{ai,1} > \pi_{ci,1} > \frac{\pi_{ai,1}}{2}. \tag{8.22}$$

It should be pointed out that the above mentioned adaptive laws of the critic and actor NNs weights are designed based on Lyapunov stability analysis and gradient descent algorithm, respectively. The detailed derivation process for these adaptive laws is shown below.

Substituting (8.17) and (8.18) into (8.9), the HJBE can be reformulated as

$$\begin{aligned} H_{i,1}(z_{i,1}, \alpha_{i,1}^{*}, \frac{dJ_{i,1}^{*}}{dz_{i,1}}) =& z_{i,1}^{2} + \frac{1}{(d_i+p_i)^2}\Big\{-\rho_{i,1}z_{i,1}^{4L-3} - \frac{z_{i,1}^{3}}{4\gamma_{i,1}} - \hat{\Theta}_{Fi,1}^{T}S_{Fi,1} \\ & - \frac{1}{2}\hat{\Theta}_{ai,1}^{T}S_{Ji,1}\Big\}^2 + \frac{1}{(d_i+p_i)^2}\Big\{2\rho_{i,1}z_{i,1}^{4L-3} + \frac{z_{i,1}^{3}}{2\gamma_{i,1}} \\ & + 2\hat{\Theta}_{Fi,1}^{T}S_{Fi,1} + \hat{\Theta}_{ci,1}^{T}S_{Ji,1}\Big\} \times \Big\{(d_i+p_i)[f_{i,1} \\ & + \frac{1}{d_i+p_i}(-\frac{z_{i,1}^{3}}{4\gamma_{i,1}} - \hat{\Theta}_{Fi,1}^{T}S_{Fi,1} - \frac{1}{2}\hat{\Theta}_{ai,1}^{T}S_{Ji,1} \\ & - \rho_{i,1}z_{i,1}^{4L-3})] - \sum_{s\in\mathcal{N}_i} a_{i,s}(x_{s,2} + f_{s,1}(x_{s,1})) - p_i\dot{y}_r\Big\} \\ & + \frac{1}{2}\frac{d^2J_{i,1}^{*}}{dz_{i,1}^{2}} \parallel (d_i+p_i)\Omega_{i,1} - \sum_{s\in\mathcal{N}_i} a_{i,s}\Omega_{s,1} \parallel^2 = 0. \end{aligned} \tag{8.23}$$

Define the Bellman residual error $\nabla_{i,1}$ as

$$\begin{aligned} \nabla_{i,1} =& H_{i,1}(z_{i,1}, \hat{\alpha}_{i,1}^{*}, \frac{d\hat{J}_{i,1}^{*}(z_{i,1})}{dz_{i,1}}) - H_{i,1}(z_{i,1}, \alpha_{i,1}^{*}, \frac{dJ_{i,1}^{*}(z_{i,1})}{dz_{i,1}}) \\ =& H_{i,1}(z_{i,1}, \hat{\alpha}_{i,1}^{*}, \frac{d\hat{J}_{i,1}^{*}(z_{i,1})}{dz_{i,1}}) \end{aligned} \tag{8.24}$$

With the above analysis, the local optimal virtual control is expected to satisfy $\nabla_{i,1} = H_{i,1}(z_{i,1}, \hat{\alpha}_{i,1}^{*}, d\hat{J}_{i,1}^{*}(z_{i,1})/dz_{i,1}) \to 0$. Therefore, supposed that the condition $H_{i,1}(z_{i,1}, \hat{\alpha}_{i,1}^{*}, d\hat{J}_{i,1}^{*}(z_{i,1})/dz_{i,1}) = 0$ holds and has the unique solution, one has

$$\frac{\partial H_{i,1}(z_{i,1}, \hat{\alpha}_{i,1}^{*}, \frac{d\hat{J}_{i,1}^{*}(z_{i,1})}{dz_{i,1}})}{\partial\hat{\Theta}_{ai,1}} = \frac{1}{2(d_i+p_i)^2}S_{Ji,1}S_{Ji,1}^{\mathrm{T}}(\hat{\Theta}_{ai,1} - \hat{\Theta}_{ci,1}) = 0 \tag{8.25}$$

Design a positive function $\lambda_{i,1} = \mathrm{Tr}\{(\hat{\Theta}_{ai,1} - \hat{\Theta}_{ci,1})^T(\hat{\Theta}_{ai,1} - \hat{\Theta}_{ci,1})\}$, and it is obvious that $\lambda_{i,1} = 0$. Since $\frac{\partial \lambda_{i,1}}{\partial \hat{\Theta}_{ai,1}} = -\frac{\partial \lambda_{i,1}}{\partial \hat{\Theta}_{ci,1}} = 2(\hat{\Theta}_{ai,1} - \hat{\Theta}_{ci,1})$, we derive that

$$\begin{aligned}\dot{\lambda}_{i,1} =&\mathrm{Tr}\left\{\frac{\partial \lambda_{i,1}}{\partial \hat{\Theta}_{ai,1}}\dot{\hat{\Theta}}_{ai,1} + \frac{\partial \lambda_{i,1}}{\partial \hat{\Theta}_{ci,1}}\dot{\hat{\Theta}}_{ci,1}\right\}\\ =&\mathrm{Tr}\left\{-\frac{\partial \lambda_{i,1}}{\partial \hat{\Theta}_{ai,1}}S_{Ji,1}S_{Ji,1}^T(\pi_{ai,1}(\hat{\Theta}_{ai,1} - \hat{\Theta}_{ci,1})\right.\\ &\left.+\pi_{ci,1}\hat{\Theta}_{ci,1}) + \pi_{ci,1}S_{Ji,1}S_{Ji,1}^T\hat{\Theta}_{ci,1}\frac{\partial \lambda_{i,1}}{\partial \hat{\Theta}_{ai,1}}\right\}\\ =&-\frac{\pi_{ai,1}}{2}\mathrm{Tr}\left\{\frac{\partial \lambda_{i,1}}{\partial \hat{\Theta}_{ai,1}}S_{Ji,1}S_{Ji,1}^T\frac{\partial \lambda_{i,1}}{\partial \hat{\Theta}_{ai,1}}\right\} \le 0. \end{aligned} \tag{8.26}$$

It can be seen that the derived adaptive control laws (8.20) and (8.21) can assure $\lambda_{i,1} = 0$, and meanwhile, (8.25) holds.

Step s ($2 \le s \le n-1$): According to (8.1) and (8.6), we have

$$dz_{i,s} =(x_{i,s+1} + f_{i,s}(\bar{x}_{i,s}) - \ell\hat{\alpha}_{i,s-1}^*)dt + \Omega_{i,s}(\bar{y})d\omega_i \tag{8.27}$$

Construct the performance function as

$$\begin{aligned}J_{i,s}^*(z_{i,s}) =& \min_{\alpha_{i,s}\in\Xi(\underline{\Omega}_{i,s})}\int_t^\infty \hbar_{i,s}(z_{i,s}(\xi), \alpha_{i,s}(\xi))d\xi\\ =& \int_t^\infty \hbar_{i,s}(z_{i,s}(\xi), \alpha_{i,s}^*(\xi))d\xi\end{aligned} \tag{8.28}$$

where $\Xi(\underline{\Omega}_{i,s})$ represents the admissible control and $\underline{\Omega}_{i,s}$ stands for a compact set containing the origin. $\hbar_{i,s}(z_{i,s}, \alpha_{i,s}) = z_{i,s}^2(t) + \alpha_{i,s}^2(z_{i,s})$ denotes the cost function.

Viewing $x_{i,s+1}$ as the optimal virtual controller $\alpha_{i,s}^*$, the HJBE is expressed as

$$\begin{aligned}H_{i,s}\left(z_{i,s}, \alpha_{i,s}^*, \frac{dJ_{i,s}^*(z_{i,s})}{dz_{i,s}}\right) =&z_{i,s}^2 + \alpha_{i,s}^{*2} + \frac{dJ_{i,s}^*(z_{i,s})}{dz_{i,s}}(f_{i,s}(\bar{x}_{i,s}) - \ell\hat{\alpha}_{i,s-1}^*\\ &+\alpha_{i,s}^*) + \frac{1}{2}\frac{d^2J_{i,s}^*(z_{i,s})}{dz_{i,s}^2}\parallel \Omega_{i,s}(\bar{y})\parallel^2 = 0\end{aligned} \tag{8.29}$$

From $\frac{\partial H_{i,s}}{\partial \alpha_{i,s}^*} = 0$, we can obtain $\alpha_{i,s}^* = -\frac{1}{2}\frac{dJ_{i,s}^*(z_{i,s})}{dz_{i,s}}$. And then, decompose $\frac{dJ_{i,s}^*(z_{i,s})}{dz_{i,s}}$ as

$$\frac{dJ_{i,s}^*(z_{i,s})}{dz_{i,s}} =2\rho_{i,s}z_{i,s}^{4L-3} + 2F_{i,s}(\chi_{i,s}) + \frac{z_{i,s}^3}{2\gamma_{i,s}} + J_{i,s}^0(X_{i,s}), \tag{8.30}$$

where $\rho_{i,s} > 0$ and $\gamma_{i,s} > 0$ represent design parameters, $F_{i,s}(\chi_{i,s}) = z_{i,s} \| \Omega_{i,s}(\bar{y}) \|^4 + f_{i,s} - \ell \hat{\alpha}^*_{i,s-1} + z_{i,s} + (1/2) z^3_{i,s}$ with $\chi_{i,s} = [\hat{\Theta}_{Fi,1}, \ldots, \hat{\Theta}_{Fi,s-1}, x_{i,s}, \hat{\Theta}_{ai,1}, \ldots, \hat{\Theta}_{ai,s-1}, z_{i,s}, \bar{y}]^T$ and $J^0_{i,s}(X_{i,s}) = -2\rho_{i,s} z^{4L-3}_{i,s} - z^3_{i,s}/2\gamma_{i,s} - 2F_{i,s}(\chi_{i,s}) + dJ^*_{i,s}(z_{i,s})/dz_{i,s}$ with $X_{i,s} = [x_{i,s}, z_{i,s}]^T$.

Substituting (8.30) into $\alpha^*_{i,s}$, we have

$$\alpha^*_{i,s} = -\rho_{i,s} z^{4L-3}_{i,s} - F_{i,s}(\chi_{i,s}) - \frac{z^3_{i,s}}{4\gamma_{i,s}} - \frac{1}{2} J^0_{i,s}(X_{i,s}). \tag{8.31}$$

Estimating the unknown terms $F_{i,s}(\chi_{i,s})$ and $J^0_{i,s}(X_{i,s})$ by NNs as follows

$$F_{i,s}(\chi_{i,s}) = \Theta^{*T}_{Fi,s} S_{Fi,s}(\chi_{i,s}) + \Lambda_{Fi,s}(\chi_{i,s}) \tag{8.32}$$

$$J^0_{i,s}(X_{i,s}) = \Theta^{*T}_{Ji,s} S_{Ji,s}(X_{i,s}) + \Lambda_{Ji,s}(X_{i,s}) \tag{8.33}$$

where $\Theta^*_{Fi,s}$ and $\Theta^*_{Ji,s}$ represent the ideal weights, $S_{Fi,s}(\chi_{i,s})$ and $S_{Ji,s}(X_{i,s})$ denote the activation function vectors, and $\Lambda_{Fi,s}(\chi_{i,s})$ and $\Lambda_{Ji,s}(X_{i,s})$ are NNs approximation errors.

Then, it follows from (8.32) and (8.33) that

$$\begin{aligned} \frac{dJ^*_{i,s}(z_{i,s})}{dz_{i,s}} =& 2\rho_{i,s} z^{4L-3}_{i,s} + \frac{z^3_{i,s}}{2\gamma_{i,s}} + \Theta^{*T}_{Ji,s} S_{Ji,s}(X_{i,s}) + \Lambda_{i,s} \\ &+ 2\Theta^{*T}_{i,s} S_{Fi,s}(\chi_{i,s}) \end{aligned} \tag{8.34}$$

$$\begin{aligned} \alpha^*_{i,s} =& -\rho_{i,s} z^{4L-3}_{i,s} - \frac{z^3_{i,s}}{4\gamma_{i,s}} - \frac{1}{2}\Theta^{*T}_{Ji,s} S_{Ji,s}(X_{i,s}) - \frac{1}{2}\Lambda_{i,s} \\ &- \Theta^{*T}_{Fi,s} S_{Fi,s}(\chi_{i,s}) \end{aligned} \tag{8.35}$$

with $\Lambda_{i,s} = 2\Lambda_{Fi,s} + \Lambda_{Ji,s}$.

Similarly with Step 1, the identifier, critic and actor NNs are given as follows:

$$\hat{F}_{i,s}(\chi_{i,s}) = \hat{\Theta}^T_{Fi,s} S_{Fi,s}(\chi_{i,s}) \tag{8.36}$$

$$\begin{aligned} \frac{d\hat{J}^*_{i,s}(z_{i,s})}{dz_{i,s}} =& 2\rho_{i,s} z^{4L-3}_{i,s} + 2\hat{\Theta}^T_{Fi,s} S_{Fi,s}(\chi_{i,s}) + \frac{z^3_{i,s}}{2\gamma_{i,s}} \\ &+ \hat{\Theta}^T_{ci,s} S_{Ji,s}(X_{i,s}) \end{aligned} \tag{8.37}$$

$$\begin{aligned} \hat{\alpha}^*_{i,s} =& -\rho_{i,s} z^{4L-3}_{i,s} - \hat{\Theta}^T_{Fi,s} S_{Fi,s}(\chi_{i,s}) - \frac{z^3_{i,s}}{4\gamma_{i,s}} \\ &- \frac{1}{2}\hat{\Theta}^T_{ai,s} S_{Ji,s}(X_{i,s}) \end{aligned} \tag{8.38}$$

where $\hat{F}_{i,s}$, $d\hat{J}^*_{i,s}/dz_{i,s}$, and $\hat{\alpha}^*_{i,s}$ are the estimation of $F_{i,s}$, $dJ^*_{i,s}/dz_{i,s}$ and $\alpha^*_{i,s}$, respectively. $\hat{\Theta}_{Fi,s}$, $\hat{\Theta}_{ci,s}$, and $\hat{\Theta}_{ai,s}$, respectively, denote the identifier, critic, and actor NNs weights, which is updated by the following laws:

$$\dot{\hat{\Theta}}_{Fi,s} = \eta_{i,s}(S_{Fi,s}(\chi_{i,s})z^3_{i,s} - \phi_{i,s}\hat{\Theta}_{Fi,s}) \tag{8.39}$$

$$\dot{\hat{\Theta}}_{ci,s} = -\pi_{ci,s}S_{Ji,s}(X_{i,s})S^T_{Ji,s}(X_{i,s})\hat{\Theta}_{ci,s} \tag{8.40}$$

$$\dot{\hat{\Theta}}_{ai,s} = -S_{Ji,s}(X_{i,s})S^T_{Ji,s}(X_{i,s})(\pi_{ai,s}(\hat{\Theta}_{ai,s} - \hat{\Theta}_{ci,s}) + \pi_{ci,s}\hat{\Theta}_{ci,s}) \tag{8.41}$$

where $\eta_{i,s}$ represents a positive definite matrix. $\phi_{i,s}$, $\pi_{ci,s}$, and $\pi_{ai,s}$ stand for the positive design parameters satisfying the following conditions:

$$\pi_{ai,s} > \frac{1}{2},\ \pi_{ai,s} > \frac{\phi_{i,s}}{2},\ \pi_{ai,s} > \pi_{ci,s} > \frac{\pi_{ai,s}}{2}. \tag{8.42}$$

Step n: The error dynamic infers from (8.1) and (8.6) that

$$\begin{aligned} dz_{i,n} =& (u_i + f_{i,n}(\bar{x}_{i,n}) + \Phi_i(t - T_0)\lambda_i(\bar{x}_{i,n}, u_i) \\ &- \ell\hat{\alpha}^*_{i,n-1})dt + \Omega_{i,n}(\bar{y})d\omega_i. \end{aligned} \tag{8.43}$$

Define the performance function as:

$$\begin{aligned} J^*_{i,n}(z_{i,n}) =& \min_{\alpha_{i,n}\in\Xi(\underline{\Omega}_{i,n})} \int_t^\infty \hbar_{i,n}(z_{i,n}(\xi), u_i(\xi))d\xi \\ =& \int_t^\infty \hbar_{i,n}(z_{i,n}(\xi), u^*_i(\xi))d\xi \end{aligned} \tag{8.44}$$

where $\Xi(\underline{\Omega}_{i,n})$ represents the admissible control and $\underline{\Omega}_{i,n}$ denotes the compact set containing the origin. $\hbar_{i,n}(z_{i,n}, u_i) = z^2_{i,n}(t) + u^2_i(z_{i,n})$ stands for the cost function.

According to (8.44), the HJBE is defined as

$$\begin{aligned} H_{i,n}(z_{i,n}, u^*_i, \frac{dJ^*_{i,n}(z_{i,n})}{dz_{i,n}}) =& z^2_{i,n} + u^{*2}_i + \frac{dJ^*_{i,n}(z_{i,n})}{dz_{i,n}}(f_{i,n}(\bar{x}_{i,n}) + u^*_i \\ &+ \Phi_i(t - T_0)\lambda_i(\bar{x}_{i,n}, u_i) - \ell\hat{\alpha}^*_{i,n-1}) \\ &+ \frac{1}{2}\frac{d^2J^*_{i,n}(z_{i,n})}{dz^2_{i,n}} \parallel \Omega_{i,n}(\bar{y}) \parallel^2 = 0 \end{aligned} \tag{8.45}$$

By solving equation $\partial H_{i,n}/\partial u^*_i = 0$, one has $u^*_i = -dJ^*_{i,n}(z_{i,n})/2dz_{i,n}$. Similar to Step 1, $dJ^*_{i,n}(z_{i,n})/dz_{i,n}$ is denoted as

$$\frac{dJ_{i,n}^*(z_{i,n})}{dz_{i,n}} = 2\rho_{i,n}z_{i,n}^{4L-3} + 2F_{i,n}(\chi_{i,n}, u_i) + \frac{z_{i,n}^3}{2\gamma_{i,n}} + J_{i,n}^0(X_{i,n}), \tag{8.46}$$

where $\rho_{i,n}$ and $\gamma_{i,n}$ stand for positive design parameters, $F_{i,n}(\chi_{i,n}, u_i) = \Gamma(\bar{x}_{i,n}, u_i) - \ell\hat{\alpha}_{i,n-1}^* + z_{i,n}\| \Omega_{i,n}(\bar{y}) \|^4 + (1/2)z_{i,n}^3 + (1/4)z_{i,n}$ with $\chi_{i,n} = [z_{i,n}, \hat{\Theta}_{Fi,1}, ..., \hat{\Theta}_{Fi,n-1}, \bar{y}, \hat{\Theta}_{ai,1}, ..., \hat{\Theta}_{ai,n-1}, u_i]^T$ being a unknown function, and $J_{i,n}^0(X_{i,n}) = -2\rho_{i,n}z_{i,n}^{4L-3} - z_{i,n}^3/2\gamma_{i,n} - 2F_{i,n}(\chi_{i,n}, u_i) + dJ_{i,n}^*(z_{i,n})/dz_{i,n}$ with $X_{i,n} = [x_{i,n}, z_{i,n}]^T$.

Distinguished from the previous $n-1$ steps, the function $F_{i,n}(\chi_{i,n}, u)$ includes the term $\Gamma(\bar{x}_{i,n}, u_i)$ in this last step, where the actual controller u_i is affected by the non-affine nonlinear faults. In what follows, the adverse effects brought by the faults are considered in the actual controller design process. Substituting (8.46) into u_i^*, we have

$$u_i^* = -\rho_{i,n}z_{i,n}^{4L-3} - F_{i,n}(\chi_{i,n}, u_i) - \frac{1}{4\gamma_{i,n}}z_{i,n}^3 - \frac{1}{2}J_{i,n}^0(X_{i,n}). \tag{8.47}$$

Using the NNs to estimate the unknown function $F_{i,n}(\chi_{i,n}, u_i)$ and $J_{i,n}^0(X_{i,n})$, we have

$$F_{i,n}(\chi_{i,n}, u_i) = \Theta_{Fi,n}^{*T}S_{Fi,n}(\chi_{i,n}, u_i) + \Lambda_{Fi,n}(\chi_{i,n}, u_i) \tag{8.48}$$

$$J_{i,n}^0(X_{i,n}) = \Theta_{Ji,n}^{*T}S_{Ji,n}(X_{i,n}) + \Lambda_{Ji,n}(X_{i,n}) \tag{8.49}$$

where $\Theta_{Fi,n}^*$ and $\Theta_{Ji,n}^*$ stand for the ideal weights, $S_{Fi,n}(\chi_{i,n}, u_i)$ and $S_{Ji,n}(X_{i,n})$ denote the activation function vectors, and $\Lambda_{Fi,n}(\chi_{i,n}, u_i)$ and $\Lambda_{Ji,n}(X_{i,n})$ are NNs approximation errors.

Then, one infers from (8.48) and (8.49) that

$$\begin{aligned}\frac{dJ_{i,n}^*}{dz_{i,n}} =& 2\rho_{i,n}z_{i,n}^{4L-3} + \frac{z_{i,n}^3}{2\gamma_{i,n}} + \Theta_{Ji,n}^{*T}S_{Ji,n}(X_{i,n}) + \Lambda_{i,n} \\ &+ 2\Theta_{Fi,n}^{*T}S_{Fi,n}(\chi_{i,n}, u_i)\end{aligned} \tag{8.50}$$

$$\begin{aligned}u_i^* =& -\rho_{i,n}z_{i,n}^{4L-3} - \frac{1}{4\gamma_{i,n}}z_{i,n}^3 - \frac{1}{2}\Theta_{Ji,n}^{*T}S_{Ji,n}(X_{i,n}) - \frac{1}{2}\Lambda_{i,n} \\ &- \Theta_{Fi,n}^{*T}S_{Fi,n}(\chi_{i,n}, u_i)\end{aligned} \tag{8.51}$$

with $\Lambda_{i,n} = 2\Lambda_{Fi,n} + \Lambda_{Ji,n}$.

To obtain an available actual control input, we apply a NN-based RL algorithm by the following forms:

$$\hat{F}_{i,n}(\chi_{i,n}, u_i) = \hat{\Theta}_{Fi,n}^T S_{Fi,n}(\chi_{i,n}, u_i) \tag{8.52}$$

$$\frac{d\hat{J}_{i,n}^*}{dz_{i,n}} = 2\rho_{i,n}z_{i,n}^{4L-3} + 2\hat{\Theta}_{Fi,n}^T S_{Fi,n}(\chi_{i,n}, u_i) + \frac{z_{i,n}^3}{2\gamma_{i,n}}$$

$$+ \hat{\Theta}_{ci,n}^T S_{Ji,n}(X_{i,n}) \tag{8.53}$$

$$\hat{u}_i^* = -\rho_{i,n} z_{i,n}^{4L-3} - \hat{\Theta}_{Fi,n}^T S_{Fi,n}(\chi_{i,n}, u_i) - \frac{z_{i,n}^3}{4\gamma_{i,n}} - \frac{1}{2}\hat{\Theta}_{ai,n}^T S_{Ji,n}(X_{i,n}) \tag{8.54}$$

where $\hat{F}_{i,n}$, $d\hat{J}_{i,n}^*/dz_{i,n}$, and $\hat{u}_i^*$ are the estimation of $F_{i,n}$, $dJ_{i,n}^*/dz_{i,n}$ and u_i^*, respectively. $\hat{\Theta}_{Fi,n}$, $\hat{\Theta}_{ai,n}$, and $\hat{\Theta}_{ci,n}$, respectively, denote the identifier, critic, and actor NNs weights, and their updating laws are constructed as follows:

$$\dot{\hat{\Theta}}_{Fi,n} = \eta_{i,n}(S_{Fi,n}(\chi_{i,n}, u_i) z_{i,n}^3 - \phi_{i,n}\hat{\Theta}_{Fi,n}) \tag{8.55}$$

$$\dot{\hat{\Theta}}_{ci,n} = -\pi_{ci,n} S_{Ji,n}(X_{i,n}) S_{Ji,n}^T(X_{i,n}) \hat{\Theta}_{ci,n} \tag{8.56}$$

$$\dot{\hat{\Theta}}_{ai,n} = -S_{Ji,n}(X_{i,n}) S_{Ji,n}^T(X_{i,n})(\pi_{ai,n}(\hat{\Theta}_{ai,n} - \hat{\Theta}_{ci,n}) + \pi_{ci,n}\hat{\Theta}_{ci,n}) \tag{8.57}$$

where $\eta_{i,n}$ denotes a positive definite matrix. $\phi_{i,n}$, $\pi_{ci,n}$, and $\pi_{ai,n}$ represent positive design parameters. For follow-up analysis, $\phi_{i,n}$, $\pi_{ci,n}$, and $\pi_{ai,n}$ should satisfy the following conditions:

$$\pi_{ai,n} > \frac{1}{2},\ \pi_{ai,n} > \frac{\phi_{i,n}}{2},\ \pi_{ai,n} > \pi_{ci,n} > \frac{\pi_{ai,n}}{2}. \tag{8.58}$$

8.3 Stability Analysis

In view of the above n step backstepping design, we can derive the theoretical results by using the following Lyapunov stability analysis method and Theorem 8.1 given later.

By means of $z_{i,2} = x_{i,2} - \hat{\alpha}_{i,1}^*$, the consensus error in (8.7) can be further reformulated as

$$dz_{i,1} = \{(d_i + p_i)(z_{i,2} + f_{i,1}(x_{i,1}) + \hat{\alpha}_{i,1}^*) - \sum_{s\in\mathcal{N}_i} a_{i,s}(x_{s,2} + f_{s,1}(x_{s,1})) - p_i\dot{y}_r\}dt - \{(d_i + p_i)\Omega_{i,1}(\bar{y}) - \sum_{s\in\mathcal{N}_i} a_{i,s}\Omega_{s,1}(\bar{y})\}d\omega_i \tag{8.59}$$

Step 1: Design the Lyapunov function as

$$V_{i,1} = \frac{1}{4}z_{i,1}^4 + \frac{1}{2}\tilde{\Theta}_{Fi,1}^T \eta_{i,1}^{-1}\tilde{\Theta}_{Fi,1} + \frac{1}{2}\tilde{\Theta}_{ci,1}^T\tilde{\Theta}_{ci,1} + \frac{1}{2}\tilde{\Theta}_{ai,1}^T\tilde{\Theta}_{ai,1},$$

where $\tilde{\Theta}_{\xi i,1} = \hat{\Theta}_{\xi i,1} - \Theta_{\xi i,1}^*(\xi = F, c, a)$ denotes the corresponding error.

From (8.1), (8.12), (8.18), (8.19), (8.20), (8.21), and (8.59), according to Itô stochastic rule (8.4), we have

$$\begin{aligned}\ell V_{i,1} =& -\rho_{i,1} z_{i,1}^{4L} - \frac{1}{4\gamma_{i,1}} z_{i,1}^6 - z_{i,1}^4 \| G \|^4 - \frac{1}{2} z_{i,1}^6 + (d_i + p_i) z_{i,1}^3 z_{i,2} \\ &+ \frac{3}{2} z_{i,1}^2 \| G \|^2 - \frac{3}{4}(d_i + p_i) z_{i,1}^4 - \phi_{i,1} \tilde{\Theta}_{Fi,1}^T \hat{\Theta}_{Fi,1} + z_{i,1}^3 \Lambda_{Fi,1} \\ &- \tilde{\Theta}_{ai,1}^T S_{Ji,1} S_{Ji,1}^T (\pi_{ai,1}(\hat{\Theta}_{ai,1} - \hat{\Theta}_{ci,1}) + \pi_{ci,1} \hat{\Theta}_{ci,1}) \\ &- \frac{1}{2} z_{i,1}^3 \hat{\Theta}_{ai,1}^T S_{Ji,1} - \pi_{ci,1} \tilde{\Theta}_{ci,1}^T S_{Ji,1} S_{Ji,1}^T \hat{\Theta}_{ci,1}, \end{aligned} \tag{8.60}$$

where $G = (d_i + p_i)\Omega_{i,1}(\bar{y}) - \sum_{s \in \mathcal{N}_i} a_{i,s} \Omega_{s,1}(\bar{y})$.

Using Young's inequality gives

$$\begin{aligned} \frac{3}{2} z_{i,1}^2 \| G \|^2 \leq& z_{i,1}^4 \| G \|^4 + \frac{9}{4} \\ z_{i,1}^3 \Lambda_{Fi,1} \leq& \frac{1}{2} z_{i,1}^6 + \frac{1}{2} \Lambda_{Fi,1}^2 \\ -\frac{1}{2} z_{i,1}^3 \hat{\Theta}_{ai,1}^T S_{Ji,1} \leq& \frac{z_{i,1}^6}{4\gamma_{i,1}} + \frac{\gamma_{i,1}}{4} (\hat{\Theta}_{ai,1}^T S_{Ji,1})^2 \end{aligned} \tag{8.61}$$

With the help of Lemma 8.3, along with $\Im_1 = z_{i,1}$, $\Im_2 = z_{i,2}$, $\flat_1 = 3$, $\flat_2 = \flat_3 = 1$, one has

$$(d_i + p_i) z_{i,1}^3 z_{i,2} \leq \frac{3}{4}(d_i + p_i) z_{i,1}^4 + \frac{1}{4}(d_i + p_i) z_{i,2}^4. \tag{8.62}$$

From (8.61) and (8.62), it yields

$$\begin{aligned}\ell V_{i,1} \leq& -\rho_{i,1} z_{i,1}^{4L} + (\pi_{ai,1} - \pi_{ci,1}) \tilde{\Theta}_{ai,1}^T S_{Ji,1} S_{Ji,1}^T \hat{\Theta}_{ci,1} \\ &+ \frac{\gamma_{i,1}}{4} \hat{\Theta}_{ai,1}^T S_{Ji,1} S_{Ji,1}^T \hat{\Theta}_{ai,1} - \phi_{i,1} \tilde{\Theta}_{Fi,1}^T \hat{\Theta}_{Fi,1} + \frac{9}{4} \\ &- \pi_{ci,1} \tilde{\Theta}_{ci,1}^T S_{Ji,1} S_{Ji,1}^T \hat{\Theta}_{ci,1} + \frac{(d_i + p_i)}{4} z_{i,2}^4 \\ &- \pi_{ai,1} \tilde{\Theta}_{ai,1}^T S_{Ji,1} S_{Ji,1}^T \hat{\Theta}_{ai,1} + \frac{1}{2} \Lambda_{Fi,1}^2. \end{aligned} \tag{8.63}$$

Based on $\tilde{\Theta}_{\xi i,1} = \hat{\Theta}_{\xi i,1} - \Theta_{\xi i,1}^*(\xi = F, c, a)$, we have

$$\begin{aligned} \tilde{\Theta}_{Fi,1}^T \hat{\Theta}_{Fi,1} =& \frac{1}{2} \hat{\Theta}_{Fi,1}^T \hat{\Theta}_{Fi,1} + \frac{1}{2} \tilde{\Theta}_{Fi,1}^T \tilde{\Theta}_{Fi,1} - \frac{1}{2} \Theta_{Fi,1}^{*T} \Theta_{Fi,1}^* \qquad (8.64) \\ \tilde{\Theta}_{ci,1}^T S_{Ji,1} S_{Ji,1}^T \hat{\Theta}_{ci,1} =& \frac{1}{2} (\tilde{\Theta}_{ci,1}^T S_{Ji,1})^2 + \frac{1}{2} (\hat{\Theta}_{ci,1}^T S_{Ji,1})^2 \end{aligned}$$

$$-\frac{1}{2}(\Theta_{Ji,1}^{*T}S_{Ji,1})^2 \tag{8.65}$$

$$\tilde{\Theta}_{ai,1}^T S_{Ji,1} S_{Ji,1}^T \hat{\Theta}_{ai,1} = \frac{1}{2}(\tilde{\Theta}_{ai,1}^T S_{Ji,1})^2 + \frac{1}{2}(\hat{\Theta}_{ai,1}^T S_{Ji,1})^2 - \frac{1}{2}(\Theta_{Ji,1}^{*T}S_{Ji,1})^2 \tag{8.66}$$

By using (8.64)–(8.66), (8.63) can be rewritten as

$$\begin{aligned}\ell V_{i,1} \le & -\rho_{i,1} z_{i,1}^{4L} - \frac{\phi_{i,1}}{2}\tilde{\Theta}_{Fi,1}^T \tilde{\Theta}_{Fi,1} + \frac{(d_i+p_i)}{4} z_{i,2}^4 + \frac{\phi_{i,1}}{2}\Theta_{Fi,1}^{*T}\Theta_{Fi,1}^* \\ & + \frac{\pi_{ai,1}+\pi_{ci,1}}{2}\Theta_{Ji,1}^{*T} S_{Ji,1} S_{Ji,1}^T \Theta_{Ji,1}^* - \frac{\pi_{ci,1}}{2}\hat{\Theta}_{ci,1}^T S_{Ji,1} S_{Ji,1}^T \hat{\Theta}_{ci,1} \\ & - \frac{\pi_{ci,1}}{2}\tilde{\Theta}_{ci,1}^T S_{Ji,1} S_{Ji,1}^T \tilde{\Theta}_{ci,1} - \frac{\pi_{ai,1}}{2}\tilde{\Theta}_{ai,1}^T S_{Ji,1} S_{Ji,1}^T \tilde{\Theta}_{ai,1} \\ & - (\frac{\pi_{ai,1}}{2} - \frac{\gamma_{i,1}}{4})\hat{\Theta}_{ai,1}^T S_{Ji,1} S_{Ji,1}^T \hat{\Theta}_{ai,1} + \frac{1}{2}\Lambda_{Fi,1}^2 + \frac{9}{4} \\ & + (\pi_{ai,1} - \pi_{ci,1})\tilde{\Theta}_{ai,1}^T S_{Ji,1} S_{Ji,1}^T \hat{\Theta}_{ci,1}. \end{aligned} \tag{8.67}$$

Applying Young's inequality, one obtains

$$\begin{aligned}&(\pi_{ai,1} - \pi_{ci,1})\tilde{\Theta}_{ai,1}^T S_{Ji,1} S_{Ji,1}^T \hat{\Theta}_{ci,1} \\ &\le \frac{\pi_{ai,1} - \pi_{ci,1}}{2}\{(\tilde{\Theta}_{ai,1}^T S_{Ji,1})^2 + (\hat{\Theta}_{ci,1}^T S_{Ji,1})^2\}. \end{aligned} \tag{8.68}$$

Combining with inequality (8.68) and (8.22), one can have

$$\begin{aligned}\ell V_{i,1} \le & -\rho_{i,1} z_{i,1}^{4L} + B_{i,1} - \frac{\pi_{ci,1}}{2}\tilde{\Theta}_{ai,1}^T S_{Ji,1} S_{Ji,1}^T \tilde{\Theta}_{ai,1} \\ & - \frac{\phi_{i,1}}{2}\tilde{\Theta}_{Fi,1}^T \tilde{\Theta}_{Fi,1} - \frac{\pi_{ci,1}}{2}\tilde{\Theta}_{ci,1}^T S_{Ji,1} S_{Ji,1}^T \tilde{\Theta}_{ci,1} \\ & + \frac{(d_i+p_i)}{4} z_{i,2}^4, \end{aligned} \tag{8.69}$$

where $B_{i,1} = (1/2)\phi_{i,1}\Theta_{Fi,1}^{*T}\Theta_{Fi,1}^* + (\pi_{ci,1}/2 + \pi_{ai,1}/2)\Theta_{Ji,1}^{*T} S_{Ji,1} S_{Ji,1}^T \Theta_{Ji,1}^* + (1/2)\Lambda_{Fi,1}^2 + 9/4$.

Defined that $\varpi_{\eta_{i,1}^{-1}}^{\max}$ denotes the maximal eigenvalue of $\eta_{i,1}^{-1}$ and $\varpi_{S_{Ji,1}}^{\min}$ represents the minimal eigenvalue of $S_{Ji,1}S_{Ji,1}^T$, we have

$$\begin{aligned}-\tilde{\Theta}_{Fi,1}^T \tilde{\Theta}_{Fi,1} &\le -\frac{1}{\varpi_{\eta_{i,1}^{-1}}^{\max}}\tilde{\Theta}_{Fi,1}^T \eta_{i,1}^{-1}\tilde{\Theta}_{Fi,1} \\ -\tilde{\Theta}_{ci,1}^T S_{Ji,1} S_{Ji,1}^T \tilde{\Theta}_{ci,1} &\le -\varpi_{S_{Ji,1}}^{\min}\tilde{\Theta}_{ci,1}^T \tilde{\Theta}_{ci,1} \\ -\tilde{\Theta}_{ai,1}^T S_{Ji,1} S_{Ji,1}^T \tilde{\Theta}_{ai,1} &\le -\varpi_{S_{Ji,1}}^{\min}\tilde{\Theta}_{ai,1}^T \tilde{\Theta}_{ai,1} \end{aligned} \tag{8.70}$$

Based on the above mentioned results, one has

$$\begin{aligned}\ell V_{i,1} \leq &- \rho_{i,1} z_{i,1}^{4L} + B_{i,1} - \frac{\phi_{i,1}}{2\varpi^{\max}_{\eta_{i,1}^{-1}}} \tilde{\Theta}_{Fi,1}^T \eta_{i,1}^{-1} \tilde{\Theta}_{Fi,1} + \frac{(d_i + p_i)}{4} z_{i,2}^4 \\ &- \frac{\pi_{ci,1}}{2} \varpi^{\min}_{S_{Ji,1}} \tilde{\Theta}_{ci,1}^T \tilde{\Theta}_{ci,1} - \frac{\pi_{ci,1}}{2} \varpi^{\min}_{S_{Ji,1}} \tilde{\Theta}_{ai,1}^T \tilde{\Theta}_{ai,1}. \end{aligned} \tag{8.71}$$

Step s ($2 \leq s \leq n-1$): Considering the Lyapunov function as

$$V_{i,s} = \frac{1}{4} z_{i,s}^4 + \frac{1}{2} \tilde{\Theta}_{Fi,s}^T \eta_{i,s}^{-1} \tilde{\Theta}_{Fi,s} + \frac{1}{2} \tilde{\Theta}_{ci,s}^T \tilde{\Theta}_{ci,s} + \frac{1}{2} \tilde{\Theta}_{ai,s}^T \tilde{\Theta}_{ai,s} + V_{i,s-1}.$$

Based on (8.27), (8.32), and (8.38)–(8.41), the infinitesimal generator of $V_{i,s}$ satisfies

$$\begin{aligned}\ell V_{i,s} =& \ell V_{i,s-1} + z_{i,s}^3 (z_{i,s+1} - \rho_{i,s} z_{i,s}^{4L-3} - \frac{1}{2} \hat{\Theta}_{ai,s}^T S_{Ji,s} - \phi_{i,s} \tilde{\Theta}_{Fi,s} \hat{\Theta}_{Fi,s} \\ &+ \Lambda_{Fi,s} - \frac{1}{4\gamma_{i,s}} z_{i,s}^3) + \frac{3}{2} z_{i,s}^2 \| \Omega_{i,s}(\bar{y}) \|^2 - z_{i,s}^4 - z_{i,s}^4 \| \Omega_{i,s}(\bar{y}) \|^4 \\ &- \frac{1}{2} z_{i,s}^6 - \tilde{\Theta}_{ai,s}^T S_{Ji,s} S_{Ji,s}^T (\pi_{ai,s}(\hat{\Theta}_{ai,s} - \hat{\Theta}_{ci,s}) + \pi_{ci,s} \hat{\Theta}_{ci,s}) \\ &- \pi_{ci,s} \tilde{\Theta}_{ci,s}^T S_{Ji,s} S_{Ji,s}^T \hat{\Theta}_{ci,s}. \end{aligned} \tag{8.72}$$

Defined that $\varpi^{\max}_{\eta_{i,s}^{-1}}$ and $\varpi^{\min}_{S_{Ji,s}}$ denote the maximal eigenvalue of $\eta_{i,s}^{-1}$ and the minimal eigenvalue of $S_{Ji,s} S_{Ji,s}^T$, respectively. Similar to the calculation process of (8.61)–(8.70), and substituting (8.42) into (8.72),we can obtain

$$\begin{aligned}\ell V_{i,s} \leq& \sum_{k=1}^{s} (-\rho_{i,k} z_{i,k}^{4L} - \frac{\phi_{i,k}}{2\varpi^{\max}_{\eta_{i,k}^{-1}}} \tilde{\Theta}_{Fi,k}^T \eta_{i,k}^{-1} \tilde{\Theta}_{Fi,k} - \frac{\pi_{ci,k}}{2} \varpi^{\min}_{S_{Ji,k}} \tilde{\Theta}_{ai,k}^T \tilde{\Theta}_{ai,k} \\ &- \frac{\pi_{ci,k}}{2} \varpi^{\min}_{S_{Ji,k}} \tilde{\Theta}_{ci,k}^T \tilde{\Theta}_{ci,k} + B_{i,k}) + \frac{z_{i,s+1}^4}{4}, \end{aligned} \tag{8.73}$$

where $B_{i,s} = (1/2)\phi_{i,s} \Theta_{Fi,s}^{*T} \Theta_{Fi,s}^* + (\pi_{ci,s}/2 + \pi_{ai,s}/2) \Theta_{Ji,s}^{*T} S_{Ji,s} S_{Ji,s}^T \Theta_{Ji,s}^* + (1/2) \Lambda_{Fi,s}^2 + 9/4$.

Step n: The Lyapunov function is chosen as

$$V_{i,n} = \frac{1}{4} z_{i,n}^4 + \frac{1}{2} \tilde{\Theta}_{Fi,n}^T \eta_{i,n}^{-1} \tilde{\Theta}_{Fi,n} + \frac{1}{2} \tilde{\Theta}_{ci,n}^T \tilde{\Theta}_{ci,n} + \frac{1}{2} \tilde{\Theta}_{ai,n}^T \tilde{\Theta}_{ai,n} + V_{i,n-1}. \tag{8.74}$$

The infinitesimal generator of $V_{i,n}$ along with (8.43) and (8.55)–(8.57), one obtains

$$\begin{aligned}\ell V_{i,n} =&\ell V_{i,n-1} + z_{i,n}^3(u_i + \Phi_i(t-T_0)\lambda_i(\bar{x}_{i,n}, u_i) + f_{i,n}(\bar{x}_{i,n}) - \ell\hat{\alpha}_{i,n-1}^*)\\ &- \pi_{ci,n}\tilde{\Theta}_{ci,n}^T S_{Ji,n} S_{Ji,n}^T \hat{\Theta}_{ci,n} + \tilde{\Theta}_{Fi,n}^T(S_{Fi,n} z_{i,n}^3 - \phi_{i,n}\hat{\Theta}_{Fi,n})\\ &- \tilde{\Theta}_{ai,n}^T S_{Ji,n} S_{Ji,n}^T(\pi_{ai,n}(\hat{\Theta}_{ai,n} - \hat{\Theta}_{ci,n}) + \pi_{ci,n}\hat{\Theta}_{ci,n})\\ &+ \frac{3}{2} z_{i,n}^2 \| \Omega_{i,n}(\bar{y}) \|^2.\end{aligned} \tag{8.75}$$

Using Young's inequality and Assumption 8.2, it yields

$$\begin{aligned}z_{i,n}^3[f_{i,n}(\bar{x}_{i,n}) + \Phi_i(t-T_0)\lambda_i(\bar{x}_{i,n}, u_i)] \le& |z_{i,n}^3||\bar{\Gamma}(\bar{x}_{i,n}, u_i)|\\ \le& \frac{1}{2} z_{i,n}^6 \bar{\Gamma}^2(\bar{x}_{i,n}, u_i) + \frac{1}{2}\\ \le& z_{i,n}^3 \Gamma(\bar{x}_{i,n}, u_i) + \frac{1}{2}\end{aligned} \tag{8.76}$$

where $\Gamma(\bar{x}_{i,n}, u_i) = \frac{1}{2} z_{i,n}^3 \bar{\Gamma}^2(\bar{x}_{i,n}, u_i)$.

Remark 8.2 Obviously, the unknown function $F_{i,n}(\chi_{i,n}, u_i)$ contains multiple signals. Enlightened by [14, 15], there will cause the algebraic loop problem, if the NN (8.16) is employed directly in the controller design progress. The filtered signal is utilized to cope with the knotty problem, which is reasonable since the majority of actuators have low-pass features.

Hence, (8.48) can be rewritten as

$$F_{i,n}(\chi_{i,n}, u_i) = \Theta_{Fi,n}^{*T} S_{Fi,n}(\chi_{i,n}, u_{fi}) + \Lambda_{Fi,n}(\chi_{i,n}, u_{fi}), \tag{8.77}$$

and u_{fi} denotes the filtered signal depicted by

$$u_{fi} = H_L(z) u_i \approx u_i, \tag{8.78}$$

where $H_L(z)$ represents Butterworth low-pass filer.

Combining (8.75)–(8.77), and similar to Step 1, we have

$$\begin{aligned}\ell V_{i,n} \le& \sum_{k=1}^{n}(-\rho_{i,k} z_{i,k}^{4L} + B_{i,k} - \frac{\phi_{i,k}}{2\varpi_{\eta_{i,k}^{-1}}^{\max}} \tilde{\Theta}_{Fi,k}^T \eta_{i,k}^{-1} \tilde{\Theta}_{Fi,k}\\ &- \frac{\pi_{ci,k}}{2} \varpi_{S_{Ji,k}}^{\min} \tilde{\Theta}_{ci,k}^T \tilde{\Theta}_{ci,k} - \frac{\pi_{ci,k}}{2} \varpi_{S_{Ji,k}}^{\min} \tilde{\Theta}_{ai,k}^T \tilde{\Theta}_{ai,k}),\end{aligned} \tag{8.79}$$

where $B_{i,n} = (1/2)\phi_{i,n}\Theta_{Fi,n}^{*T}\Theta_{Fi,n}^* + (\pi_{ci,n}/2 + \pi_{ai,n}/2)\Theta_{Ji,n}^{*T} S_{Ji,n} S_{Ji,n}^T \Theta_{Ji,n}^* + (1/2)\Lambda_{Fi,n}^2 + 11/4$.

Let $\psi_n = \min\{\rho_{i,k}, \frac{\phi_{i,k}}{2\varpi^{\max}_{\eta^{-1}_{i,k}}}, \frac{\pi_{ci,k}}{2}\varpi^{\min}_{S_{Ji,k}}, \frac{\pi_{ci,k}}{2}\varpi^{\min}_{S_{Ji,k}}\}$, which yields

$$\begin{aligned}\ell V_{i,n} \leq & -\psi_n \sum_{k=1}^{n}(z_{i,k}^{4L} + \tilde{\Theta}_{ai,k}^T\tilde{\Theta}_{ai,k} + \tilde{\Theta}_{ci,k}^T\tilde{\Theta}_{ci,k} \\ & + \tilde{\Theta}_{Fi,k}^T\eta_{i,k}^{-1}\tilde{\Theta}_{Fi,k}) + \sum_{k=1}^{n} B_{i,k}.\end{aligned} \tag{8.80}$$

From Lemma 8.2, one obtains

$$-\psi_n \sum_{k=1}^{n} z_{i,k}^{4L} \leq -4^L\psi_n(\sum_{k=1}^{n}\frac{1}{4}z_{i,k}^4)^L. \tag{8.81}$$

To deal with this term $\sum_{k=1}^{n}\tilde{\Theta}_{ai,k}^T\tilde{\Theta}_{ai,k}$, and employing Lemma 8.3, along with $\Im_1 = 1, \Im_2 = \sum_{k=1}^{n}\tilde{\Theta}_{ai,k}^T\tilde{\Theta}_{ai,k}, \flat_1 = 1 - L, \flat_2 = L, \flat_3 = L^{\frac{L}{1-L}}$, one has

$$2^L(\sum_{k=1}^{n}\frac{1}{2}\tilde{\Theta}_{ai,k}^T\tilde{\Theta}_{ai,k})^L \leq (1-L)L^{\frac{L}{1-L}} + \sum_{k=1}^{n}\tilde{\Theta}_{ai,k}^T\tilde{\Theta}_{ai,k}. \tag{8.82}$$

Similarly, one has

$$\begin{aligned}& 2^L(\sum_{k=1}^{n}\frac{1}{2}\tilde{\Theta}_{ci,k}^T\tilde{\Theta}_{ci,k})^L + 2^L(\sum_{k=1}^{n}\frac{1}{2}\tilde{\Theta}_{Fi,k}^T\eta_{i,k}^{-1}\tilde{\Theta}_{Fi,k})^L \\ \leq & 2(1-L)L^{\frac{L}{1-L}} + \sum_{k=1}^{n}\tilde{\Theta}_{ci,k}^T\tilde{\Theta}_{ci,k} + \sum_{k=1}^{n}\tilde{\Theta}_{Fi,k}^T\eta_{i,k}^{-1}\tilde{\Theta}_{Fi,k}.\end{aligned} \tag{8.83}$$

From (8.80)–(8.83), we have

$$\begin{aligned}\ell V_{i,n} \leq & -2^L\psi_n(\sum_{k=1}^{n}\frac{1}{2}\tilde{\Theta}_{Fi,k}^T\eta_{i,k}^{-1}\tilde{\Theta}_{Fi,k})^L - 2^L\psi_n(\sum_{k=1}^{n}\frac{1}{2}\tilde{\Theta}_{ai,k}^T\tilde{\Theta}_{ai,k})^L \\ & -2^L\psi_n(\sum_{k=1}^{n}\frac{1}{2}\tilde{\Theta}_{ci,k}^T\tilde{\Theta}_{ci,k})^L - 4^L\psi_n(\sum_{k=1}^{n}\frac{1}{4}z_{i,k}^4)^L + \sum_{k=1}^{n}B_{i,k} \\ & + 3(1-L)L^{\frac{L}{1-L}}.\end{aligned} \tag{8.84}$$

Letting $A_i = \min\{4^L\psi_n, 2^L\psi_n\}$ and $B_i = \sum_{k=1}^{n}B_{i,k} + 3(1-L)L^{\frac{L}{1-L}}$, one has

$$\ell V_{i,n} \leq -A_i V_{i,n}^L + B_i. \tag{8.85}$$

Theorem 8.1 *Consider the stochastic nonlinear MASs (8.1) under Assumptions 8.1-8.2, the optimal controller (8.18), (8.38), and (8.54), the updating laws of the identifier, critic, actor NNs (8.19), (8.39), (8.55), (8.20), (8.40), (8.56), (8.21), (8.41), and (8.57), and the design parameters satisfy (8.22), (8.42), and (8.58), the presented approach ensures that the whole state variables are SGFSP and the outputs of each follower can track the output of the leader accurately.*

Proof. Selecting the overall Lyapunov function as $V = \sum_{i=1}^{N} V_{i,n}$, and utilizing Lemma 8.1, then we can derive that

$$\ell V \leq -AV^L + B,$$

where $A = \min\{A_i, i = 1, 2, ..., N\}$, $B = \sum_{i=1}^{N} B_i$.

Let $T^* = \frac{1}{(1-k)\vartheta A}[\omega^{1-k}(\chi(0)) - (\frac{B}{(1-\vartheta)A})^{\frac{1-\vartheta}{\vartheta}}]$, with $0 < \vartheta < 1$ being a constant, and $\chi(0) = [z(0), \tilde{\Theta}_F(0), \tilde{\Theta}_c(0), \tilde{\Theta}_a(0)]^T$ with initial conditions $z(0) = [z_{i,1}(0), ..., z_{i,n}(0)]^T$, $\tilde{\Theta}_F(0) = [\tilde{\Theta}_{Fi,1}(0), ..., \tilde{\Theta}_{Fi,n}(0)]^T$, $\tilde{\Theta}_c(0) = [\tilde{\Theta}_{ci,1}(0), ..., \tilde{\Theta}_{ci,n}(0)]^T$ and $\tilde{\Theta}_a(0) = [\tilde{\Theta}_{ai,1}(0), ..., \tilde{\Theta}_{ai,n}(0)]^T$. Hence, according to Lemma 8.1, the inequality $E[V^L(z, \tilde{\Theta}_F, \tilde{\Theta}_c, \tilde{\Theta}_a)] \leq \frac{B}{(1-\vartheta)A}$ holds for $t \geq T^*$, which implies that all signals in closed-loop systems are SGFSP. Furthermore, from the definition of V, for $\forall t \geq T^*$, we can obtain that

$$E[(\frac{1}{4} \mid z_{i,1} \mid^4)^L] \leq \frac{B}{(1-\vartheta)A},$$

consequently, one can conclude that the consensus error will converge to the origin in finite time by adjusting the design parameters appropriately. Finally, the proof of Theorem 8.1 is completed.

8.4 Simulation

In this section, we shall demonstrate the feasibility of the presented control strategy through the following two simulation examples.

Example 8.1 Consider the following stochastic nonlinear MASs composed of one leader and three followers, which can be formulated as

$$\begin{cases} dx_{i,1} = (x_{i,2} + 0.5\cos(t) + 0.1x_{i,1}\sin(x_{i,1}))dt + (0.1\cos(t)\sin(x_{i,1}))dw \\ dx_{i,2} = (u_i + x_{i,2}^2\sin(x_{i,1}) + \Phi_i(t - T_0)\lambda_i(x, u_i))dt \\ \qquad\quad + (0.1\sin(t)\cos(x_{i,1}x_{i,2}))dw \\ \quad y_r = 0.5\sin(t) + 0.2 \end{cases}$$

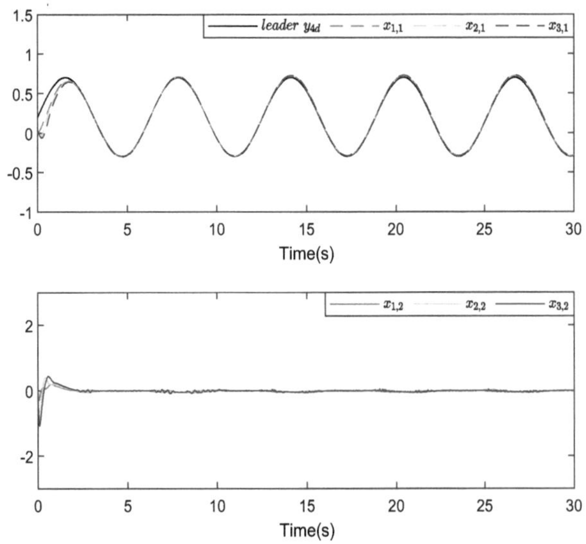

Fig. 8.1 Trajectories of $x_{i,s}(i = 1, 2, 3; s = 1, 2)$ and leader y_r

The fault function, the time profile of failure, and the Butterworth low-pass filter are defined as

$$\Phi_i(t - T_0) = \begin{cases} 0, & t < 10 \\ 1 - e^{-16(t-10)}, & t \geq 10 \end{cases}$$

$$\lambda_i(x, u_i) = 8(x_{i,1}x_{i,2} + \sin(u_i)) + 6$$

$$H_L(z) = \frac{1}{z^2 + 1.414z + 1}$$

Besides, the adjacency matrix $\mathcal{A}$ can be described as

$$\mathcal{A} = \begin{bmatrix} 0 & 0 & 1 \\ 1 & 0 & 0 \\ 0 & 1 & 0 \end{bmatrix}.$$

Moreover, the initial values are defined as $x_{i,s}(0) = 0$; $\Theta_{Fi,1}(0) = \Theta_{ci,1}(0) = \Theta_{ai,1}(0) = 0.4(i = 1, 2, 3; s = 1, 2)$. Meanwhile, the design parameters are selected as $L = \frac{25}{27}$; $\pi_{ci,s} = 13$; $\pi_{ai,s} = 15$; $\eta_{i,s} = 10$; $\phi_{i,s} = 2$; $\rho_{i,1} = 3$, $\rho_{i,2} = 8$; $\gamma_{i,1} = 4$, $\gamma_{i,2} = 8$.

Simulation results are plotted in Figs. 8.1, 8.2, 8.3and 8.4. From Fig. 8.1, it is well verified that the three followers can be aligned with the leader to achieve consensus, and it also depicts that the second states of systems are all bounded. Figure 8.2 is given to demonstrate the consensus error can converge to zero. The control input signals with faults are depicted in Fig. 8.3. The trajectories of identifier, critic, and actor NNs weights are drawn in Fig. 8.4, respectively.

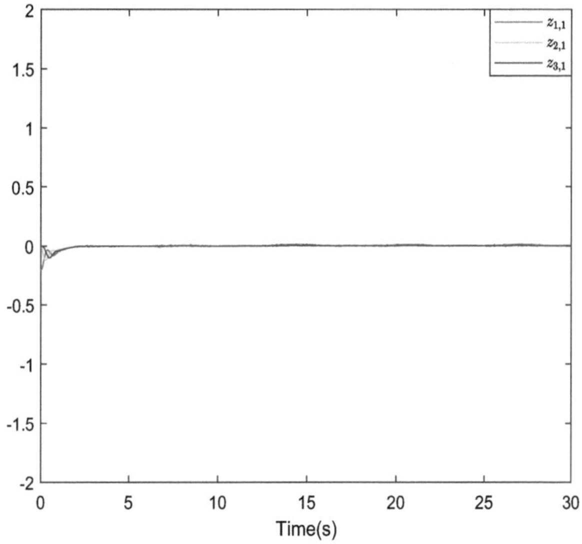

Fig. 8.2 Trajectories of Consensus errors $z_{i,1}(i = 1, 2, 3)$

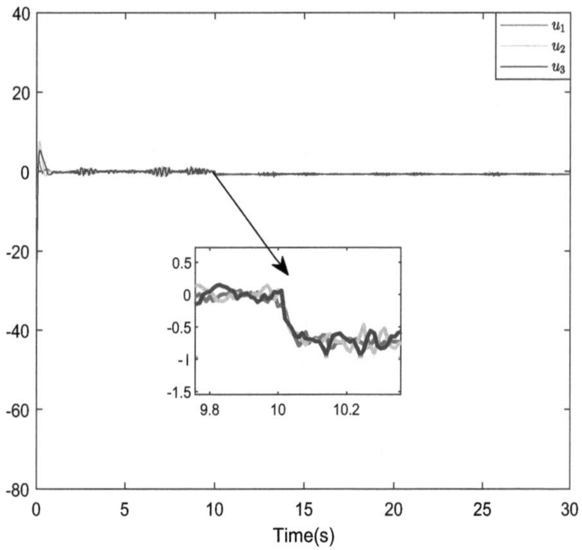

Fig. 8.3 Trajectories of the actual control signals $u_i(i = 1, 2, 3)$

Example 8.2 Next, we further consider the simulation with single-link robots to illustrate the applicability of the presented strategy. The single-link manipulator systems are modeled as:

$$R_i \ddot{\varsigma}_i + \Upsilon_i g \delta_i \sin \varsigma_i = u_i,$$

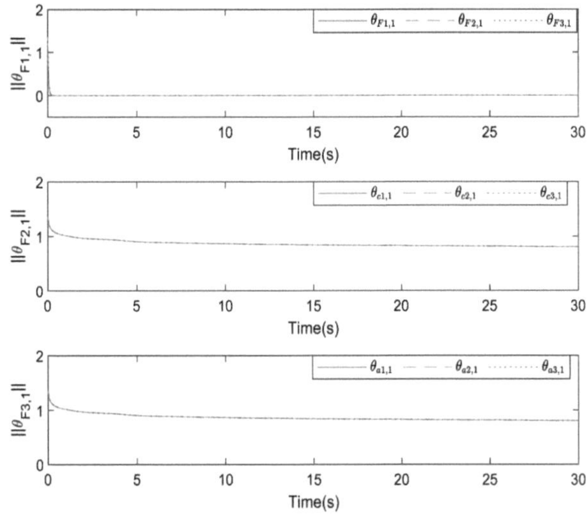

Fig. 8.4 Trajectories of $\| \hat{\Theta}_{Fi,1} \|$, $\| \hat{\Theta}_{ci,1} \|$ and $\| \hat{\Theta}_{ai,1} \|$ $(i = 1, 2, 3)$

where R_i denotes the moment of inertia. u_i, δ_i, g and Υ_i refer to the control torque, the length, the gravitational acceleration and the mass of link, respectively. ς_i represents the angle position. Let $x_{i,1} = \varsigma_i$, $x_{i,2} = \dot{\varsigma}_i$, the systems can be represented as

$$\begin{cases} dx_{i,1} = x_{i,2}dt + o_{i,1}d\omega_i \\ dx_{i,2} = \left(\frac{1}{R_i}u_i - \frac{\Upsilon_i}{R_i}g\delta_i \sin x_{i,1}\right)dt + o_{i,2}d\omega_i. \end{cases}$$

The design parameters are given by: $R_i = 1$; $\Upsilon_i = \delta_i = 0.5$; $g = 10$; $o_{i,1} = 0.1\sin(x_{i,1})$; $o_{i,2} = 0.1\cos(x_{i,1}x_{i,2})$; $L = \frac{37}{39}$; $\phi_{i,s} = 0.2$; $\eta_{i,s} = \pi_{ci,s} = 2$; $\pi_{ai,s} = 4$; $\rho_{i,1} = 10.5$, $\rho_{i,2} = 23$; $\gamma_{i,1} = 1$, $\gamma_{i,2} = 2$ $(i = 1, 2, 3; s = 1, 2)$. And other parameters are the same as defined in the numerical simulation. From Figs. 8.5-8.6, we can obtain the simulation results. Figure 8.5 describes that three followers will track the leader exactly. According to the Fig. 8.5, the consensus error is bounded. The curves of the control signals can be seen in Fig. 8.6.

8.5 Conclusion

8.5.1 Contribution

In this paper, we delve into the finite-time adaptive optimal control problem for nonlinear MASs under the simultaneous influence of stochastic disturbances and non-affine nonlinear faults, a scenario of great practical significance as real-world systems often face such complex and combined challenges. To address this intricate problem,

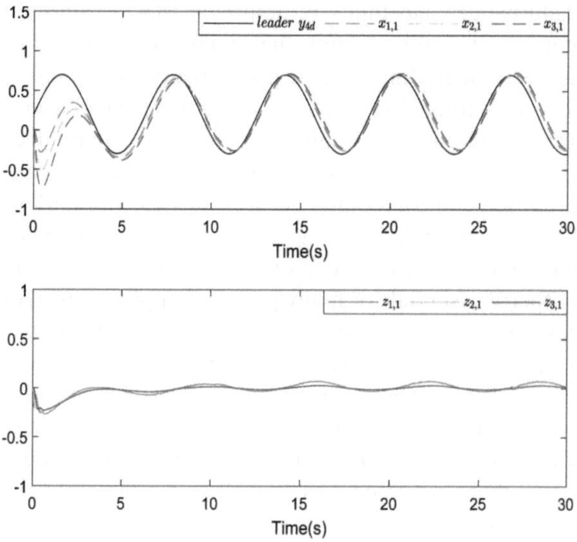

Fig. 8.5 Trajectories of $x_{i,1}$, leaders y_r and $z_{i,1} (i = 1, 2, 3)$

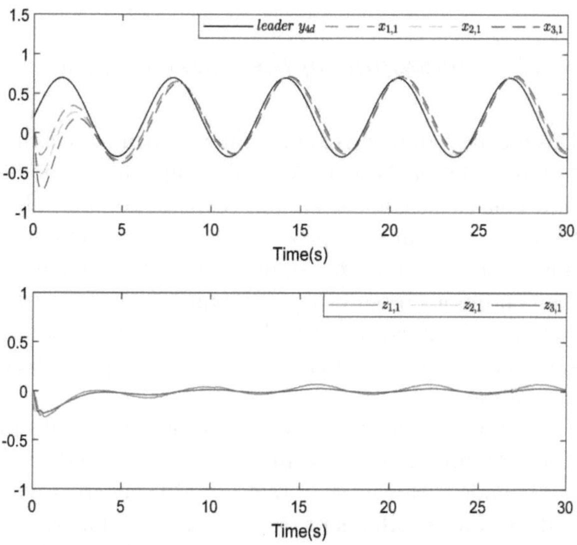

Fig. 8.6 Trajectories of the actual control signals $u_i (i = 1, 2, 3)$

we adopt a novel approach by integrating optimal control theory with RL algorithms, successfully establishing an I-C-A architecture. This architecture is a significant innovation, leveraging the strengths of optimal control theory in achieving desired system performance through optimizing control inputs based on system models and performance criteria, while also harnessing the adaptability of RL algorithms that can learn from environmental interactions and adapt to changing conditions, which is crucial for handling the uncertainties and complexities introduced by stochastic disturbances and faults in nonlinear MASs. Subsequently, to mitigate the adverse impacts of nonlinear faults that can introduce high-frequency noise and disturbances, severely degrading system performance and stability, we introduce a Butterworth low-pass filter due to its excellent frequency-selective characteristics, effectively smoothing out high-frequency components to reduce fault-induced negative effects. It is rigorously verified through theoretical analysis and extensive simulations that the proposed control method ensures all state variables of the controlled system exhibit SGFSP, a remarkable achievement guaranteeing system robustness and convergence within a finite time frame even in the presence of stochastic elements and faults, a property essential for many practical applications requiring quick and stable system responses. Finally, the effectiveness of the proposed control strategy is thoroughly validated through a series of carefully designed simulation examples covering a wide range of scenarios, including different levels of stochastic disturbances and various types of non-affine nonlinear faults, with the simulation results clearly demonstrating the strategy's superior performance in maintaining system stability and achieving optimal control objectives, as well as its adaptability to different fault and disturbance conditions, showcasing its robustness and versatility.

8.5.2 Limitations of the Current Work

Despite the promising results achieved in the research, several limitations still persist.

Computational complexity: The implementation of the optimized adaptive finite-time consensus control strategy involves complex algorithms and computations. For instance, the real-time processing of stochastic nonlinear dynamics and non-affine nonlinear faults requires significant computational resources. In large-scale MASs, the continuous monitoring and adjustment of control inputs based on the system's state and the presence of faults can lead to high computational overhead. This may limit the practical application of the strategy in systems with limited computational capabilities.

Assumption dependency: The proposed control strategy relies on several assumptions. Assumption 8.1 assumes that the first and second derivatives of disturbances are bounded. However, in real - world scenarios with abrupt disturbances, such as sudden sensor failures or external shocks, this assumption may not hold. Assumption 8.2 requires the reference signal and its derivatives to be bounded by certain constants and functions. In dynamic environments where the reference signal may change rapidly or unpredictably, meeting this assumption can be challenging.

Parameter sensitivity: The performance of the control strategy is highly sensitive to the design parameters. Parameters such as the time-varying gains and the neural network parameters play a crucial role in the system's stability and tracking performance. If inappropriate parameter values are chosen, it may result in reduced tracking precision or even system instability. For example, if the adaptation gains are set too high, the system may exhibit oscillatory behavior, while too low values may lead to slow convergence.

These limitations highlight the need for further research to enhance the robustness, applicability, and ease of implementation of the optimized adaptive finite-time consensus control strategy for stochastic nonlinear MASs with non-affine nonlinear faults.

8.5.3 Future Directions

The future research could be pursued in the following directions:

Robustness enhancements: Future work should place emphasis on relaxing the assumption of bounded disturbances. This can be achieved by developing more sophisticated observers and control laws that are capable of effectively handling unbounded or state-dependent uncertainties. By doing so, the control strategy can be made more robust and applicable to a wider range of real-world scenarios where disturbances may not be strictly bounded.

Advanced filtering techniques: To improve the tracking accuracy of the MASs, more advanced command filtering techniques with explicit error compensation mechanisms could be investigated. These techniques can help mitigate the impact of filtering errors, which may otherwise degrade the system's performance. For example, adaptive filtering methods that can adjust their parameters based on the system's state and the characteristics of the disturbances could be explored.

Adaptive event-triggering Schemes: Exploring dynamic or adaptive event-triggering thresholds is a promising direction. Such schemes can provide a better balance between communication efficiency and control performance across a wider range of operating conditions. For instance, the ETM that can adjust the triggering threshold based on the system's current state, the level of disturbances, or the communication load can be developed to optimize the overall system performance.

Expanded system complexities: As originally proposed, future research will focus on extending the current methods to output-feedback control for non-affine MASs that are also subject to unknown states and disturbances. Additionally, other practical considerations such as switching topologies, actuator faults, unknown hysteresis, and anti-consensus techniques will also be explored. This will make the control strategy more comprehensive and applicable to complex real-world MASs.

Experimental validation: Testing the proposed control strategy on physical multi-agent platforms, such as mobile robots or UAVs, would be essential. This will help evaluate its performance and robustness under real - world conditions and communication constraints. By conducting experimental validation, the practicality and effectiveness of the control strategy can be verified, and any necessary adjustments or improvements can be identified.

References

1. Yin Z, Lin Y, Zhang Y, Qian Y, Shu F, Li J (2022) Collaborative multiagent reinforcement learning aided resource all location for UAV anti-jamming communication. IEEE Internet Things J 9(23):23995–24008
2. Mahelaet OP et al (2022) Comprehensive overview of multi-agent systems for controlling smart grids. CSEE J Power Energy Syst 8(1):115–131
3. Zargarzadeh H, Dierks T, Jagannathan S (2015) Optimal control of nonlinear continuous-time systems in strict-feedback form. IEEE Trans Neural Netw Learn Syst 26(10):2535–2549
4. Sun K, Sui S, Tong S (2018) Fuzzy adaptive decentralized optimal control for strict feedback nonlinear large-scale systems. IEEE Trans Cybern 48(4):1326–1339
5. Li Y, Sun K, Tong S (2019) Observer-based adaptive fuzzy fault-tolerant optimal control for SISO nonlinear systems. IEEE Trans Cybern 49(2):649–661
6. Bhat SP, Bernstein DS (1998) Continuous finite-time stabilization of the translational and rotational double integrators. IEEE Trans Autom Control 43(5):678–682
7. Du H, He Y, Cheng Y (2014) Finite-time synchronization of a class of second-order nonlinear multi-agent systems using output feedback control. IEEE Trans Circuits Syst I Regul Pap 61(6):1778–1788
8. Sui S, Chen CLP, Tong S (2019) Fuzzy adaptive finite-time control design for nontriangular stochastic nonlinear systems. IEEE Trans Fuzzy Syst 27(1):172–184
9. Li Y, Li Y-X, Tong S (2023) Event-based finite-time control for non linear multiagent systems with asymptotic tracking. IEEE Trans Autom Control 68(6):3790–3797
10. Xu Y, Wu Z-G, Pan Y-J (2023) Observer-based dynamic event triggered adaptive control of distributed networked systems with application to ground vehicles. IEEE Trans Ind Electronics 70(4):4148–4157
11. Liu Y, Zhang H, Sun J, Wang Y (2024) Event-triggered adaptive finite time containment control for fractional-order nonlinear multiagent systems. IEEE Trans Cybern 54(2):1250–1260
12. Wang X, Rui X, Huang T, Kurths J (2024) Event-triggered adaptive containment control for heterogeneous stochastic nonlinear multiagent systems. IEEE Trans Neural Netw Learn Syst 35(6):8524–8534
13. Wang X, Zhou Y, Huang T, Chakrabarti P (2022) Event-triggered adaptive fault-tolerant control for a class of nonlinear multiagent systems with sensor and actuator faults. IEEE Trans Circuits Syst I Regul Pap 69(10):4203–4214
14. Li Y-X, Yang G-H (2016) Fuzzy adaptive output feedback fault-tolerant tracking control of a class of uncertain nonlinear systems with nonaffine nonlinear faults. IEEE Trans Fuzzy Syst 24(1):223–234
15. Sun K, Liu L, Qiu J, Feng G (2021) Fuzzy adaptive finite-time fault-tolerant control for strict-feedback nonlinear systems. IEEE Trans Fuzzy Syst 29(4):786–796
16. Sui S, Chen CLP, Tong S (2019) Fuzzy adaptive finite-time control design for nontriangular stochastic nonlinear systems. IEEE Trans Fuzzy Syst 27(1):172–184
17. Wen G, Ge SS, Tu F (2018) Optimized backstepping for tracking control of strict-feedback systems. IEEE Trans Neural Netw Learn Syst 29(8):3850–3862

FSC
www.fsc.org
MIX
Papier aus verantwortungsvollen Quellen
Paper from responsible sources
FSC® C105338

If you have any concerns about our products,
you can contact us on
ProductSafety@springernature.com

In case Publisher is established outside the EU,
the EU authorized representative is:
Springer Nature Customer Service Center GmbH
Europaplatz 3, 69115 Heidelberg, Germany

Printed by Libri Plureos GmbH
in Hamburg, Germany